DEVELOPMENTAL GENE EXPRESSION REGULATION

DEVELOPMENTAL GENE EXPRESSION REGULATION

NATHAN C. KURZFIELD

EDITOR

Nova Biomedical Books
New York

Library of Congress Cataloging-in-Publication Data
Developmental gene expression regulation / [edited by] Nathan C. Kurzfield.
    p. ; cm.
  Includes bibliographical references and index.
  ISBN 978-1-60692-794-6 (hardcover)
 1. Developmental genetics. 2. Genetic regulation. I. Kurzfield, Nathan C.
 [DNLM: 1. Gene Expression Regulation, Developmental. QU 475 D489 2009]
 QH453.D475 2009
 571.8'5--dc22
                                2008053222

Published by Nova Science Publishers, Inc. ✦ New York

Contents

Preface

Developmental Gene Expression Regulation consists of any of the processes by which nuclear, cytoplasmic, or intercellular factors influence the differential control of gene action during the developmental stages of an organism. This new book presents the latest research in this field from around the world.

Chapter I - Developmental gene regulation is to elucidate the mechanisms of spatio-temporal gene expression in organisms during development and disease occurring. This chapter is focusing on the regulation of early developmental gene expression based on the latest progress of vertebrate developmental studies. The fate of germ cells in extra-embryonic ectoderm is determined during PGC formation by predetermined germ plasm in the oocyte, from which VASA, the DEAD box family protein of ATP-dependent RNA helicase is identified as a regulator in germline cell specification, spermatogenesis, RNA splicing and post-translational degradation, and cell growth. The regulation of germline cell growth needs the multi-functional growth factor LIF, maintaining the pluripotency of ES cells in vitro and up-regulating its expression during implantation suggested the involvement of LIF in the event, which was further supported by direct evidence from gene knockout. The gonad in the early fetal life as one tissue, indifferent and indistinguishable by morphology, has two fates, making it a unique regulatory model of gene expression. Gene regulation and their interactions of many genes involved in this process including SRY, SOX9, WT1, FGF9, WNT4, DAX1 and DHH, and their regulating roles and interaction during sexual development will be discussed, particularly regulatory roles in alternative splicing and signal transduction pathway in gonadal development.

Chapter II - Genetic control of proliferation, morphogenesis, and differentiation during development of multicellular organisms is crucial for the proper formation of adults. Regulation of developmental gene expression, however, goes beyond the developmental stages of an organism. The rate at which organisms age for example, is also regulated by many developmental genes.

While it is intuitive to consider aging a simple byproduct of accumulated wear and tear on the organism, it nevertheless has been shown that altering expression of single genes can extend life span significantly; demonstrating that the aging process can be genetically influenced. Many genes involved in developmental processes modulate, later in life, adult aging. For instance, altered expression of genes affecting endocrine signaling, stress responses, metabolism, and growth during developmental stages can increase the life span of model organisms. Furthermore, the study of these genes has revealed evolutionarily conserved pathways for the modulation of aging. In contrast to the precise genetic regulation that occurs during development, life span, while genetically regulated, is not so tightly controlled. For example, there are significant differences in aging phenotype even in monozygotic human twins. In this chapter, I will discuss whether aging is mainly due to an organism's post-embryonic developmental process or a haphazard process, and I will describe and compare the regulation of genetic pathways involved in both development and aging. In doing so, I will consider what these two fields of science can learn from each other to progress our understanding of the regulation of these genes.

Chapter III - Establishment of specific cell fates requires orchestrated interaction of an array of transcription factors and signaling pathways which incorporate major developmental roles. The transition of proliferating precursor or immature cells into a certain cell type and terminal differentiation has been studied in great detail on various neuronal cell types, due to the large degree of diversity, complex functional roles, and intrinsic properties of cells in the nervous system. In the peripheral nervous system sensory specificity is established by the choice of an immature, but committed, postmitotic progenitor to express a specific sensory receptor gene. Findings of sensory receptor gene regulation stemming form the olfactory system and visual system and in mouse and fruit fly provide insight in how this highly complex process is regulated and genetically controlled. After the initial decision to become a sensory neuron the cell then decides which receptor gene to express and subsequently maintain the expression of this given receptor gene. Genetic mechanisms for this choice depend upon transcriptional regulators which are expressed in subtype of sensory neurons, thus may provide a combinatorial code to orchestrate the expression of a specific receptor gene: For instance in the fly visual system, where an array of six *rhodopsin* genes can be expressed a combinatorial code of transcription factors is required for the sensory receptor gene regulation. Interestingly during larval and adult stages the regulation of *rhodopsins* makes use of distinct developmental genetic program. Regulatory regions of *rhodopsins* display a bipartite architecture with a proximal domain required for general PR expression and a distal domain encoding subtype specificity. In the human retina cone cells can express L and M opsin genes, which are located in close proximity on the chromosome. Regulation of L and M opsin depends upon "locus control regions" (LCR) a common long range *cis*-acting element regulating both genes. Interestingly, in the mouse retina *Opsins* are not clustered and co-expression of two *Opsins* genes occurs. In the mouse olfactory system, an array of over 1200 Odorant receptor (OR) genes can be expressed. OR genes are often arranged in clusters along the chromosome and seem to depend on distant and local *cis*-acting elements. Interestingly only one of the two parental copies of an OR gene is expressed, resulting in monoallelic expression of the gene. Taken together the choice of a sensory neuron to adopt a specific sensory specificity depends upon the complex interaction of *cis*- and *trans* acting factors. Even though generally only one form of sensory receptor gene is expressed, various mechanisms may be acting to achieve a similar outcome of sensory receptor gene regulation. Moreover recent findings reveal that sensory receptor genes can be co-expressed and that this co-expression is genetically controlled, thus adding an additional layer of complexity in the regulatory properties of sensory receptor genes.

Chapter IV - This chapter deals with an organotypic culture system to examine transcriptional events contributing to cell survival of the organ of Corti (OC), the modiolus (MOD) and the stria vascularis (SV) of newborn rats. mRNA profiling using Affymetrix gene chips was carried out in tissue obtained immediately after preparation and after 24 h in culture. The probe sets of 45 genes were subjected to a cluster analysis. A number of identified genes represented three major processes associated with the preparation of the cultures: mechanical injury and inflammation, hypoxia and excitotoxicity. The inflammatory response ontology was represented by inflammatory cytokines Interleukin-1beta (Il-1b), Interleukin 6 (Il-6), TNF-alpha converting enzyme (Tace) and Intercellular adhesion molecule (Icam). The hypoxia response was represented by the increase of Hypoxia-inducible factor-1 alpha (Hif-1a),

Glucose transporter 1 (Glut1) and Glucose transporter 3 (Glut3). The excitotoxic damaging process included changes in the glutamate transporters and NMDAR receptors. We identified Tace expression as a novel gene in the inner ear with a potentially important role in inner ear injury. The MOD region belongs to the most vulnerable regions of the ear characterized by a particularly high increase of Il-1b, Il-6 and Hif-1alpha mRNA expression.

Cell survival in culture is maintained by a complex regulation of pro-death genes on the one hand and protective genes on the other. In general, genes encoding proteins involved in triggering or executing cell death are downregulated and genes encoding protective acting proteins are upregulated. We found caspase 2, caspase 6 and calpain downregulated and the mitochondrial superoxide dismutase Sod2, the heat shock proteins Hsp27 and Hsp 70 and the insulin like growth factor binding proteins Igfbp3 and Igfbp5 upregulated. For two genes (Tace, Bax) we observed a differential response of coding and non-coding sequences. These data provide new insights into the role of the various members of the pro-death and pro-survival genes in protecting inner ear cells from injury-induced damage during the developing period.

Chapter V - The TGF-β superfamily consists of numerous members, including TGF-β proper, bone morphogenetic proteins (BMP) and growth differentiation factors (GDF). All TGF-β are dimeric cytokines present a biological active carboxy terminal domain of 110–140 amino acids following proteolysis.

Bone morphogenetic proteins (BMPs), first identified for their involvement in vertebrate bone formation, are now widely recognized as key factors in the regulation of many fundamental developmental processes in all deuterostomes. The active gradient established by BMP secreted ligands is one of the essential factors responsible for generating the positional information that underlies developmental patterning, including the regeneration of lost parts.

Myostatin or GDF-8, a recently discovered GDF subfamily member, acts as a negative regulator in maintaining the mammalian proper muscle mass during both embryogenesis and post-natal muscle development. Unlike most other members of the BMP/GDF superfamily, mammalian myostatin is secreted as a latent complex, usually linked to regulatory proteins, and its mature dimer produces an effect almost exclusively on muscle tissue.

The present chapter deals with the importance of the TGF-β family of growth factors in relation to regulatory spheres through the animal kingdom, focusing in particular on the expression of BMP molecules during regeneration and myostatin in "non-canonical" animal models; it also focuses on the regulative actions of myostatin during the development of vertebrates and in different experimental conditions, including *in vitro* chick co-culture and endurance training. Data available in literature indicate that there is substantial scope for future research in the area of TGF-β /myostatin linked to development.

"There is grandeur in this view of life, with its several powers, having been originally breathed by the Creator into a few forms or into one; and that, whilst this planet has gone cycling on according to the fixed law of gravity, from so simple a beginning endless forms most beautiful and most wonderful have been, and are being evolved."

Charles Darwin, The Origin of Species.

Chapter VI - Alpha-foetoprotein (AFP) is a well known diagnostic biomarker used in medicine to detect foetal developmental anomalies such as neural tube defects or Down's syndrome, or to follow the development of tumors such as hepatocellular carcinomas. However, the role of AFP goes way further than that. AFP is involved at least in rodents in the correct differentiation of the female brain, through its estrogen binding capacity. This chapter present an overview of what is known about the regulation of the *Afp* gene, describes the phenotype of the AFP knock-out (AFP KO) mouse and offers an overview of other mouse models available to study estrogen function.

Being in the right place, at the right time, is a key factor for alpha-foetoprotein (AFP). Firstly, because it is involved in major events occurring during narrow time-windows, such as sexual differentiation of the female brain. Secondly, because AFP is expressed in an onco-foetal way.

Chapter VII - Neurosecretory cells play critical roles in different and specific physiological and behavioral processes via spatiotemporal regulation of neurohormone secretion. In insects, diapause and metamorphosis are induced by neuropeptide hormones secreted from a few neurosecretory cells that project intrinsic axons to intrinsic neurohemal sites. Diapause hormone (DH) is responsible for induction of embryonic diapause in *Bombyx mori*. The diapause hormone-pheromone biosynthesis activating neuropeptide gene, *DH-PBAN,* is expressed exclusively in seven pairs of DH-PBAN-producing neurosecretory cells (DHPCs) on the terminally differentiated processes of the subesophageal ganglion (SG). On the other hand, prothoracicotropic hormone (PTTH) plays a central role in controlling molting and metamorphosis in *Bombyx mori* by stimulating the prothoracic glands to synthesize and release the molting hormone, ecdysone. The *PTTH* gene is constantly expressed during larval-pupal development, and the peptide is produced exclusively in two pairs of lateral PTTH-producing neurosecretory cells (PTPCs) in the brain. To help reveal the regulatory mechanisms of cell-specific expression of *DH-PBAN* and *PTTH*, we identified *cis*-regulatory elements that regulate expression in DHPCs and PTPCs, respectively, using a recombinant baculovirus (AcNPV)-mediated gene transfer system and a gel-mobility shift assay. Interestingly, *Bombyx mori* Pitx (BmPitx), a bicoid-like homeobox transcription factor, binds the 5'-upstream sequence of both *DH-PBAN* and *PTTH* and activates gene expression. This article describes the regulatory mechanisms of cell-specific expression of the neuropeptide hormone genes involved in diapause and metamorphosis in *Bombyx*.

Chapter VIII - Regulation of gene expression during embryogenesis and development is a crucial clue for a normal anatomy and physiology. In fact, very little is known regarding factors that influence and regulate developmental gene expression. Similarly, there is little information available concerning the effects of a coordinate expression of a group of functionally related genes.

The analysis of temporal patterns of gene expression in embryos is essential for the understanding of the molecular mechanisms that control development. This scientific field has been innovated by the combined use of experimental high-throughput methods, such as DNA microarrays, and bioinformatic methods that take advantage of the completion of the human genome sequence, along with the genomes of related species. Microarray analysis, in fact, provides a large amount of data -at molecular level- that once acquired, must be functionally integrated in order to find common patterns within a defined group of biological samples. Following the enormous number of data obtained from these experiments, a new type of comparative embryology is now emerging, and it is based on the comparison of gene expression patterns. The sequencing of several new genomes, the increasing computational power and new bioinformatic algorithms cooperate to overcome some of the intrinsic difficulties in the study of gene regulation, thus permitting, for example, to identify regulation sites located far away from the genes. Recent bioinformatic methods applied to gene regulation are reviewed that either follow the "single species, many genes" approach or the "single gene, many species" one.

In this chapter we would review the new application of DNA microarray and bioinformatics to define a new combinatorial approach for analysis of gene expression during development.

Chapter IX - The developing central neural circuits are genetically controlled and initiated by developmental signals. Recent progress in molecular and cellular developmental biology provides evidence of how the brain is feminized or masculinized during the critical developmental period. Research into the development of brain architecture requires experimenting with animals, specifically, interfering with normal development and with environmental conditions. Drosophilae, sea urchins, and metazoans are simple invertebrates used for standard research models. Recently, the teleosts, bony fish with biological and genomic complexity found in the higher vertebrates, have become important models for developmental and molecular neurobiology studies. As in mammals, sexual dimorphic genetic expression is found in the developing brain of teleosts. The cellular and synaptic organization of brain architecture is determined by the genomic program and triggered by environmental cues such as photoperiod and temperature. This review highlights some of the methodological issues related to current findings about the gene expression regulation involved in the complex process of neural development, particularly in brain-sex differentiation.

Chapter X - *SRY/Sry*, a single-copy gene on the Y-chromosome, was identified to play the critical role in initiating testicular differentiation during gonadal development in humans and mice two decades ago. Nonetheless, neither the regulation of *Sry* expression nor the mode of SRY action during gonadal differentiation is well understood. The B6.Y^{TIR} mouse carries a Y-chromosome originally from a *Mus musculus domesticus* mouse caught in Tirano, Italy (Y^{TIR}) and the X-chromosome and autosomes from the C57BL/6J (B6) inbred mouse strain, which belongs to *Mus musculus molossinus*. It has been demonstrated that the SRY protein is expressed normally both in pattern and onset, yet, B6.Y^{TIR} mice develop only ovaries or ovotestes. Therefore, this mouse model provides an opportunity to study the mechanism of SRY action during gonadal sex determination. We hypothesize that the testis determining pathway in the B6.Y^{TIR} gonad is impaired by at least two mechanisms that act synergistically. First, *Sry* transcript levels from the Y^{TIR}-chromosome are reduced on the B6 genetic background. Second, polymorphisms of *Sry* sequences lead to inefficient biological activity of the SRY protein encoded on the Y^{TIR}-chromosome. Both dysfunctions are requisite to impairing testicular differentiation.

Chapter XI - Fibroblast growth factors (FGFs) constitute a large family of signaling polypeptides that play critical roles in development. During morphogenesis, FGFs are involved in cell proliferation, differentiation and migration; however, in adults these proteins function as homeostatic factors. FGFs mediate their functions through a cell surface receptor, the fibroblast growth factor receptors (FGFRs), which are a member of the tyrosine kinase superfamily. Both the *FGF* and *FGFR* gene families are identified in multicellular organisms but not in unicellular ones and have expanded greatly during evolution. FGF gene organization is highly conserved among vertebrates. In human and mouse, the FGF gene family consists of 22 members; however, in zebrafish (*Danio rerio*) there are 27 identified *fgf* members. Japanese medaka *(Oryzias latipes),* like zebrafish, is a small aquarium fish used as a model organism in vertebrate development. During evolution, these two fish species (zebrafish and Japanese medaka) were separated from their last common ancestor about 110 million years ago. The medaka genome is only half (800 Mb) of the zebrafish genome (1700 Mb). We have searched medaka genome data bases and identified 28 *fgf* genes in this species of which nine are paralogs. We have done a phylogenetic and conserved gene location (synteny) analysis of the identified *fgf* genes of medaka and analyzed the evolutionary relationships of these genes with human *FGF* gene families.

Chapter XII - Connection between dynamics of yolk lipovitellin degradation and specific features of germ genes expression was studied in early development of intergenetic reciprocal F1 hybrids of the bream, roach and blue bream.

According to the modern view lipovitellin is the main protein of oocyte and embryo yolk. Its main function is reserve, nutritional and structural. But moreover lipovitellin is active component of internal fluid embryo invironment in wich embryo cells and germ genes are developing and expressing. There is a view that synchronous expression of parental alleles of genetic loci is related, as a rule, to kindred fish crossing, and asynchronous expression – to remote fish crossing. But when we analysed character of expression of some loci of intergeneric F1 hybrids (*6-Pgd, Ldh-B, β–Est-1, 2, 3, Aat-1, Me-1, 2* and others) with different expression time in embryogenesis, we show that character of loci expression is related to stage of development also.

As objects we used zygotes, embryos, larvas and frysof bream, roach, blue bream and intergeneric reciprocal F1 hybrids. Identification and analysis of enzymes activity and lipoviteelline properties were performed using methods of disk-, gradient and SDS-electrophoresis in polyacrylamide gel.

There are some arguments for regulation function of lipovitellin : first argument is connected with different expressions character (synchronous and asynchronous) of germ genes in early and late stages of embryogenesis; second - with different distribution of isoenzymes activity in early and late stages of embryogenesis; and third – with biocatalytic activity of lipovitelline because of oogenesis isoenzymes connected with lipovitelline by weak connection.

When first locus expression was timed to the early stages (blastodisk – gastrula), the gene parental alleles were activated asynchronously according to the maternal types. When the first expression was timed to later stages (yolk sac resorption), parenteral alleles were activated synchronously. In early development lipovitellin and oogenesis enzymes form the maternal (by origin) metabolic medium, which preferential activation of maternal alleles. At later developmental stages, when the yolk reserves are partially or fully resorbed and maternal proteins don' t play importance role, and germ proteins form new (germ) internal fluid invironment, in wich the embryonic genes are activated synchronously.

Short Communication - Regulation of embryonic axis patterning by Hox genes has been shown to be widely conserved among metazoans. In *Drosophila melanogaster* the Hox gene *abdominal-A* (*abd*-A) is important for the development of the legless abdomen. In contrast to the clear tagmata-correlated activity in insects the analysis of Hox expression patterns during crustacean development has turned out to be more diverse. While in the branchiopod brine shrimp *Artemia franciscana* the posterior genes show a more ancestral overlapping arrangement, this is not the case for the malacostracan isopod *Porcellio scaber*. In this more modern species *abd-A* is mainly restricted to the developing pleon. Here we present the cloning and expression pattern of the *abd-A* gene from the freshwater crustacean *Asellus aquaticus*. In contrast to the related isopod *Porcellio scaber*, *Asellus aquaticus* differs in the regulation of posterior segment patterning. While *Porcellio scaber* displays distinct and separate segments in the pleon, posterior segments are partially fused in *Asellus aquaticus* to yield a pleotelson. The *abd-A* signal was significantly reduced or absent in the pleotelson of *Asellus aquaticus*, while *abd-A* was expressed in the free segments of *Porcellio scaber*. An additional correlation between *abd-A* gene expression and patterning in these two species was found in that the orientation of walking legs was towards the posterior pole in segments expressing *abd-A*. *Asellus aquaticus* thus may provide a highly interesting and novel arthropod model organism to study evolution of segment identity and patterning.

In: Development Gene Expresión Regulation
Editor: Nathan C. Kurzfield

ISBN: 978-60692-794-6
©2009 Nova Science Publishers, Inc.

Chapter I

Gene Regulation and Early Developmental Gene Expression in Vertebrate

Hongshi Yu and Shuliang Cui

Department of Zoology, The University of Melbourne
Royal Parade, Parkville, Victoria 3010, Australia

Abstract

Developmental gene regulation is to elucidate the mechanisms of spatio-temporal gene expression in organisms during development and disease occurring. This chapter is focusing on the regulation of early developmental gene expression based on the latest progress of vertebrate developmental studies. The fate of germ cells in extra-embryonic ectoderm is determined during PGC formation by predetermined germ plasm in the oocyte, from which VASA, the DEAD box family protein of ATP-dependent RNA helicase is identified as a regulator in germline cell specification, spermatogenesis, RNA splicing and post-translational degradation, and cell growth. The regulation of germline cell growth needs the multi-functional growth factor LIF, maintaining the pluripotency of ES cells in vitro and up-regulating its expression during implantation suggested the involvement of LIF in the event, which was further supported by direct evidence from gene knockout. The gonad in the early fetal life as one tissue, indifferent and indistinguishable by morphology, has two fates, making it a unique regulatory model of gene expression. Gene regulation and their interactions of many genes involved in this process including SRY, SOX9, WT1, FGF9, WNT4, DAX1 and DHH, and their regulating roles and interaction during sexual development will be discussed, particularly regulatory roles in alternative splicing and signal transduction pathway in gonadal development.

Introduction

Molecular mechanism of gene expression and regulation became a fast growing field in modern genetics after the discovery of DNA as genetic material and unveiling of the genetic codes in all living organisms. Gene expression is to decode genetic information from DNA to protein or RNA, including transcription to form a primary transcript (pre-mRNA), conversion of pre-mRNA into mature mRNA and translation to synthesize the protein. Thus any step of gene expression may be modulated, from the DNA-RNA transcription step to post-translational modification of a protein. Regulation at transcriptional level is the basic modulation of gene expression, which decides when transcription occurs and how much RNA is created. Transcription factors play a central role in activation or repression of transcription. Some transcription factors bind directly to the DNA molecule (sequence of promoter region); others bind to other transcription factors. Thus, protein-DNA interactions and protein-protein interactions regulate gene activity activating or blocking the process of transcription. Post-transcriptional regulation determines isoforms of transcripts by alternative splicing and the stability of mature mRNA by capping at 5' end and addition of poly (A) tail at 3' end. The modulation of gene expression by small non-coding RNAs is a recently discovered level of gene regulation. Small non-coding RNAs are kinds of small RNAs with regulatory function without being translated into protein including microRNAs (miRNA), short interfering RNAs (siRNAs) and Piwi-interacting RNAs (piRNAs). These small non-coding RNAs play important roles in animal development by controlling translation or stability of mRNAs (Stefani and Slack, 2008). In addition, epigenetic events participate in regulation of gene expression. DNA methylation and histone modifications are two major areas of epigenetics. DNA methylation is the process of adding methyl groups to specific cytosine residues in the promoter regions of DNA and triggers heritable gene silencing. It is involved in the regulation of imprinted gene expression and X-chromosome inactivation. Histone modifications including methylation, acetylation, phostphorylation, ubiquitination and ADP-ribosylaiton cause profound changes in local chromatin structure and further control the accessibility of the chromatin and transcriptional activities inside a cell. Therefore, regulation of gene expression occurs at multiple levels and produces complicated networks.

Development begins with the fusion of two gametes, the sperm and egg cell to form a zygote described as fertilization in vertebrates. Embryogenesis follows fertilization to produce a complex, multicellular organism. How does this process regulate and orchestrate during ontogenesis is one of the greatest mysteries of life, and represents a fundamental challenge in developmental biology. During embryogenesis, a fertilized egg divides by cleavage to form a ball of cells called morula at 16-cell stage, and then develops a cavity named blastocyst, the first structure in which any cell specialization occurs. By embryonic day 4.0 (E4.0) in mice, and between 5 to 7 days post-fertilization in humans, the blastocyst, composed of trophectoderm, blastocoet and inner cell mass (ICM), reaches the uterus. At this stage embryonic stem (ES) cells can be derived from the ICM of the blastocyst. 1- 2 days later in human and 0.5 day later in mouse the blastocyst implants in the uterine wall. As development proceeds, blastocyst forms gastrula at between about E6.5 and E8.0 in mouse that has the three primary germ layers of cells, endoderm, mesoderm, and ectoderm. And then organogenesis occurs from three different layers to develop all kinds of tissues and organs. In this process, it is of interest that the primordial germ cells (PGCs) do not arise within the genital ridge or the mesonephros but migrate from an entirely separate source. Therefore understanding formation, specification and migration of PGCs contribute to the understanding this intriguing event and mechanisms of gene regulation. Moreover sex determination and differentiation is unique event during organogenesis, two fates in one tissue depending on status of specific gene expression and regulation. Therefore we will discuss the two fundamental decisions in an early life, germ cell specification and sex determination, to illustrate the molecular mechanism of gene expression and regulation in early development in this chapter.

Gene Regulation in Germ Cell Specification

A new life begins since the formation of a gamete, the fusion of living germ cells produced in parental generation either by sexual reproduction or asexual reproduction through fertilization. All phenotypes including the physical appearance and behaviour present in the adulthood are derived from the single cell by development, a series of phases of growth and modifications described as morphogenesis. In the development of a new individual, the single-celled zygote turns into an organism compose of multiple cells and cell types specialized as different tissues and organs with different biological functions. Germ cells play central roles in generation of new lives, they are the founder cells of the gametes carrying genetic make-ups into the future generations. Germline stem cells are formed before they migrate into the gonads and have the properties of self-renewal and pluripotency. These cells are coordinately undergoing proliferation and differentiation to ensure the success of an individual growth and development. Discovery of genes and their regulators involved in the establishment of the germline, the migration of germ cells to the gonads and the cellular microenvironment is one of the major tasks of developmental genetics.

Germ cells arise as a cell population of PGCs during gametogenesis. The germ cells migrate to the gonad to form accessory cells with somatic cells, and then give rise to gametes in the future (Saffman and Lasko, 1999; Wylie, 1999; Wylie, 2000). The germ cell lineage is potentially immortal and is controlled by a special developmental program different from that of the somatic cell lineage. Mammalian germ cells are specified by germ cell-specific cytoplasmic determinants in the germplasm of the fertilized egg (Eddy, 1975), which is associated with changes of chromosome (Beams and Kessel, 1974). The fate of germ cells was determined by those gerplasm determinants in germ cell precursors. The germplasm provides the germ cell determinants and the microenvironment essentially leading to gametogensis (Noce et al., 2001).

VASA: A Germplasmic Determinant for Germ Cell Formation and Migration

The molecular characterization of the germplasm to search for germ cell determinants started with *Drosophila melanogaster* (Rongo et al., 1995). During oogenesis, the polar granules form the mitochondrial clouds consisting of RNAs and proteins (Kobayashi et al., 1993). Those germplasms were analyzed for their role in the germ cell determination, including Oskar, Vasa, Nanos and Tudor, amongst which *vasa* gene, encoding a DEAD-family protein of ATP-dependent RNA helicase, is well characterized (Hay et al., 1988; Lasko and Ashburner, 1988; Liang et al., 1994). The germ cell formation in Drosophila needs expressed VASA and females with homozygous *vasa* gene mutation failed to develop posterior structures and pole cells (Ashburner et al., 1990). The germ cell lineage in early development requires *vasa* gene expression, which activates transcription factors by binding to the downstream target RNAs involved in germ cell establishment to regulate gene translation (Dahanukar and Wharton, 1996; Gavis et al., 1996; Styhler et al., 1998; Tomancak et al., 1998).

Homologous genes coding for VASA protein have been identified in other animals (Table 1), including *C. elegans*, *Xenopus*, zebrafish, mice, humans, chickens, trout, and rats and marsupials (Castrillon et al., 2000; Cui et al., in press; Fujiwara et al., 1994; Komiya et al., 1994; Komiya and Tanigawa, 1995; Olsen et al., 1997; Roussell and Bennett, 1993; Tsunekawa et al., 2000; Yoon et al., 1997; Yoshizaki et al., 2000) according to the structural conservation. *Vasa* genes are expressed in germ line cells and therefore used as specific molecular marker for germ cell profiles (Fujiwara et al., 1994; Lee et al., 2005). Functional studies of VASA showed the *vasa* genes are required for germ cell formation in mammals. Male mice with a targeted *vasa* homolog (Mvh) are abnormal in spermatogenesis and sterile although homozygous females are fertile (Tanaka et al., 2000). Human infertile male patients with no VASA immunogenicity cannot produce mature sperm (Castrillon et al., 2000). These studies also show a high similarity in DNA sequences and amino acids sequences across species. By sequence alignment and structural comparison, VASAs from different species show a typical constitutional structure of 10 conserved motifs in 2 domains (Linder, 2006) as a member of the DEAD box superfamily proteins, an ATP-dependant RNA helicases and RNA-dependant ATPases (Cordin et al., 2006; Linder et al., 1989).

Germ plasm is specified as regions of cytoplasm of eggs and early embryos in many species, which is derived from some embryonic cells and contains RNAs and proteins as electron dense masses of granules and fibrils. *Vasa* gene encodes a germ plasmic RNA-binding protein of the DEAD box family. Functional VASA protein is required for the provision of normal germ plasm in cytoplasm for the formation of germ cells. The maternally derived *vasa* mRNA is uniformly distributed in oocytes but the protein becomes localized to the germ plasm. Homologues of *vasa* have showed to be expressed in germ plasm in many species (Braat et al., 1999; Olsen et al., 1997; Shibata et al., 1999; Yoon et al., 1997) and its localization is controlled by maternal signals (Pelegri et al., 1999). VASA protein interacts with transcription factors, such as the initiation factor diF2 in Drosophila (Carrera et al., 2000), to regulate gene expression at the translational level.

Table 1. Vasa gene expression patterns in different representative species

Species	Gene name	Transcripts	Proteins	Functions	References
Fruit fly	vasa	Early embryos, germ cells in ovary and testis.	Germ cells in ovary and testis.	Early embryogenesis and oogenesis.	Hay et al., 1988; Lasko et al., 1988; Liang et al., 1994; Markussen et al., 1995.
Nematode	glh	Early cleavage stages in germline cells.	Germline blastomeres and gonads.	Germ cell proliferation and gametogenesis	Kuznicki et al., 2000; Gruidi et al., 1996.
Zebrafish	vasa	Early embryos, testis and ovary.	Germinal vesicles in early oogenesis		Braat et al., 1999; Yoon et al., 1997; Weiginger et al., 1999.
Frog	XVLG1	PGC specific expression in gonads	PGCs, Ooocyte and eggs	Survival of PGCs.	Komiya et al., 1994; Ikenishi et al., 1996; Ikenishi and Tanaka, 2000.
Chicken	Cvh	Testis	Embryos, PGCs, testis and ovary.		Tsunekawa et al., 2000
Possum (marsupial)	Tvvh	Fetus, newborn,adult testis and ovary	Testis and ovary.		Cui et al., in press.
Mouse	Mvh	Testis	Germ cells, testis and ovary.	PGC growth in spermatogenesis	Carrera et al., 2000; Tanaka et al., 2000.
Pig	Pvh	Early embryo, fetus, testis and ovary	PGCs, oocyte and spermatocyte.		Lee et al., 2005.
Human	VASA	Fetus and adult gonads	PGCs, testis and ovary		Castrillon et al., 2000.

In the presence of VASA in the germplasm, other RNA-binding proteins, such as Nanos, HMG CoA reductase and Oct4 are also involved in the germ cell formation and migration. Nanos initiates the gene expression for germ cell specification and control germ cell specific genes expression for the cell growth otherwise holds the expression at low levels (Asaoka et al., 1998) and is needed for germ cell migration (Kobayashi et al., 1996). The zygotically expressed Columbus, the lipid-metabolizing enzyme HMG CoA reductase, generate signals to direct the migrating germ cells to the destination, the competent gonad (Van Doren et al., 1998). The green fluorescent protein (GFP) expression driven by *Oct4* gene promoter

(Anderson et al., 1999) confirms that *Oct4* is only expressed in the germ cells and allows to trace the germ cell migration (Anderson et al., 2000; Anderson et al., 1999; Bendel-Stenzel et al., 2000) as well as their associated proteins, such as β1-integrin and E-cadherin essential for germ cell colonization and compaction into the genital ridges (Houston and King, 2000; Maegawa et al., 1999). The mitochondrial ribosomal RNAs (rRNAs) in the germ plasm also showed their role in germ cell formation (Iida and Kobayashi, 1998). The coordinative regulation of expression levels of those chemoattractants and repellants secures the germ cells migration and colonization in the destination gonad. Those migration signals generated in the gonadal mesoderm has to be recognized by its receptors expressed in the germ cells, such as the maternally expressed Nanos. The mouse germ cell migration pattern was based on the expression of alkaline phosphatase (AP) and polysaccharides expressed on the surfaces of germ cells detected by antibodies. The germ cell migration pattern in zebrafish is established based on genetic mutation (Pelegri et al., 1999) of Vasa signal in the germ plasm (Weidinger et al., 1999; Yoon et al., 1997). Since bird germ cells are formed around the blood island, which would be the starting point for germ cell migration driven by positive chemoattraction of the gonad (Kuwana and Rogulska, 1999). Further investigations into the molecules as attractive mediators involved in germ cell migration are needed for a good understanding of the important biological process.

Stem Cells and ES Cells

A stem cell is a cell capable of self replication indefinitely and differentiation to give rise to different cell types. The fertilized egg is totipotent since it has the potential to generate all cells and tissues through division and differentiation for an organism. Stem cells are pluripotent to cells derived from all embryonic germ layers, from which all cells of the body are developed. The pluripotent stem cells have the potential to give rise to any type of cell while unipotent stem cell, derived from adult tissues, can renew itself and specialize into cell types of their source tissue (Slack, 2000). Both types of stem cells are capable of self-renewal and differentiation to generate cells for tissues, but pluripotent stem cells are versatile to have the potency of generating a new life, defined as the embryonic stem cell obtained from early blastocyst embryos.

The germinal PGCs migrate and proliferate in the genital ridge (Tam and Snow, 1981), from which germ cells have been isolated at about the time of their migration (Durcova-Hills et al., 2001; Matsui et al., 1992; Resnick et al., 1992; Shamblott et al., 1998) as embryonic germ (EG) cells. The EG cells show similar features to ES cells (Donovan and de Miguel, 2003). Because the changes of somatic status of imprinted genes as they migrate and maturate (Yamazaki et al., 2003), EG cells from different phases of migration are somewhat different (Hajkova et al., 2002) and, as showed in mouse PGCs at E10.5, methylation erasure occurred and imprinted genes expressed (Geijsen et al., 2004; Yamazaki et al., 2003), suggesting that ES cells should be derived from Germ cells earlier than the described stages.

The first stem cell started when the egg is fertilized by a sperm, which give rise to all types of cells in tissues and organs in later embryonic development. Cell lines with properties resembling the differentiation potential of the early embryo are referred as ES cells that are derived from early blastocyst before implantation. ES cells are propagated *in vitro* indefinitely without losing their stem cell properties. Under the differentiation conditions, ES cells start to give rise to wide range of cell types, as demonstrated in the mouse ES cell line (Evans and Kaufman, 1981; Martin, 1981) and human ES cells (Thomson et al., 1998). ES cells are experimentally isolated from the ICM of the blastocyst embryo in mammals. ES cells retain their developmental potency after passages of cell culture and are able to integrate into the ICM of an early embryo to generate an individual animal. ES cells have two important characteristics: they propagate in extended culture without differentiation, the self-renewal, and these cells contribute all cells and tissue in the next generation, including the pluripotent germ cells (Capecchi, 1989). ES cell line was initially established with cells isolated from the ICM by cell culture in the presence of mitotically inactivated fibroblast feeder layer cells (Evans and Kaufman, 1981; Martin, 1981), which produce the multifunctional cytokine leukemia inhibitory factor (LIF) as a stem cell regulator essential for suppression of ES cell differentiation (Gearing et al., 1987; Smith et al., 1988). Then LIF is used in the culture medium to replace the feeder layer cells in the establishment and the maintenance of ES cell lines (Smith et al., 1988) to avoid fibroblast contaminants in the ES cell population. The combination of LIF and the bone morphogenic proteins (BMPs) were found to enhance the capacity of self-renewal and pluripotency of ES cells cultured in serum-free media (Ying et al., 2003).

Cytokine LIF In ES Cell Culture

The derivation and propagation of the early mouse ES cells required feeder layer cells adopted from the recipe for embryonal carcinoma cell culture (Evans and Kaufman 1981; Martin 1981). The feeder cells secrete growth factors into the culture medium to keep the ES cells undifferentiated. The LIF was identified by analysis of the conditioned media (CM) as a growth factor. And, in later experiments, purified LIF can be used in the ES cell culture medium to replace the fibroblast feeder layer cells (Smith et al., 1988; Williams et al., 1988). In the attempts to establish and maintain ES cell lines from various species and its critical role in embryo implantation (Nachtigall et al., 1996; Stewart et al., 1992), genes encoding LIF have been cloned and characterized in many mammalian species in addition to the mouse, including human, cow, pig, sheep, rat, mink and marsupials (Cui et al., 2001; Gearing et al., 1987; Gearing et al., 1988; Piedrahita et al., 1997; Song et al., 1998; Willson et al., 1992; Yamamori et al., 1989), suggested a structural and functional conservation across a wide range of species. The organization of the LIF genes and the alternative transcription in the mouse LIF has been characterized (Rathjen et al., 1990; Stahl et al., 1990; Willson et al., 1992).

LIF is a multifunctional growth factor that has been extensively studied in many mammalian species, particularly in the mouse. Mouse LIF was originally characterized by its ability to suppress the proliferation of cells in the murine myeloid leukaemia cell line M1, by inducing their irreversible differentiation to macrophage cells (Metcalf et al., 1988), and its ability to suppress the differentiation of totipotent mouse embryonic stem cells (Smith et al., 1988; Williams et al., 1988). LIF is essential to keep the pluripotency of the mouse embryonic cells by inhibiting their differentiation. Further LIF studies have showed a diverse set of biological activities on different target cell types *in vitro* and on different adult and embryonic tissues *in vivo* (Gough et al., 1989; Hilton et al., 1991; Thompson and Majithia, 1998). The cellular responses to LIF are initiated through binding of the LIF molecule to a specific receptor on the cell surface (Nichols et al., 1996).

LIF belongs to the family of interleukin 6 (IL-6)-type cytokines (Heinrich et al., 1998; Hibi and Hirano, 1998; Rose-John, 2002). IL-6 family cytokines share a common receptor component gp130 for activation of transcription of the downstream genes through the JAK-STAT pathways (Brivanlou and Darnell, 2002; Heinrich et al., 2003; Levy and Darnell, 2002; O'Shea et al., 2002). The bound complexes enter the nucleus, bind to DNA sequence of genes and activate transcriptions. However, LIF action is not necessary *in vivo* since LIF-null mice are fertile and produce normal embryos although failed to implant (Stewart et al., 1992). But normal expression of the LIF receptor (LIFR) and the receptor partner gp130 are critical because the newborn mice die shortly after birth without LIFR (Li et al., 1995; Ware et al., 1995) and early embryos die without gp130 (Yoshida et al., 1996). The LIF-STAT pathway is not operational in human ES cells (Daheron et al., 2004). The search of a LIF-like growth factor in human ES cell culture leads to the discovery of the pluripotency sustaining factor Nanog found in the germ cells in early embryos (Chambers et al., 2003; Silva et al., 2006). This divergent homeodomain protein stimulates propagation of ES cells without differentiation. The Nanog transcripts are present in mouse and human ES cell lines, but not present in the differentiated cells. Nanog defines the ES cells through the activation of STAT3 transcription pathways (Chambers et al., 2003).

LIF activates the LIF-STAT signaling pathway mainly through the signal transducer and activator of transcription 3 (STAT3) mediated by LIF receptor complexes (Niwa et al., 1998; Takeda et al., 1997). BMP acts through the BMP-SMAD pathway to enhance self-renewal of mouse ES cells in collaboration with the cytokine gp130, a component of LIF receptor complexes (Ying et al., 2003), implies that BMP reacts with the cytokine protein in the receptor complexes to activate an alternative pathways for downstream gene transcription in the process. In LIF-free media, BMP induces differentiation of mouse ES cells, probably because the missing LIF function cannot be compensated by BMP without LIF receptor complex.

Factors Maintain Self-Renewal and Pluripotency of Mouse ES Cells

Blastocyst feeder layer cells support the ES cell culture by secreting the cytokine LIF to sustain ES cell self-renewal (Smith et al., 1988; Williams et al., 1988). LIF inhibits the mouse ES cell differentiation by activation of STAT3, which can also be activated by other growth factors efficiently for the cell growth (Matsuda et al., 1999). LIF strengthens its capacity for the growth of ES cells by the activation of other signals involved in ES cell growth, such as extracellular receptor kinases (ERKs) (Burdon et al., 1999) and phosphoinositide 3-kinase (PI3K) (Paling et al., 2004), both of which promote differentiations. The coordinated activation of those transduction signal pathways controls the division of the undifferentiated ES cell (Fig.1).

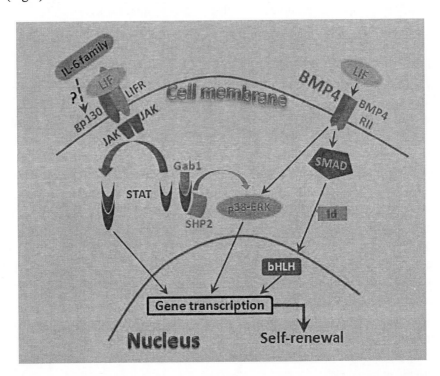

Figure 1. Signal transduction pathways for ES cell self-renewal. LIF and BMP4 signaling are required for ES cells to retain their capacity of self-renewal and their pluripotency. The cytokine LIF binds to its receptors (LIFR) to form a low affinity receptor complex that recruit the common signaling subunit gp130, leading to the formation of hetero-dimerized high affinity receptor complexes. The dimerization activates JAK family of kinases which further activate the STAT family of transcription factors to initiate gene transcription for the ES cell growth. The LIF signal transduction is strengthened by activating other signal. The activated STAT transcription factor binds to the adaptors SHP2 and Gab1 to gain its ability of activating the ERK pathway important for self-renewal of ES cells. The receptor complex component LIFR is shared by some of the IL-6 cytokine family members, and the gp130 is shared by all cytokines in the family. Whether these cytokines can compensate LIF function through activating STAT pathway is to be clarified. Binding of BMP4 to its receptors (BMP4 RII), in the presence of LIF, activates SMAD pathway, initiating transcription of Id genes, which in turn to promote expression of the negative factor bHLH that regulate genes transcription for self-renewal of ES cells. BMP4 also support ES cell renewal by sharing the p38-ERK pathway, the alternative pathway of LIF-STAT.

LIF binds to its receptor on the cell surface to form a primary LIF-LIFR ligand, which induces the dimerization with gp130 receptors to become a high affinitive receptor complex. This receptor complex stimulates tyrosine phosphorylation, which actives the transcription factor STAT3 with the mediation of JAK kinases (Lutticken et al., 1994). The activated STAT3 also bind to the adapters SHP2 and Gab1 to stimulate Ras-Erk mitogen-activated protein kinase pathways (Takahashi-Tezuka et al., 1998). STAT3 plays a critical role in the signal transduction by connections to both subsequent activations for ES cell self-renewal (Niwa et al., 1998). ES cells cultured with LIF differentiate when normal STAT3 expression is interrupted. The ES cell self-renewal is promoted by suppression of the SHP2/Erk signaling pathways (Burdon et al., 1999), probably because of its down-regulation of JAK-

STAT signaling (Symes et al., 1997). The cytokine LIF has actions on a variety of cell types (Kishimoto et al., 1994) to stimulate cell differentiation through the mediation of STAT3, which is controversial to its action in ES cells, implying that the LIF-STAT3 action in ES cells selectively activate the transduction signals in favor repression of cell differentiation. The lack of the evidence of direct linkage between normal expansion of the epiblast earl and the requirement of the factors, such as LIF, gp130 and STAT3 in embryogenesis, suggests either the expansion of epiblast is autonomous or controlled by other signaling pathway.

LIF lends its support to BMP4 for a full capacity of repressing ES cell differentiation. BMP4 is a signaling protein of the TGF-β superfamily involved in the inhibition of differentiation of the mouse ES cells (Ying et al., 2003). LIF in the culture medium with serum can suppress differentiation of the mouse ES cells, but LIF fails to maintain the undifferentiated state on withdrawal of the serum, suggesting the involvement of other components in serum for ES cells. BMP4 in serum-free culture medium stimulates ES cells to differentiate, but remain them undifferentiated in the presence of LIF. In the absence of both factors, the ES cells proceed to neural differentiation. For self-renewal of the mouse ES cells, BMP4 binds to SMAD proteins to form a complex that further activates of Id genes to express the negative bHLH factor (Fig.1), which can also be initiated by serum components through multiple pathways. ES cells propagate properly in culture media with expression of Id genes but no added BMP4 (Ying et al., 2003). BMP4 also support the ES cells self-renewal through binding to ERK-MARK of the p38 pathway (Qi et al., 2004). However, BMP4 promotes differentiation of human ES cells to trophoblast cells in culture (Xu et al., 2002), implying that the growth of human ES cells is controlled by factors or mechanism different to the mouse cells.

Transcription Factors Specify Human ES Cells

LIF and BMP4 are not supportive to the human ES cell culture for self-renewal. It becomes clearer that the molecular mechanisms controlling the mouse ES cell self-renewal and pluripotency is different from that in human. In searches for factors involved in human ES cell growth in culture found evidence alternative self-renewal pathways for the human ES cell growth, which is not dependent of LIF and STAT3, agreeing the finding that LIF is not essential for mouse germ cell pluripotency *in vivo* (Berger and Sturm, 1997; Dani et al., 1998).

The initiation and maintenance of human ES cells require feeder cells in the medium and cannot be replaced by either human or mouse LIF. The feeders for human ES cell culture cannot be substituted by mouse blastocyst feeder layer cells either because mouse LIF failed to act on human cells (Layton et al., 1994) or the LIF receptor is expressed in different levels of effectiveness in human ES cell lines (Ginis et al., 2004). It seemed that STAT signal transduction pathway does not suspend the differentiation of human ES cells (Daheron et al., 2004; Humphrey et al., 2004). There is even no active STAT3 found in undifferentiated human ES cells. The feeders for human ES cell culture can be substituted by conditioned medium (Xu et al., 2001). These early observation suggested that human ES cell self-renewal is controlled by some other regulatory mechanisms.

Oct4, an atypical homeodomain protein, plays an important role in ES cell specification. Oct4 was first identified and cloned among the important transcription factors. Oct4 belongs to the POU transcription factor family, which has an octamer recognition sequence in the promoters and enhancers shared in cell-specific genes expressed ubiquitously. The expression of Oct4 is always associated with germline development in the mouse, exclusively found in ICM of early blastocyst, germ cells and the ES cells (Rosner et al., 1990). Oct4 expression is restricted to pluripotent lineages (Palmieri et al., 1994; Yeom et al., 1996). It keeps expressing in germ cells from as early as 4- to 8-cell stages stage in the mouse embryos until the epiblast stage when they start to differentiate. Without Oct4 expression, the mutated mouse ES cells differentiate and the embryo dies at implantation because these embryos are composed of trophoblast with no ICM, from which all germ layers, the mesoderm, ectoderm and endoderm derive (Nichols et al., 1998). The ES cells with down-regulated Oct4 gene expression differentiate to just trophoblast cells in the mouse (Niwa et al., 2000) and in human (Matin et al., 2004). Over-expressed Oct4 in mouse ES cells leads to a fate of extraembryonic endoderm differentiation. Oct4 is an activator and a repressor for gene transcription in conjunction with co-factors, such as the adeno virus E1A (Scholer et al., 1991), Sox-2 (Yuan et al., 1995), Foxd3 (Guo et al., 2002), and HMG-1 (Butteroni et al., 2000). Oct4 activates the target genes when a quantitative balance to its co-factors reached since lower or higher levels of Oct4 failed to activate their targets by forming no active transcriptional complex (Ben-Shushan et al., 1998). Oct4 regulates the process in collaboration with its co-factors. Oct4 plays its role in the ES cell renewal and differentiation by adjusting its expression level to maintain a constant ratio to its co-factors.

Co-existing with Oct4 found in many cell targets, the SRY-related HMG box family protein Sox2 is also involved in the development of ICM (Avilion et al., 2003). Sox2 binds to many gene targets with Oct4 and stimulate the formation of the ICM (Avilion et al., 2003). Sox2 encodes transcription factors with a single HMG DNA-binding domain expressed in the pluripotent lineages of early embryos. It is also expressed in the multipotential cells of the extraembryonic ectoderm (Avilion et al., 2003) and involved in the maintenance of neural progenitors of the central nervous system (Graham et al., 2003). The differentiation of ES cells shows down-regulated Sox2 expression. Sox2-null embryos develop to blastocyst stage, probably driven by maternal Sox2 protein, but fail soon after the implantation (Avilion et al., 2003) with abnormal cellular structure. Sox2-null blastocysts fail to differentiate into ICM in attempts to generate ES cells, suggesting its role in maintaining the pluripotent cells in the earlier embryonic stages.

Nanog, atypical homeodomain protein related to the Nkx subfamily, characterized as a pluripotency factor involved in self-renewal of ES cells (Chambers et al., 2003; Mitsui et al., 2003). Nanog is expressed in the germline cells including the inner cells of the morula and blastocyst, early germ cells, ES cells, embryonic germ cells, and embryonic carcinoma cells in the mouse but it is not expressed in differentiated cells (Chambers et al., 2003; Mitsui et al., 2003). The mouse ES cells with over-expressed Nanog self-renew and maintains the pluripotency without LIF, but their self-renewal capacity is reduced, indicating that those cells are still dependant to LIF for the full renewal capacity. Nanog strengthen the functions in germ cell self-renewal working together with LIF. It seems that the actions of Nanog are mediated rather through other pathways than STAT3 for the LIF (Chambers et al., 2003).

Cells with Nanog expression can propagate in serum-free media without BMP (Ying et al. 2003), but still need Oct4 activity (Chambers et al., 2003). Both Nanog and Oct4 are needed for the self-renewal of mouse ES cells, otherwise they differentiate to extraembryonic endoderm (Mitsui et al., 2003). Embryos without functional Nanog (gene knockout) can develop up to the blastocyst stage and the ICM differentiates completely. Nanog contains two transcription-activating domains (Pan and Pei, 2005) that positively regulate ES cell-specific genes. Nanog is required for the maintenance of pluripotency of the ICM next to Oct4. Therefore Oct4 inhibits ES cell differentiation, Nanog also inhibit ES cell differentiation and maintains its pluripotency. Human Nanog is expressed in ES cells, embryonic cells and embryonic germ cells, and tumor cells (Mitsui et al., 2003). The mouse ES cells with human Nanog expression can partially compensate their need for LIF (Chambers et al., 2003).

Forced expression of Nanog in ES cells lifts the requirement for LIF to maintain their pluripotency, bypassed the LIF-STAT pathway, suggesting that Nanog is a major regulator of the pluripotency (Chambers et al., 2003; Daheron et al., 2004; Mitsui et al., 2003). Target genes bound by Nanog, Oct4 and Sox2 have been recently characterized (Boyer et al., 2005; Loh et al., 2006) and those factors work together on their target genes, forming a regulatory circuit for the maintenance of pluripotency of ES cells (Fig.2). The mouse cell fusion experiment showed more cells with ES cell phenotype than cells with somatic phenotype, those ES cell nuclear proteins are able to reprogram differentiated cells back to stemness (Cowan et al., 2005).

The factors involved in establishing and maintaining pluripotency, and cell reprogramming have been further analyzed by a functional genomics approach (Ivanova et al., 2006) in an attempt to identify novel factors required for self-renewal in mouse ES cells model. By using the microarray techniques, they screened expression of genes in the ES cells and compared their expression levels at the different developmental stages. 65 out of 901 DNA-binding proteins were found rapidly down-regulated when the ES cells differentiate (Ivanova et al., 2006). Using a combination of green fluorescent protein (GFP) labeling and RNA interference (RNAi) techniques to detect the expression decreases in the marked differentiating cells in culture, 6 genes, including 3 previously defined *Nanog, Oct4* and *Sox2* plus *Esrrb, Tcl1* and *Tbx3* genes, were responsible for self-renewal, agreeing with results from previous observations. Further analysis showed all those 6 factors are involved in differentiation suppression either in a distinct program or in collaboration with others (Ivanova et al., 2006). The RNAi techniques revealed hundreds of genes (474) were up or down-regulated by knockdown of *Nanog, Oct4* or *Sox2*, and those genes were not affected by knockdown of *Esrrb, Tbx3* or *Tcl1*. The genes up-regulated by *Esrrb, Tbx3* or *Tcl1* knockdowns were not affected by *Nanog, Oct4* or *Sox2* either. These results confirmed the previously described transcriptional pathways in ES cell self-renewal and the maintenance of their pluripotency. Is there a possible second pathways activated by Esrrb, Tbx3 and Tcl1 blocking the differentiation process? The effect of Esrrb, Tbx3 and Tcl1 in gene transcription can be mended by Nanog, at least indicating levels of communicating mechanism between Nanog and the other factors. Nanog would be a master regulator in the systems.

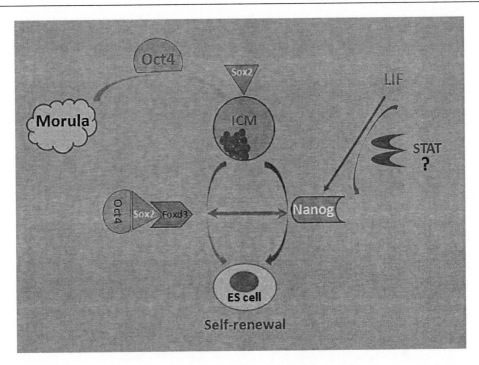

Figure 2. Regulatory network formed by Nanog, Oct4 and Sox in ES cell pluripotency and self-renewal. The transcription factor Nanog, Oct4, Sox2 are identified in the intrinsic regulatory circuit in the maintenance of ES cell pluripotency and self-renewal. Expressing with germline development, Oct4 initiates germ cells with pluripotency and stimulates the formation of the inner cell mass (ICM) of blastocysts by balancing expression levels with co-factors, e.g. Sox2 essential for ICM formation. Nanog, Oct4 and Sox2 bind to the promoter region of same gene as well as their own promoters to form auto-regulation loops to maintain the ES cell properties. Oct4 stimulates Nanog expression by binding to Nanog promoter, Foxd3 and Nanog ensure Oct4 not over-expressed. These factors act together to control the expression of their target genes for their roles in ES cell pluripotency. The cytokine LIF strengthens the capacity of the Nanog-Oct4-Sox2 networks for ES cell self-renewal, either by activating STAT or joining the Nanog pathway, which is to be studied.

Gene Regulation of Sex Determination in Vertebrate

Sexual development is genetically and hormonally controlled processes comprising three main sequential processes. The first phase is the establishment of chromosomal or genetic sex at fertilization, which represents the genetic sex determination with no morphologic indication of sexes. Early gonadal development is characterized by the formation of a bipotential structure, composed of PGCs and somatic precursor cells, identical in males and females, and independent of genetic sex. In the mouse the genital ridge is visible at E10.0, in chicken at about 4 days of incubation, and in human at about 4 weeks of gestation. The second stage is to determine the gonadal fates into testes or ovaries depending on differentiation of somatic cell lineages precisely controlled by some curial transcription factors and signals (Brennan and Capel, 2004; Capel, 2000; Park and Jameson, 2005; Wilhelm et al., 2007b), and gonads are distinguishable morphologically. It occurs at E11.5 in mouse, 5.5 d in chicken and about 6 weeks in human. And the third phase is the hormonal control of proper development into functional gonads with internal and external sexual duct systems. The testis is composed of germ cells and somatic cells including Sertoli cells, peritubular myoid cells, Leydig cells and other interstitial cells; while ovary also comprises germ cells and somatic cells including granulosa cells, theca cells and other interstitial cells (Fig.3).

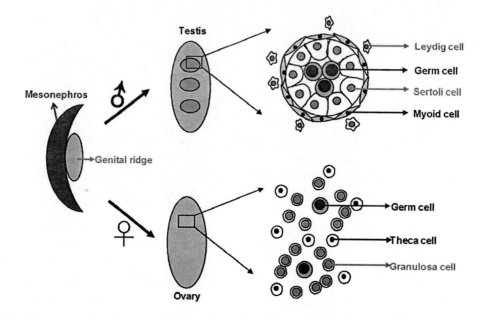

Figure 3. Gonadal development in mammals. During the early development, the gonad is undifferentiated, composed of genital ridge and mesonephros. As development proceeds, gonads have different fates based on the differentiation of somatic cell lineages in the genital ridge. In male pathway supporting cells differentiate into Sertoli cells, and then recruit other cells form interstitial cells. With the proliferation and differentiation of germ cells and somatic cells, the germ cells are surrounded by the Sertoli cells and peritubular myoid cells to form the testis cords. Interstitial cells, such as Leydig cells, distribute on the space between testis cords. In the female pathway, the development of ovary is later than testis. Ovary shows no great change morphologically compared with testis at the same stage. Granulosa cells surrounding germ cell becomes all kinds of ovary nests. Theca cells can be found between the ovary nests.

The urogenital ridge arises as paired structures within the intermediate mesoderm during the early development (Capel, 2000), and then the gonad emerges on the ventro-medial surface of the urogenital central region called mesonephros by thickening of the coelomic epithelium. Adjacent to the gonads are two simple ducts, the Müllerian and Wolffian ducts, the anlagen of female (oviduct, uterus and upper part of the vagina) and male (epididymis, vas deferens and seminal vesicles) reproductive tracts, respectively. At the same time, PGCs specified in the epiblast migrate and reside in the developing gonads between E10.0 and E11.0 in mouse. Up to this stage, the embryo is initially sexually indifferent. With the proliferation of somatic cell lineages and germ cells, the fate decision to a testis or an ovary is triggered by a certain hierarchy of transcription factors or signals, resulting in the differentiation of supporting cell precursors, and the secretion of a hormone. In male pathway, supporting cells differentiate into Sertoli cell precursors, which secrete the Müllerian inhibiting substance (MIS) and induce regression of the Müllerian duct. Then the interstitial cell lineage in the testis differentiates into Leydig cell precursors, leading to production of testosterone, thereby promoting development of Wolffian duct derivatives. Meanwhile, Sertoli cells polarize and aggregate around germ cells, thereby causing the reorganization of the gonad into two compartments: the testis cords, composed of Sertoli cells and germ cells; and the interstitial space between cords, mainly composed of Leydig cells. In female pathway, the absence of MIS permits continued development of the Müllerian duct, whereas the absence of testosterone leads to degeneration of the Wolffian duct. Owing to ovary begins to differentiate later than testis, the first cellular event, germ cells entering into meiosis, occurs at around E13.5. Additionally, the migration of mesenchymal cells and mesonephric cells to the gonads also plays an important role in proper gonadal development.

Dynamic Expression of Factors Involved In Sexual Development

There are massive crucial factors directing sexual development (table 2). These factors include genes expressed by different somatic cells and germ cells, signal molecules binding to receptors on the cell surface and thereby activating/inactivating signaling systems to regulate gene expression, specialized membrane junctions that connect cells and allow them to communicate, and genomic imprinting. Finding these factors involved in this process is the first step to unveil the mechanism of sex determination and differentiation. Among these factors, the most important discovery is the male sex-determining gene *SRY/Sry* (in order to simply genes described in the text, we use the lowercase except the first letter for mouse and the uppercase for human and other species), sex determining gene on Y chromosome, in mammals. XY mice with *Sry* expression develop testes and XX mice without *Sry* expression develop ovaries in wild type mice. Mutation analyses show that XY mice with no functional or enough *Sry* expression develop ovaries while the addition of *Sry* to XX mice initiates the male pathway (Koopman et al., 1991; Koopman et al., 1990). In addition, *Sry* has a narrow expression window from E10.5 to E12.5 when sex determination occurs in mice (Hacker et al., 1995). All of these data demonstrate that *Sry* is the master gene to initiate testicular pathway. In addition, there are lots of genes which are the upstream or downstream of *Sry* to regulate sexual development.

In mammals, several genes including the Wilms tumor suppressor gene *Wt1*, steroidogenic factor 1 (*Sf1*, also known as *Ad4BP* or *NR5A1*), Lim homeobox gene *Lhx9*, empty-spiracles homeobox gene 2 (*Emx2*) and the member of the polycomb group *M33* have been implicated in the development of indifferent gonad prior to sexual differentiation. Mutation study suggests a role of *Wt1* in specifying the coelomic epithelium cells in the developing urogenital ridge and ensuring their survival (Hammes et al., 2001; Nordenskjold et al., 1995). Disruption of mice *Sf1* (–/–) results in complete adrenal and gonadal agenesis, male-to-female sex reversal and persistence of Müllerian structures in males (Luo et al., 1994; Sadovsky et al., 1995). In mice lacking *Lhx9* function, germ cells migrate normally, but somatic cells of the genital ridge fail to proliferate and a discrete gonad fails to form (Birk et al., 2000). In *Emx2* mutants, the thickening of the coelomic epithelium, character of the first stage of the gonadal development, is not prominent, and the Müllerian duct never forms (Miyamoto et al., 1997). Deficient *M33* induces male to female sex reversal (Katoh-Fukui et al., 1998). Therefore, transcription factors play important roles in specifying cells and tissues fates prior to the sex determination.

Table 2. genes involved in sexual development

genes	product	Expression pattern	functions	References
Early gonadal development				
Wt1	Transcription factor	genital ridge; coelomic epithelium cells; Sertoli cells; germ cells	survival of genital ridge	Hammes et al., 2001; Nordenskjold et al., 1995
Sf1	Nuclear receptor	coelomic epithelium cells; Sertoli cells; Leydig cells	survival of bipotential gonad and proliferation cells	Luo et al., 1994; Sadovsky et al., 1995
Lhx9	Transcription factor	early genital ridge	somatic cell proliferation	Birk et al., 2000
Emx2	Transcription factor	coelomic epithelium cells and mesenchyme cells	proliferation of somatic cell in genital ridge	Miyamoto et al., 1997
M33	Transcription factor	No answer	formation of genital ridge	Katoh-Fukui et al., 1998
Sex determination and differentiation				
Sry	Transcription factor	Genital ridge; Sertoli cells	male sex-determining factor; somatic cell differentiation	Koopman et al., 1991; Sinclair et al., 1990
Sox9	Transcription factor	genital ridge; Sertoli cells	Sertoli cell proliferation and differentiation	Barrionuevo et al., 2006; Kent et al., 1996; Vidal et al., 2001
Amh	Hormone	Sertoli cells	regression of the Müllerian duct	Behringer et al., 1994; Josso et al., 1998; Mishina et al., 1996
Dmrt1	Transcription factor	Sertoli cells; germ cells	Testicular development; male-determining gene in Medaka	Matsuda et al., 2002; Raymond et al., 1999; Smith et al., 1999
Dax1	Nuclear receptor	Sertoli cells; Leydig cells	Testis development	Swain et al., 1996; Meeks et al., 2003
Wnt4	Signaling molecule	mesonephric mesenchyme; Leydig cells	Ovary development; testicular development	Jeays-Ward et al., 2004; Vainio et al., 1999

Table 2. (continued)

genes	product	Expression pattern	functions	References
Early gonadal development				
Fgf9	Signaling molecule	Sertoli cells	Sertoli cell proliferation	Colvin et al., 2001;
Dhh	Signaling molecule	Sertoli cells	Early testis formation; spermatogenesis	Bitgood et al., 1996; Yao et al., 2002
Gata4	Transcription factor	Sertoli cells; Leydig cells	Testis development	Tevosian et al., 2002; Viger et al., 1998

Some genes expressed in gonads in both sexes before sex determination change their expression pattern in sex-specific manner after sex determination, either up- or down-regulate to exert effects on the sexual development. *Sox9*, Sry-related gene 9, is expressed in the genital ridge of both XY and XX embryos in the early stage at about E10.5 (Kent et al., 1996), which is a little later than *Sry* expression. And by E11.5 it is up-regulated in the gonads in the XY embryos but not in the XX embryos (Kent et al., 1996; Morais da Silva et al., 1996). Interestingly, it is expressed in the cytoplasmic compartment of somatic cells in undifferentiated gonads in both male and female. At the time of sexual differentiation, *Sox9* moves into the nucleus of male pre-Sertoli cells, whereas it remains in the cytoplasm and then its expression decreases in female (de Santa Barbara et al., 2000; Malki et al., 2005a). Mutation of *Sox9* (-/-) led to sex reversal of male to female in XY embryos (Barrionuevo et al., 2006) and ectopic expression of *Sox9* resulted in female to male sex reversal in XX embryos (Vidal et al., 2001), indicating a critical role of *Sox9* in male development (Qin and Bishop, 2005). *Dmrt1*, the doublesex and mab-3 related transcription factor 1, is expressed only in the gonads examined so far in many species(Matsuda et al., 2002; Raymond et al., 1999; Smith et al., 1999), implicating its conserved role in gonad differentiation and function (Ferguson-Smith, 2007). Moreover, *Dmrt1* expression profiles exhibited a dynamic, sexually dimorphic pattern that signified a role in testis differentiation during gonadogenesis (Lei et al., 2007). Deficient *Dmrt1* causes testicular defects (Raymond et al., 2000) or male to female sex reversal (Matsuda et al., 2002), further confirming its role in testicular development. *DAX1* (dosage-sensitive sex reversal, adrenal hypoplasia congenita critical region on the X chromosome, gene 1) was originally considered as the candidate for ovary-determining gene based on sex reversal data in humans (Muscatelli et al., 1994; Swain et al., 1996). However, gene knock out and transgenic studies in mice show that *Dax1* is not required for normal ovarian development (Yu et al., 1998). Null mutation of *Dax1* has little effect on the formation of the ovary; in contrast, it causes severe testicular dysgenesis (Jeffs et al., 2001; Meeks et al., 2003a; Meeks et al., 2003b). This evidence strongly disagreed the proposed role of *Dax1* in ovarian development, but supports its role in testicular differentiation. Anti-Müllerian hormone (*Amh*), also called Müllerian inhibiting substance (*Mis*), is a distant member of the transforming growth factor-β family, regressing the development of Müllerian duct in testicular pathway (Behringer et al., 1994; Josso et al., 1998; Mishina et al., 1996). The expression pattern of *AMH/Amh* varies in different species (De Santa Barbara et al.,

1998; de Santa Barbara et al., 2000; Oreal et al., 2002; Oreal et al., 1998; Western et al., 1999), implying different mechanisms in different species for the regulation of sexual development (Takada et al., 2005). The deletion of fibroblast growth factor 9 (*Fgf9*) led to male to female sex reversal, but did not affect the ovarian development (Colvin et al., 2001). *Fgf9* appears to act as downstream gene of *Sry* to stimulate mesenchymal proliferation, mesonephric cell migration, germ cell survival and Sertoli cell differentiation in the embryonic testis (Colvin et al., 2001; DiNapoli et al., 2006). In addition, *Fgf9* expression pattern exhibits dynamics during gonadal development in the mouse (microarray data) (Nef et al., 2005; Ottolenghi et al., 2007) and the marsupial, tammar wallaby (unpublished data by real time PCR, Chung and Yu et al). Briefly, it shows higher level of expression in testis than that in ovary with dramatically increases in testis during the window of sex determination. The zinc-finger transcription factor *Gata4*, binding to the consensus motif GATA, is expressed in somatic cells but not in germ cells of XX and XY mouse embryos as early as E11.5, maintained in the testis development, and markedly down-regulated in the ovary on E13.5 (Viger et al., 1998). It has the similar temporal and spatial localization throughout gonadogenesis in porcine (McCoard et al., 2001), suggesting the conserved role during sexual development (Bielinska et al., 2007; Miyamoto et al., 2008). Moreover, interaction between *Gata4* and *Fog2* (Friend of Gata2) is required for normal gonadal development (Manuylov et al., 2007; Tevosian et al., 2002).

Gene Regulation in Sexual Development

Sex determination and differentiation is the molecular and cellular events of differentiation, migration, proliferation and communication to orchestrate the development of the dimorphic gonads. In mammals, testicular development is initiated by the action of SRY protein with a high-mobility group (HMG)-box DNA binding domain characteristic of the SOX family of transcription factors (Koopman et al., 1991; Sinclair et al., 1990), which in conjunction with other factors activates the complex hierarchical networks of transcriptional regulation to determine the plasticity of gonadal fates (Parker et al., 1999). Sex determination is the somatic cell fate decision, and germ cells just migrate from outside of gonads and continue to proliferate before E13.5 in mice. The axis of testicular development is the *Sry-Sox9* expression to promote differentiation and proliferation (Fig.4). The questions remaining in male sex determination are how the switch of Sry/Sox9 expression is turned on or off, and its activation of following events. Furthermore, the female pathway remains to be elucidated.

It has become increasingly evident in recent years that development is under the epigenetic control. Epigenetics is the study of heritable states of gene activity by modifications of DNA and chromatin other than changes in DNA sequence itself. The chromatin structure is determined by the components of DNA and modified by the epigenetic events. The conformation of chromatin makes it possible to access the appropriate transcription factors, initiate transcription and participate in gene regulation and cellular differentiation. SRY, for example, binds to the minor groove of the DNA and further bends the chromatin remodeling to switch on/off gene expression (Pontiggia et al., 1994). In addition, SOX9 and transcription factor p300 cooperatively interacts with chromatin and activates transcription via regulation of chromatin modification (Furumatsu et al., 2005). Therefore, any modification of chromatin would affect these transcription factors to access the chromatin leading to regulating gene expression.

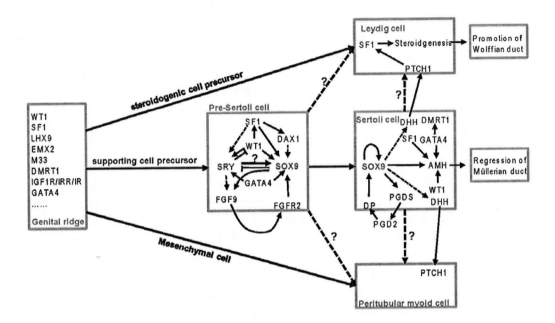

Figure 4. Mechanism of sex determination in early testicular development. Sex determination is the process of somatic cell decision from undifferentiated precursors to differentiated cells. Before sex determination occurs, lots of genes crucial sexual development have been expressed, for example, WT1, SF1, LHX9, EMX2, M33 and DMRT1. During sex determination, SRY and SOX9 play the central roles in initiating testicular pathway. SRY induces SOX9 expression by direct or indirect action whereas SOX9 expression inhibits SRY expression by the feedback loop. WT1 and GATA4 are the upstream genes to promote SRY expression. WT1, SF1 and GATA4 also affect SOX9 gene expression by binding to SOX9 promoter region or enhancer region. SOX9 regulates FGF9 gene expression while FGF9 also gives the feedback to SOX9 by its receptor FGFR2. Similarly, SOX9 modulates PGDS expression while PDGS feedbacks to SOX9 via the receptor DP. Moreover, SOX9 expression is also regulated by its autocrine loops. The product of AMH induces the regression of Mullerian duct; testosterone produced by steroidgenesis promotes wolffian duct development. Both of them play important roles in male internal and external ducts development. Lots of genes, SOX9, SF1, GATA4 and WT1 take participate in regulation of AMH expression at transcriptional level via binding to its promoter region. Meanwhile, DHH gene is expressed in Sertoli cells, functioning by its receptor PTCH1 expressed in Leydig cells and peritubular myoid cells to communicate with other somatic cells in developing gonads. The SF1 has a critical role in steroidgenesis and development, regulated by PTCH1 and WT1. In addition, proliferation and differentiation of Sertoli cells recruit differentiation and aggregation of Leydig cells and peritubular myoid cells. However, it still remains largely unknown how to interact between them in this process. Solid lines or arrows represent pathway confirmed by experiments, and dash lines or arrows stands for unknown pathway or hypothesis. AMH, anti-Müllerian hormone; DAX1, dosage-sensitive sex reversal, adrenal hypoplasia congenita critical region on the X chromosome, gene 1; DMRT1, the doublesex and mab-3 related transcription factor 1; DHH, desert hedgehog; DP, prostaglandin D2 receptor; EMX2, empty-spiracles homeobox gene 2; FGF9, fibroblast growth factor 9; FGFR2, FGF 9 receptor 2; GATA4, GATA binding protein 4; IGF1R, insulin-like growth factor 1 receptor; IR, insulin receptor; IRR, insulin receptor-related receptor; LHX9, Lim homeobox gene 9; M33, the member of the polycomb group 33; PGD2, prostaglandin D2; PGDS, prostaglandin D synthase gene; PTCH1, patched, *Drosophila*, homolog of 1; SF1, steroidogenic factor 1; SOX9, SRY (sex determining region Y) - box 9; SRY, sex determining region Y; WT1, Wilms tumor suppressor gene 1.

Interactions among transcription factors produce complex networks to control timing and level of gene expression at a certain space during sex determination and differentiation. WT1 plays a central role in the development of several organs, especially in urogenesis (Roberts, 2005). It appears to transactivate and regulate *SRY/Sry* gene expression (Hossain and Saunders, 2001; Matsuzawa-Watanabe et al., 2003; Shimamura et al., 1997), and the feedback from *Sry* down-regulates *Wt1* expression in the murine embryonic mesonephros-derived M15 cell line (Ito et al., 2006). WT1 also transactivates the target gene expression via binding to sites in the corresponding promoter region, such as *Sf1* (Wilhelm and Englert, 2002), *Sox9* (Gao et al., 2006), *Dax1* (Kim et al., 1999), *Amh* (Hossain and Saunders, 2003; Shimamura et al., 1997). In addition, *Wt1* and *Gata4* cooperatively regulates *Sry* and *Amh* during sex determination and differentiation (Miyamoto et al., 2008). The SF1 has a critical role in steroidogenesis and development, mutation of *Sf1* led to male to female sex reversal (Ozisik et al., 2003). It participates in gene regulation in the gonadal development. SF1 and SOX8/SOX9 bind to the promoter region of *Amh* to regulate its expression (De Santa Barbara et al., 1998; Schepers et al., 2003); and *Dax1* expression is dependent on the *Sf1* in the developing gonads (Hoyle et al., 2002); A recent publication showed a synergistic action of *Sry* and *Sf1* on a *Sox9* specific enhancer promote *Sox9* expression, which in turn to activate the testicular pathway (Sekido and Lovell-Badge, 2008). SOX9 induces the Sertoli cell differentiation and subsequent testis cord formation (Chaboissier et al., 2004; Qin and Bishop, 2005; Vidal et al., 2001), playing a central role in testicular development. It regulates prostaglandin D synthase gene (*PGDS*) at transcriptional level to ensure sufficient differentiation of Sertoli cells in male pathway (Wilhelm et al., 2007a; Wilhelm et al., 2005), whereas PDGS produces Prostaglandin D2 that induces nuclear import of SOX9 via its cAMP-PKA phosphorylation (Malki et al., 2005b; Wilhelm et al., 2007b). The human *SOX9* proximal promoter is regulated by the cyclic-AMP response element binding protein (CREB), transcription factor Sp1 (Piera-Velazquez et al., 2007) and two CCAAT-binding factors (Colter et al., 2005). *SRY* induces *SOX9* expression via direct or indirect regulatory mechanism while *SOX9* inhibits *SRY* expression and maintains its own expression in an auto-regulatory loop (Wilhelm et al., 2005). *Sox9* is also coordinated with factors, *Sry*, *Dax1* and *Tda1* to differentiate pre-Sertoli cells into Sertoli cells during sex determination (Bouma et al., 2005).

Alternative Splicing to Regulate the Gonadal Fates

Splicing is one of the steps in charge of removing introns from pre-mRNA to mature mRNA. It occurs co-transcriptionally or after the precursor transcript is released from the chromatin to the nucleoplasm (Minvielle-Sebastia and Keller, 1999). In higher eukaryotic cells, the same pre-mRNA can produce more than one mature mRNA by alternative splicing via variable splicing sites. It exists widely in eukaryotic genomes such that more than half of the genes in humans are alternatively spliced (Johnson et al., 2003; Modrek and Lee, 2002). This provides additional dimensions to the regulation of gene expression as different mRNA isoforms may have distinct half-lives, following different export pathways and interacting with different factors for intracellular targeting to regulate translation (Black, 2003). How alternative splicing is regulated and what is the function of individual mRNA isoforms are becoming the central questions to gain insight into the mechanism of this process.

The *WT1* gene is the best paradigm in mammals to elicit its function by alternative splicing in organogenesis, especially in sexual development. It may encode up to 24 protein isoforms sharing four C-terminal C_2H_2 zinc fingers and an N-terminal proline/glutamine-rich regulatory region. Two splice variants, with (+) or without (-) three amino acids Lys-Thr-Ser (KTS), among them are highly conserved over 450 million years in vertebrates, suggesting that they perform some distinct functions. The WT1 (-KTS) shows very high DNA binding affinity while very low in WT1 (+KTS) isform; The WT1 (-KTS) displays transcriptional activation and co-activator with SF1 on *AMH* promoter as well as retention of cell survival, but similar function in WT1 (+KTS); The WT1 (+KTS) has high activity of co-localisation with splicing speckles and strong ability to bind to the splice factor U2AF65 whereas the WT1 (-KTS) is very low or weak (Hastie, 2001). Taken together, a role for -KTS protein is in transcription and the +KTS protein is in RNA processing.

In genitourinary development +KTS and -KTS have overlapping and distinct roles (Hastie, 2001). Homozygous null WT1 mice lack kidney, adrenal glands, spleen and gonad, and die at the midgestation because of failure of cardiac abnormality (Hammes et al., 2001; Herzer et al., 1999; Moore et al., 1999). Mutant mice who just remove WT1 (+KTS) or WT1 (-KTS) can survive through to birth (Hammes et al., 2001), demonstrated that both forms of WT1 perform identical functions, initiating development of the heart, early nephrogenesis, spleen and the adrenal. However, the difference between the two sets of mice (mutation of +KTS or -KTS) becomes obvious at the later stages of genitourinary development. XY Frasier homozygous mice lacking WT1 (+KTS) isoform develop along the female pathway (Hammes et al., 2001). Furthermore, a dramatic reduction of *Sry* expression and *Wt1* expression in gonads implied that WT1 (+KTS) may be one of the crucial factors operating upstream of the Sry (Hammes et al., 2001; Hastie, 2001). Homozygous mice lacking WT1 (-KTS) produced much smaller gonads, consisting mainly of undifferentiated mesenchyme and few differentiated tissue, suggested the role of WT1 (-KTS) in inhibition of apoptosis (Hammes et al., 2001; Hastie, 2001). However, the roles of WT1 (-KTS) in male pathway remains to be clarified.

Dmrt1 is a highly conserved gene containing the DM domain among phyla (Raymond et al., 1998; Smith et al., 1999). Dmrt1 plays a crucial role in sexual development (De Grandi et al., 2000; Moniot et al., 2000; Raymond et al., 1999; Raymond et al., 2000), and is thought to be a master gene in testicular development in Medaka (Matsuda et al., 2002; Nanda et al., 2002). However, the mechanism of the Dmrt1 in sexual development remains unknown. Interestingly, recent studies showed that there are wide phenomena of alternative splicing in this gene during sexual development in distinct spices including human (Cheng et al., 2006), mouse (Lu et al., 2007), chicken (Zhao et al., 2007), zebrafish (Guo et al., 2005), rice field eel (Huang et al., 2005) and Medaka (personal communication, Zhou et al). A high conserved region in 5'-terminus, and that splicing occurs at 3'-terminus are shared common characteristics among these isoforms. Furthermore, different expression level exists in different isoforms during sexual development, such as Chicken *Dmrt1* and rice field eel

Dmrt1, suggesting that alternative splicing plays pivotal roles in gene regulation and organogenesis.

In Drosophila sex lethal gene (*SXL*), transformer gene (*TRA*) and doublesex gene (*DSX*) are the most important genes during sex determination (Fig.5). By alternative splicing, they have different transcripts functioning in different pathway (Baker, 1989; Herbert and Rich, 1999; Schutt and Nothiger, 2000). *SXL* is the key gene sitting on top of the sex determination pathway to initiate the downstream gene expression and directs the female pathway. *SXL* in female is produced by splicing the exon 2 which contains stop code, and encoding the protein with biological activity; whereas *SXL* in male is terminated because of stop codes in exon 2. *TRA* is also produced by female specific splicing to jump over the stop code-containing exon 2. In the same way, *TRA* in male is inactive because of the stop code in exon 2. Due to *TRA* and *TRA-2* gene expression, *DSX* gene in female consists of exon1-4 to switch on the female pathway whereas *DSX* in male is produced to skip the exon 4 via alternative splicing to produce the protein for the male pathway.\

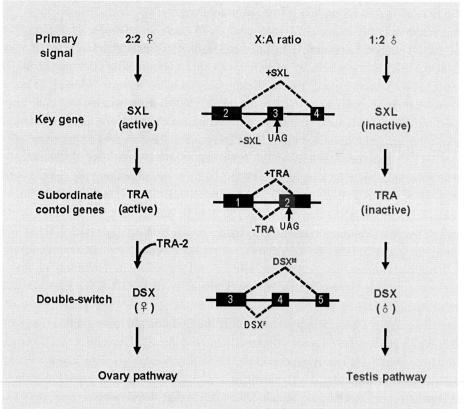

Figure 5. Sex determination in *Drosophila* via alternative splicing cascade. The sex is determined genetically by the ratio of X chromosome to Autosome (2:2 for female and 1:2 for male) in fly. This ratio directs the SLX activity in the early development by alternative splicing of exon3 which includes stop code. Then the active SLX determines the strategy of splicing in exon 2 and produces active TRA. TRA and TRA-2 cooperates to produce female double sex gene which include exon4, and initiate the ovary pathway. No SLX activity results in no TRA gene expression, thereby producing male double sex gene with no exon 4 by alternative splicing, ultimately initiating the male pathway.

Signaling Networks and Crosstalk in Gene Regulation

Signaling pathway refers to extracellular signaling molecules or stimuli binding to the cell surface receptors that trigger the intracellular cascade events resulting in regulation of gene expression. Normally, intercellular communication relies on the creation of specific signals by a signaling cell. The signals are registered by a target cell and are thereupon transmitted and processed further with the help of intracellular signal chains. In higher organisms intercellular signaling pathways play crucial roles in coordinating and regulating cell division. The pathways ensure that cells divide synchronously and, if necessary, arrest the cell division and enter a resting state. Cellular communication assumes great importance in the differentiation and development of an organism. The development of an organism is based on genetic programs that always utilize inter- and intracellular signaling pathways. Signal molecules produced by one type of cells influence and change the function and morphology of other cells in the organism. The proteins of a signaling chain can receive signals from more than one upstream signaling partner and transmit signals to more than one downstream effectors. This property allows a branching of signaling pathways and the formation of signaling networks.

During the organogenesis, signaling networks plays a key role in precise regulation of processes of the cell proliferation, differentiation, migration and death. The extracellular signaling factor FGF9 exerts an influence on somatic cell proliferation and differentiation (Colvin et al., 2001). And further analysis suggested that the FGF9 signaling is mediated by the receptor FGFR2 during male sex determination (Kim et al., 2007). Another important signal molecule WNT4 plays a significant role in ovarian development, and a distinct role in testicular development (Jeays-Ward et al., 2004; Vainio et al., 1999). The coordination and antagonism between FGF9 and WNT4 regulate mammalian sex determination (Kim et al., 2006) by forming the signaling networks. And the balance of these signaling molecules directs the morphogenesis of testis and ovary (DiNapoli and Capel, 2008). Moreover, members of other three signaling pathways have also been shown to play a role in early testis formation: platelet-derived growth factor receptor alpha (Pdgfra) (Brennan et al., 2003), Desert hedgehog (Dhh) (Yao et al., 2002) and the insulin growth factor receptors (Nef et al., 2003).

Besides these, lots of signals assume the importance of cellular activities or fates, including endocrine signaling acting the target cells via the long distance, paracrine signaling functioning by the medium range and autocrine signaling communicating the same cell type. In the hypothalamus-pituitary-gonads axis, the endocrine signaling factors FSH and LH plays key roles in proper maturation of germ cells. During early development of gonads, the precursors of Sertoli cells proliferate via the autocrine signaling mediators, such as SOX9 (Wilhelm et al., 2007b) and FGF9 (Polanco and Koopman, 2007). Furthermore, Sertoli cells establish the male pathway to reach the threshold number via autocrine signaling. And it also influences other cell types by paracrine signaling, Dhh and the receptor patch1 (Yao et al., 2002) as an example during the Leydig cell differentiation.

Conclusion

The identification and characterization of germ plasmic components in germ cells is a significant step towards a better understanding of germ cell formation. By resolving the germ plasm into molecules, the interactions of those molecules and their roles in germ cell determination are studied. The established characteristics of the germ cells also provide molecular markers for specifying germ plasms and for the identification of germ cells in species without germ plasm. Amongst those molecules as components of germ plasm, as defined in Drosophila "pole plasm" including Oskar, Vasa, Nanos and Tudor, the Vasa, a member of the DEAD-family protein of ATP-dependent RNA helicase, has been well characterized and clearly showed its involvement in the formation and maintenance of germ cells, which stimulated the identification and characterization of *vasa* in many other species. It is striking that gene expression and protein distribution of VASA follows the germline lineages in all species investigated. As a protein of DEAD box family, the VASA is thought to bind target mRNAs to modulate the initiation of their transcription and translation. The functional motifs within 2 domains of the DEAD box protein are well conserved across a broad range of living organisms, suggesting conserved fundamental functions in germ cell specification. It seems that the VASA is a master regulator of germ cell formation. Further studies on VASA-related proteins will make great contributions to a better understanding of the molecular mechanisms of specification and differentiation of the germ cells across species.

ES cells self-renew with pluripotency transiently in early developing embryos and then proceed to further development. ES cells differentiate unless the differentiation is inhibited. The establishment and maintenance of ES cells *in vitro* require growth factors secreted by the feeder layer cells, which can be replaced by the cytokine LIF. The LIF is also found to improve the pluripotent lineage *in vivo*. The serum in the ES cell culture medium can be substituted by the growth factor BMP4 to maintain the pluripotency by activating the p38 pathway mediated by ERK-MARK receptors. However, these factors failed to promote the human ES cell culture, leading to the discovery of other important factors, such as Oct4, Sox2 and Nanog. The germline expressed Oct4 plays a similar role in both mouse and human, regulates levels of gene expression in conjunction with its co-factors, including the HMG protein Sox2, by quantitative balancing. The involvement of Sox2 in the ES cell growth is showed by its role in early embryonic development. These genes have similar expression patterns with different features between mouse and human, strongly suggesting that they play regulatory roles by initiating different signal pathways. The Nanog is only expressed in pluripotent cells otherwise they differentiate to endoderm lineages. The Nanog affects ES cell self-renewal neither through the STAT3 nor BMP4 pathways independent LIF, making it a major regulator via an alternative signaling pathway. There is no hard evidence for Nanog to act through STAT activation although STAT3 and T (Brachyury) binding sites found upstream of the translation start site. However, Nanog maintains the pluripotency of mouse ES cells, indicating that Nanog might be a downstream regulator of LIF-STAT3 pathway. Nanog, in conjunction with the initiation factors Oct4 and Sox2, binds to the target genes to form an intrinsic regulatory circuit for self-renewal of the ES cell.

Although morphologically distinct males and females are observed throughout the animal kingdom, the gonads are derived from the common primordia that have the potential ability to form testes or ovaries. This situation presents a unique opportunity to study molecular mechanisms and regulatory networks in organogenesis. The discovery of *Sry* in 1990 is the milestone for understanding the mechanism of sex determination and differentiation at molecular level in mammals. However, there is no *Sry* homology gene in non-mammals. Accordingly, another issue is to look for a testis determinant in non-mammals. Unfortunately, till now there is no universal gene like *Sry* except *Dmrt1* gene in Medaka. Nevertheless, substantial progress has been made in identifying genes and regulatory networks that drives male-specific differentiation. *Sox9*, for instance, is a downstream gene of *Sry* critical for Sertoli cell differentiation. *Amh* is another gene in charge of regression of Müllerian duct in male development. Signal molecules, FGF9, DHH and PDGH, promote Sertoli cells differentiation and recruit other cells in the XY gonads to establish testis pathway. Alternative spliced WT1, WT1 (+KTS) regulating the timing and level of *Sry* expression and WT1 (-KTS) maintaining the survival of gonads, plays an important role in gonadal development. Therefore sex determination and differentiation is a complex interplay between transcription factors, secreted signaling molecules, hormones and receptors to regulate gene expression at multiple levels including transcriptional and post-transcriptional level. And they are cooperated to facilitate the proper testicular development. In contrast, least is known about the ovarian pathway, which has been regarded a default or passive pathway although more and more studies suggest an active female pathway. It still remains unknown about genes or factors controlling the female development and whether a master gene, as Sry in the male pathway, plays a commanding role in the process. All of these provide new challenges in future studies.

SRY is known as the key factor to initiate differentiation of Sertoli cells and then activates the downstream gene *Sox9* expression in the male pathway; it still remains to be clarified the direct upstream genes that switch on *Sry* gene expression, and whether *Sox9* is a direct target of SRY. Similarly, the regulation and interaction between most of transcription factors or signaling molecules identified so far remains largely unknown in organogenesis, which needs to be studied in the future. Fortunately, there are new coming techniques to be employed in modern biology. The first one is the tissue- or cell- culture to mimic the *in vivo* microenvironment of organisms. The gonad can be cultured several days under various conditions to be used in functional studies. Secondly, expression screens such as microarray analyses may discover more candidate genes that show sex-specific expression patterns. Moreover, bioinformatics provides possibility to do globing genomic analyses and find the evolutionary pathways and functional elements to regulate gene expression, which would accelerate studies on mechanisms of the early development. Furthermore, chromatin immunoprecipitation helps to characterize transcriptional regulation of target genes, especially to find target genes of SRY for recent antibody to mouse SRY. Small non-coding RNAs are common and effective modulators of gene expression, affecting on mRNA degradation and translation as well as transcription rates by altering chromatin structure. Therefore RNAi (RNA interfere), mimicking the mircoRNA *in vivo*, is new technique engaged in functional study. Therefore, all of above tools will contribute to identifying the new players involved in sexual development, establishing networks of gene interactions, and ultimately integrating all of knowledge to gain insight into mechanisms of sex determination and differentiation in organogenesis.

To study the regulation of gene expression, we found another remarkable characteristic of conservation. Firstly, genes, their structure and expression patterns, are highly conserved, resulting in a conserved role in development. *Vasa* homology, as a sample, was found from fruit fly to human, and maintained the conserved role of germ cell specification. Secondly, we can find new genes in different species by conservation, establish their regulatory pathways and further understand the mechanisms of their biological actions. *Sry* was firstly cloned in human, functional studies of *Sry* were then investigated in the mouse and finally applied in sex determination in human, which made a better understanding of gene evolutionary pathway and evolutionary events. The analysis of *Sox3*, *Sox9* and *Sry* genes in different species suggest that *Sox3* is the common ancestor of *Sry*, and the sex chromosome was derived from the autosome.

In conclusion, the development of multicellular organisms from a single cell (the fertilized eggs) is a perfectly accurate process of regulated gene expression. Study of germ cell specification and sex determination contributes greatly to the understanding of the molecular mechanisms of early development, precise interactions between transcription factors, signaling molecules, hormones, non-coding RNAs and epigenetic events at a certain time and space. However, we will still have a long way to go to understand these processes because new developed molecular biology techniques would identify more unknown players to expand the regulatory networks, which brings new concepts towards a better understanding of this fundamental biological process and will definitely be the future direction of researches in developmental molecular biology.

References

Anderson, R., Copeland, T. K., Scholer, H., Heasman, J. and Wylie, C. (2000). The onset of germ cell migration in the mouse embryo. *Mech Dev* 91, 61-8.

Anderson, R., Fassler, R., Georges-Labouesse, E., Hynes, R. O., Bader, B. L., Kreidberg, J. A., Schaible, K., Heasman, J. and Wylie, C. (1999). Mouse primordial germ cells lacking beta1 integrins enter the germline but fail to migrate normally to the gonads. *Development* 126, 1655-64.

Asaoka, M., Sano, H., Obara, Y. and Kobayashi, S. (1998). Maternal Nanos regulates zygotic gene expression in germline progenitors of Drosophila melanogaster. *Mech Dev* 78, 153-8.

Ashburner, M., Thompson, P., Roote, J., Lasko, P. F., Grau, Y., el Messal, M., Roth, S. and Simpson, P. (1990). The genetics of a small autosomal region of Drosophila melanogaster containing the structural gene for alcohol dehydrogenase. VII. Characterization of the region around the snail and cactus loci. *Genetics* 126, 679-94.

Avilion, A. A., Nicolis, S. K., Pevny, L. H., Perez, L., Vivian, N. and Lovell-Badge, R. (2003). Multipotent cell lineages in early mouse development depend on SOX2 function. *Genes Dev* 17, 126-40.

Baker, B. S. (1989). Sex in flies: the splice of life. *Nature* 340, 521-4.

Barrionuevo, F., Bagheri-Fam, S., Klattig, J., Kist, R., Taketo, M. M., Englert, C. and Scherer, G. (2006). Homozygous inactivation of Sox9 causes complete XY sex reversal in mice. *Biol Reprod* 74, 195-201.

Beams, H. W. and Kessel, R. G. (1974). The problem of germ cell determinants. *Int Rev Cytol* 39, 413-79.

Behringer, R. R., Finegold, M. J. and Cate, R. L. (1994). Mullerian-inhibiting substance function during mammalian sexual development. *Cell* 79, 415-25.

Ben-Shushan, E., Thompson, J. R., Gudas, L. J. and Bergman, Y. (1998). Rex-1, a gene encoding a transcription factor expressed in the early embryo, is regulated via Oct-3/4 and Oct-6 binding to an octamer site and a novel protein, Rox-1, binding to an adjacent site. *Mol Cell Biol* 18, 1866-78.

Bendel-Stenzel, M. R., Gomperts, M., Anderson, R., Heasman, J. and Wylie, C. (2000). The role of cadherins during primordial germ cell migration and early gonad formation in the mouse. *Mech Dev* 91, 143-52.

Berger, C. N. and Sturm, K. S. (1997). Self renewal of embryonic stem cells in the absence of feeder cells and exogenous leukaemia inhibitory factor. *Growth Factors* 14, 145-59.

Bielinska, M., Seehra, A., Toppari, J., Heikinheimo, M. and Wilson, D. B. (2007). GATA-4 is required for sex steroidogenic cell development in the fetal mouse. *Dev Dyn* 236, 203-13.

Birk, O. S., Casiano, D. E., Wassif, C. A., Cogliati, T., Zhao, L., Zhao, Y., Grinberg, A., Huang, S., Kreidberg, J. A., Parker, K. L. et al. (2000). The LIM homeobox gene Lhx9 is essential for mouse gonad formation. *Nature* 403, 909-13.

Black, D. L. (2003). Mechanisms of alternative pre-messenger RNA splicing. *Annu Rev Biochem* 72, 291-336.

Bouma, G. J., Albrecht, K. H., Washburn, L. L., Recknagel, A. K., Churchill, G. A. and Eicher, E. M. (2005). Gonadal sex reversal in mutant Dax1 XY mice: a failure to upregulate Sox9 in pre-Sertoli cells. *Development* 132, 3045-54.

Boyer, L. A., Lee, T. I., Cole, M. F., Johnstone, S. E., Levine, S. S., Zucker, J. P., Guenther, M. G., Kumar, R. M., Murray, H. L., Jenner, R. G. et al. (2005). Core transcriptional regulatory circuitry in human embryonic stem cells. *Cell* 122, 947-56.

Braat, A. K., Zandbergen, T., van de Water, S., Goos, H. J. and Zivkovic, D. (1999). Characterization of zebrafish primordial germ cells: morphology and early distribution of vasa RNA. *Dev Dyn* 216, 153-67.

Brennan, J. and Capel, B. (2004). One tissue, two fates: molecular genetic events that underlie testis versus ovary development. *Nat Rev Genet* 5, 509-21.

Brennan, J., Tilmann, C. and Capel, B. (2003). Pdgfr-alpha mediates testis cord organization and fetal Leydig cell development in the XY gonad. *Genes Dev* 17, 800-10.

Brivanlou, A. H. and Darnell, J. E., Jr. (2002). Signal transduction and the control of gene expression. *Science* 295, 813-8.

Burdon, T., Stracey, C., Chambers, I., Nichols, J. and Smith, A. (1999). Suppression of SHP-2 and ERK signalling promotes self-renewal of mouse embryonic stem cells. *Dev Biol* 210, 30-43.

Butteroni, C., De Felici, M., Scholer, H. R. and Pesce, M. (2000). Phage display screening reveals an association between germline-specific transcription factor Oct-4 and multiple cellular proteins. *J Mol Biol* 304, 529-40.

Capecchi, M. R. (1989). Altering the genome by homologous recombination. *Science* 244, 1288-92.

Capel, B. (2000). The battle of the sexes. *Mech Dev* 92, 89-103.

Carrera, P., Johnstone, O., Nakamura, A., Casanova, J., Jackle, H. and Lasko, P. (2000). VASA mediates translation through interaction with a Drosophila yIF2 homolog. *Mol Cell* 5, 181-7.

Castrillon, D. H., Quade, B. J., Wang, T. Y., Quigley, C. and Crum, C. P. (2000). The human VASA gene is specifically expressed in the germ cell lineage. *Proc Natl Acad Sci U S A* 97, 9585-90.

Chaboissier, M. C., Kobayashi, A., Vidal, V. I., Lutzkendorf, S., van de Kant, H. J., Wegner, M., de Rooij, D. G., Behringer, R. R. and Schedl, A. (2004). Functional analysis of Sox8 and Sox9 during sex determination in the mouse. *Development* 131, 1891-901.

Chambers, I., Colby, D., Robertson, M., Nichols, J., Lee, S., Tweedie, S. and Smith, A. (2003). Functional expression cloning of Nanog, a pluripotency sustaining factor in embryonic stem cells. *Cell* 113, 643-55.

Cheng, H. H., Ying, M., Tian, Y. H., Guo, Y., McElreavey, K. and Zhou, R. J. (2006). Transcriptional diversity of DMRT1 (dsx- and mab3-related transcription factor 1) in human testis. *Cell Res* 16, 389-93.

Colter, D. C., Piera-Velazquez, S., Hawkins, D. F., Whitecavage, M. K., Jimenez, S. A. and Stokes, D. G. (2005). Regulation of the human Sox9 promoter by the CCAAT-binding factor. *Matrix Biol* 24, 185-97.

Colvin, J. S., Green, R. P., Schmahl, J., Capel, B. and Ornitz, D. M. (2001). Male-to-female sex reversal in mice lacking fibroblast growth factor 9. *Cell* 104, 875-89.

Cordin, O., Banroques, J., Tanner, N. K. and Linder, P. (2006). The DEAD-box protein family of RNA helicases. *Gene* 367, 17-37.

Cowan, C. A., Atienza, J., Melton, D. A. and Eggan, K. (2005). Nuclear reprogramming of somatic cells after fusion with human embryonic stem cells. *Science* 309, 1369-73.

Cui, S., Hope, R. M., Rathjen, J., Voyle, R. B. and Rathjen, P. D. (2001). Structure, sequence and function of a marsupial LIF gene: conservation of IL-6 family cytokines. *Cytogenet Cell Genet* 92, 271-8.

Cui, S., Nanayakkara, K. and Selwood, L. (in press). A marsupial homolog of vasa (Tvvh): Conserved DEAD box family protein from the Brushtail possum.

Dahanukar, A. and Wharton, R. P. (1996). The Nanos gradient in Drosophila embryos is generated by translational regulation. *Genes Dev* 10, 2610-20.

Daheron, L., Opitz, S. L., Zaehres, H., Lensch, M. W., Andrews, P. W., Itskovitz-Eldor, J. and Daley, G. Q. (2004). LIF/STAT3 signaling fails to maintain self-renewal of human embryonic stem cells. *Stem Cells* 22, 770-8.

Dani, C., Chambers, I., Johnstone, S., Robertson, M., Ebrahimi, B., Saito, M., Taga, T., Li, M., Burdon, T., Nichols, J. et al. (1998). Paracrine induction of stem cell renewal by LIF-deficient cells: a new ES cell regulatory pathway. *Dev Biol* 203, 149-62.

De Grandi, A., Calvari, V., Bertini, V., Bulfone, A., Peverali, G., Camerino, G., Borsani, G. and Guioli, S. (2000). The expression pattern of a mouse doublesex-related gene is consistent with a role in gonadal differentiation. *Mech Dev* 90, 323-6.

De Santa Barbara, P., Bonneaud, N., Boizet, B., Desclozeaux, M., Moniot, B., Sudbeck, P., Scherer, G., Poulat, F. and Berta, P. (1998). Direct interaction of SRY-related protein SOX9 and steroidogenic factor 1 regulates transcription of the human anti-Mullerian hormone gene. *Mol Cell Biol* 18, 6653-65.

de Santa Barbara, P., Moniot, B., Poulat, F. and Berta, P. (2000). Expression and subcellular localization of SF-1, SOX9, WT1, and AMH proteins during early human testicular development. *Dev Dyn* 217, 293-8.

DiNapoli, L., Batchvarov, J. and Capel, B. (2006). FGF9 promotes survival of germ cells in the fetal testis. *Development* 133, 1519-27.

DiNapoli, L. and Capel, B. (2008). SRY and the standoff in sex determination. *Mol Endocrinol* 22, 1-9.

Donovan, P. J. and de Miguel, M. P. (2003). Turning germ cells into stem cells. *Curr Opin Genet Dev* 13, 463-71.

Durcova-Hills, G., Ainscough, J. and McLaren, A. (2001). Pluripotential stem cells derived from migrating primordial germ cells. *Differentiation* 68, 220-6.

Eddy, E. M. (1975). Germ plasm and the differentiation of the germ cell line. *Int Rev Cytol* 43, 229-80.

Evans, M. J. and Kaufman, M. H. (1981). Establishment in culture of pluripotential cells from mouse embryos. *Nature* 292, 154-6.

Ferguson-Smith, M. (2007). The evolution of sex chromosomes and sex determination in vertebrates and the key role of DMRT1. *Sex Dev* 1, 2-11.

Fujiwara, Y., Komiya, T., Kawabata, H., Sato, M., Fujimoto, H., Furusawa, M. and Noce, T. (1994). Isolation of a DEAD-family protein gene that encodes a murine homolog of Drosophila vasa and its specific expression in germ cell lineage. *Proc Natl Acad Sci U S A* 91, 12258-62.

Furumatsu, T., Tsuda, M., Yoshida, K., Taniguchi, N., Ito, T., Hashimoto, M., Ito, T. and Asahara, H. (2005). Sox9 and p300 cooperatively regulate chromatin-mediated transcription. *J Biol Chem* 280, 35203-8.

Gao, F., Maiti, S., Alam, N., Zhang, Z., Deng, J. M., Behringer, R. R., Lecureuil, C., Guillou, F. and Huff, V. (2006). The Wilms tumor gene, Wt1, is required for Sox9 expression and maintenance of tubular architecture in the developing testis. *Proc Natl Acad Sci U S A* 103, 11987-92.

Gavis, E. R., Lunsford, L., Bergsten, S. E. and Lehmann, R. (1996). A conserved 90 nucleotide element mediates translational repression of nanos RNA. *Development* 122, 2791-800.

Gearing, D. P., Gough, N. M., King, J. A., Hilton, D. J., Nicola, N. A., Simpson, R. J., Nice, E. C., Kelso, A. and Metcalf, D. (1987). Molecular cloning and expression of cDNA encoding a murine myeloid leukaemia inhibitory factor (LIF). *EMBO J* 6, 3995-4002.

Gearing, D. P., King, J. A. and Gough, N. M. (1988). Complete sequence of murine myeloid leukaemia inhibitory factor (LIF). *Nucleic Acids Res* 16, 9857.

Geijsen, N., Horoschak, M., Kim, K., Gribnau, J., Eggan, K. and Daley, G. Q. (2004). Derivation of embryonic germ cells and male gametes from embryonic stem cells. *Nature* 427, 148-54.

Ginis, I., Luo, Y., Miura, T., Thies, S., Brandenberger, R., Gerecht-Nir, S., Amit, M., Hoke, A., Carpenter, M. K., Itskovitz-Eldor, J. et al. (2004). Differences between human and mouse embryonic stem cells. *Dev Biol* 269, 360-80.

Gough, N. M., Williams, R. L., Hilton, D. J., Pease, S., Willson, T. A., Stahl, J., Gearing, D. P., Nicola, N. A. and Metcalf, D. (1989). LIF: a molecule with divergent actions on myeloid leukaemic cells and embryonic stem cells. *Reprod Fertil Dev* 1, 281-8.

Graham, V., Khudyakov, J., Ellis, P. and Pevny, L. (2003). SOX2 functions to maintain neural progenitor identity. *Neuron* 39, 749-65.

Guo, Y., Cheng, H., Huang, X., Gao, S., Yu, H. and Zhou, R. (2005). Gene structure, multiple alternative splicing, and expression in gonads of zebrafish Dmrt1. *Biochem Biophys Res Commun* 330, 950-7.

Guo, Y., Costa, R., Ramsey, H., Starnes, T., Vance, G., Robertson, K., Kelley, M., Reinbold, R., Scholer, H. and Hromas, R. (2002). The embryonic stem cell transcription factors Oct-4 and FoxD3 interact to regulate endodermal-specific promoter expression. *Proc Natl Acad Sci U S A* 99, 3663-7.

Hacker, A., Capel, B., Goodfellow, P. and Lovell-Badge, R. (1995). Expression of Sry, the mouse sex determining gene. *Development* 121, 1603-14.

Hajkova, P., Erhardt, S., Lane, N., Haaf, T., El-Maarri, O., Reik, W., Walter, J. and Surani, M. A. (2002). Epigenetic reprogramming in mouse primordial germ cells. *Mech Dev* 117, 15-23.

Hammes, A., Guo, J. K., Lutsch, G., Leheste, J. R., Landrock, D., Ziegler, U., Gubler, M. C. and Schedl, A. (2001). Two splice variants of the Wilms' tumor 1 gene have distinct functions during sex determination and nephron formation. *Cell* 106, 319-29.

Hastie, N. D. (2001). Life, sex, and WT1 isoforms--three amino acids can make all the difference. *Cell* 106, 391-4.

Hay, B., Jan, L. Y. and Jan, Y. N. (1988). A protein component of Drosophila polar granules is encoded by vasa and has extensive sequence similarity to ATP-dependent helicases. *Cell* 55, 577-87.

Heinrich, P. C., Behrmann, I., Haan, S., Hermanns, H. M., Muller-Newen, G. and Schaper, F. (2003). Principles of interleukin (IL)-6-type cytokine signalling and its regulation. *Biochem J* 374, 1-20.

Heinrich, P. C., Behrmann, I., Muller-Newen, G., Schaper, F. and Graeve, L. (1998). Interleukin-6-type cytokine signalling through the gp130/Jak/STAT pathway. *Biochem J* 334 (Pt 2), 297-314.

Herbert, A. and Rich, A. (1999). RNA processing and the evolution of eukaryotes. *Nat Genet* 21, 265-9.

Herzer, U., Crocoll, A., Barton, D., Howells, N. and Englert, C. (1999). The Wilms tumor suppressor gene wt1 is required for development of the spleen. *Curr Biol* 9, 837-40.

Hibi, M. and Hirano, T. (1998). Signal transduction through cytokine receptors. *Int Rev Immunol* 17, 75-102.

Hilton, D. J., Nicola, N. A., Waring, P. M. and Metcalf, D. (1991). Clearance and fate of leukemia-inhibitory factor (LIF) after injection into mice. *J Cell Physiol* 148, 430-9.

Hossain, A. and Saunders, G. F. (2001). The human sex-determining gene SRY is a direct target of WT1. *J Biol Chem* 276, 16817-23.

Hossain, A. and Saunders, G. F. (2003). Role of Wilms tumor 1 (WT1) in the transcriptional regulation of the Mullerian-inhibiting substance promoter. *Biol Reprod* 69, 1808-14.

Houston, D. W. and King, M. L. (2000). Germ plasm and molecular determinants of germ cell fate. *Curr Top Dev Biol* 50, 155-81.

Hoyle, C., Narvaez, V., Alldus, G., Lovell-Badge, R. and Swain, A. (2002). Dax1 expression is dependent on steroidogenic factor 1 in the developing gonad. *Mol Endocrinol* 16, 747-56.

Huang, X., Guo, Y., Shui, Y., Gao, S., Yu, H., Cheng, H. and Zhou, R. (2005). Multiple alternative splicing and differential expression of dmrt1 during gonad transformation of the rice field eel. *Biol Reprod* 73, 1017-24.

Humphrey, R. K., Beattie, G. M., Lopez, A. D., Bucay, N., King, C. C., Firpo, M. T., Rose-John, S. and Hayek, A. (2004). Maintenance of pluripotency in human embryonic stem cells is STAT3 independent. *Stem Cells* 22, 522-30.

Iida, T. and Kobayashi, S. (1998). Essential role of mitochondrially encoded large rRNA for germ-line formation in Drosophila embryos. *Proc Natl Acad Sci U S A* 95, 11274-8.

Ito, M., Miyagishi, M., Murata, C., Kawasaki, H., Baba, T., Tachi, C. and Taira, K. (2006). Down-regulation of endogenous Wt1 expression by Sry transgene in the murine embryonic mesonephros-derived M15 cell line. *J Reprod Dev* 52, 415-27.

Ivanova, N., Dobrin, R., Lu, R., Kotenko, I., Levorse, J., DeCoste, C., Schafer, X., Lun, Y. and Lemischka, I. R. (2006). Dissecting self-renewal in stem cells with RNA interference. *Nature* 442, 533-8.

Jeays-Ward, K., Dandonneau, M. and Swain, A. (2004). Wnt4 is required for proper male as well as female sexual development. *Dev Biol* 276, 431-40.

Jeffs, B., Ito, M., Yu, R. N., Martinson, F. A., Wang, Z. J., Doglio, L. T. and Jameson, J. L. (2001). Sertoli cell-specific rescue of fertility, but not testicular pathology, in Dax1 (Ahch)-deficient male mice. *Endocrinology* 142, 2481-8.

Johnson, J. M., Castle, J., Garrett-Engele, P., Kan, Z., Loerch, P. M., Armour, C. D., Santos, R., Schadt, E. E., Stoughton, R. and Shoemaker, D. D. (2003). Genome-wide survey of human alternative pre-mRNA splicing with exon junction microarrays. *Science* 302, 2141-4.

Josso, N., Racine, C., di Clemente, N., Rey, R. and Xavier, F. (1998). The role of anti-Mullerian hormone in gonadal development. *Mol Cell Endocrinol* 145, 3-7.

Katoh-Fukui, Y., Tsuchiya, R., Shiroishi, T., Nakahara, Y., Hashimoto, N., Noguchi, K. and Higashinakagawa, T. (1998). Male-to-female sex reversal in M33 mutant mice. *Nature* 393, 688-92.

Kent, J., Wheatley, S. C., Andrews, J. E., Sinclair, A. H. and Koopman, P. (1996). A male-specific role for SOX9 in vertebrate sex determination. *Development* 122, 2813-22.

Kim, J., Prawitt, D., Bardeesy, N., Torban, E., Vicaner, C., Goodyer, P., Zabel, B. and Pelletier, J. (1999). The Wilms' tumor suppressor gene (wt1) product regulates Dax-1 gene expression during gonadal differentiation. *Mol Cell Biol* 19, 2289-99.

Kim, Y., Bingham, N., Sekido, R., Parker, K. L., Lovell-Badge, R. and Capel, B. (2007). Fibroblast growth factor receptor 2 regulates proliferation and Sertoli differentiation during male sex determination. *Proc Natl Acad Sci U S A* 104, 16558-63.

Kim, Y., Kobayashi, A., Sekido, R., DiNapoli, L., Brennan, J., Chaboissier, M. C., Poulat, F., Behringer, R. R., Lovell-Badge, R. and Capel, B. (2006). Fgf9 and Wnt4 act as antagonistic signals to regulate mammalian sex determination. *PLoS Biol* 4, e187.

Kishimoto, T., Taga, T. and Akira, S. (1994). Cytokine signal transduction. *Cell* 76, 253-62.

Kobayashi, S., Amikura, R. and Okada, M. (1993). Presence of mitochondrial large ribosomal RNA outside mitochondria in germ plasm of Drosophila melanogaster. *Science* 260, 1521-4.

Kobayashi, S., Yamada, M., Asaoka, M. and Kitamura, T. (1996). Essential role of the posterior morphogen nanos for germline development in Drosophila. *Nature* 380, 708-11.

Komiya, T., Itoh, K., Ikenishi, K. and Furusawa, M. (1994). Isolation and characterization of a novel gene of the DEAD box protein family which is specifically expressed in germ cells of Xenopus laevis. *Dev Biol* 162, 354-63.

Komiya, T. and Tanigawa, Y. (1995). Cloning of a gene of the DEAD box protein family which is specifically expressed in germ cells in rats. *Biochem Biophys Res Commun* 207, 405-10.

Koopman, P., Gubbay, J., Vivian, N., Goodfellow, P. and Lovell-Badge, R. (1991). Male development of chromosomally female mice transgenic for Sry. *Nature* 351, 117-21.

Koopman, P., Munsterberg, A., Capel, B., Vivian, N. and Lovell-Badge, R. (1990). Expression of a candidate sex-determining gene during mouse testis differentiation. *Nature* 348, 450-2.

Kuwana, T. and Rogulska, T. (1999). Migratory mechanisms of chick primordial germ cells toward gonadal anlage. *Cell Mol Biol (Noisy-le-grand)* 45, 725-36.

Lasko, P. F. and Ashburner, M. (1988). The product of the Drosophila gene vasa is very similar to eukaryotic initiation factor-4A. *Nature* 335, 611-7.

Layton, M. J., Owczarek, C. M., Metcalf, D., Clark, R. L., Smith, D. K., Treutlein, H. R. and Nicola, N. A. (1994). Conversion of the biological specificity of murine to human leukemia inhibitory factor by replacing 6 amino acid residues. *J Biol Chem* 269, 29891-6.

Lee, G. S., Kim, H. S., Lee, S. H., Kang, M. S., Kim, D. Y., Lee, C. K., Kang, S. K., Lee, B. C. and Hwang, W. S. (2005). Characterization of pig vasa homolog gene and specific expression in germ cell lineage. *Mol Reprod Dev* 72, 320-8.

Lei, N., Hornbaker, K. I., Rice, D. A., Karpova, T., Agbor, V. A. and Heckert, L. L. (2007). Sex-specific differences in mouse DMRT1 expression are both cell type- and stage-dependent during gonad development. *Biol Reprod* 77, 466-75.

Levy, D. E. and Darnell, J. E., Jr. (2002). Stats: transcriptional control and biological impact. *Nat Rev Mol Cell Biol* 3, 651-62.

Li, M., Sendtner, M. and Smith, A. (1995). Essential function of LIF receptor in motor neurons. *Nature* 378, 724-7.

Liang, L., Diehl-Jones, W. and Lasko, P. (1994). Localization of vasa protein to the Drosophila pole plasm is independent of its RNA-binding and helicase activities. *Development* 120, 1201-11.

Linder, P. (2006). Dead-box proteins: a family affair--active and passive players in RNP-remodeling. *Nucleic Acids Res* 34, 4168-80.

Linder, P., Lasko, P. F., Ashburner, M., Leroy, P., Nielsen, P. J., Nishi, K., Schnier, J. and Slonimski, P. P. (1989). Birth of the D-E-A-D box. *Nature* 337, 121-2.

Loh, Y. H., Wu, Q., Chew, J. L., Vega, V. B., Zhang, W., Chen, X., Bourque, G., George, J., Leong, B., Liu, J. et al. (2006). The Oct4 and Nanog transcription network regulates pluripotency in mouse embryonic stem cells. *Nat Genet* 38, 431-40.

Lu, H., Huang, X., Zhang, L., Guo, Y., Cheng, H. and Zhou, R. (2007). Multiple alternative splicing of mouse Dmrt1 during gonadal differentiation. *Biochem Biophys Res Commun* 352, 630-4.

Luo, X., Ikeda, Y. and Parker, K. L. (1994). A cell-specific nuclear receptor is essential for adrenal and gonadal development and sexual differentiation. *Cell* 77, 481-90.

Lutticken, C., Wegenka, U. M., Yuan, J., Buschmann, J., Schindler, C., Ziemiecki, A., Harpur, A. G., Wilks, A. F., Yasukawa, K., Taga, T. et al. (1994). Association of transcription factor APRF and protein kinase Jak1 with the interleukin-6 signal transducer gp130. *Science* 263, 89-92.

Maegawa, S., Yasuda, K. and Inoue, K. (1999). Maternal mRNA localization of zebrafish DAZ-like gene. *Mech Dev* 81, 223-6.

Malki, S., Berta, P., Poulat, F. and Boizet-Bonhoure, B. (2005a). Cytoplasmic retention of the sex-determining factor SOX9 via the microtubule network. *Exp Cell Res* 309, 468-75.

Malki, S., Nef, S., Notarnicola, C., Thevenet, L., Gasca, S., Mejean, C., Berta, P., Poulat, F. and Boizet-Bonhoure, B. (2005b). Prostaglandin D2 induces nuclear import of the sex-determining factor SOX9 via its cAMP-PKA phosphorylation. *Embo J* 24, 1798-809.

Manuylov, N. L., Fujiwara, Y., Adameyko, II, Poulat, F. and Tevosian, S. G. (2007). The regulation of Sox9 gene expression by the GATA4/FOG2 transcriptional complex in dominant XX sex reversal mouse models. *Dev Biol* 307, 356-67.

Martin, G. R. (1981). Isolation of a pluripotent cell line from early mouse embryos cultured in medium conditioned by teratocarcinoma stem cells. *Proc Natl Acad Sci U S A* 78, 7634-8.

Matin, M. M., Walsh, J. R., Gokhale, P. J., Draper, J. S., Bahrami, A. R., Morton, I., Moore, H. D. and Andrews, P. W. (2004). Specific knockdown of Oct4 and beta2-microglobulin expression by RNA interference in human embryonic stem cells and embryonic carcinoma cells. *Stem Cells* 22, 659-68.

Matsuda, M., Nagahama, Y., Shinomiya, A., Sato, T., Matsuda, C., Kobayashi, T., Morrey, C. E., Shibata, N., Asakawa, S., Shimizu, N. et al. (2002). DMY is a Y-specific DM-domain gene required for male development in the medaka fish. *Nature* 417, 559-63.

Matsuda, T., Nakamura, T., Nakao, K., Arai, T., Katsuki, M., Heike, T. and Yokota, T. (1999). STAT3 activation is sufficient to maintain an undifferentiated state of mouse embryonic stem cells. *EMBO J* 18, 4261-9.

Matsui, Y., Zsebo, K. and Hogan, B. L. (1992). Derivation of pluripotential embryonic stem cells from murine primordial germ cells in culture. *Cell* 70, 841-7.

Matsuzawa-Watanabe, Y., Inoue, J. and Semba, K. (2003). Transcriptional activity of testis-determining factor SRY is modulated by the Wilms' tumor 1 gene product, WT1. *Oncogene* 22, 7900-4.

McCoard, S. A., Wise, T. H., Fahrenkrug, S. C. and Ford, J. J. (2001). Temporal and spatial localization patterns of Gata4 during porcine gonadogenesis. *Biol Reprod* 65, 366-74.

Meeks, J. J., Crawford, S. E., Russell, T. A., Morohashi, K., Weiss, J. and Jameson, J. L. (2003a). Dax1 regulates testis cord organization during gonadal differentiation. *Development* 130, 1029-36.

Meeks, J. J., Weiss, J. and Jameson, J. L. (2003b). Dax1 is required for testis determination. *Nat Genet* 34, 32-3.

Metcalf, D., Hilton, D. J. and Nicola, N. A. (1988). Clonal analysis of the actions of the murine leukemia inhibitory factor on leukemic and normal murine hemopoietic cells. *Leukemia* 2, 216-21.

Minvielle-Sebastia, L. and Keller, W. (1999). mRNA polyadenylation and its coupling to other RNA processing reactions and to transcription. *Curr Opin Cell Biol* 11, 352-7.

Mishina, Y., Rey, R., Finegold, M. J., Matzuk, M. M., Josso, N., Cate, R. L. and Behringer, R. R. (1996). Genetic analysis of the Mullerian-inhibiting substance signal transduction pathway in mammalian sexual differentiation. *Genes Dev* 10, 2577-87.

Mitsui, K., Tokuzawa, Y., Itoh, H., Segawa, K., Murakami, M., Takahashi, K., Maruyama, M., Maeda, M. and Yamanaka, S. (2003). The homeoprotein Nanog is required for maintenance of pluripotency in mouse epiblast and ES cells. *Cell* 113, 631-42.

Miyamoto, N., Yoshida, M., Kuratani, S., Matsuo, I. and Aizawa, S. (1997). Defects of urogenital development in mice lacking Emx2. *Development* 124, 1653-64.

Miyamoto, Y., Taniguchi, H., Hamel, F., Silversides, D. W. and Viger, R. S. (2008). A GATA4/WT1 cooperation regulates transcription of genes required for mammalian sex determination and differentiation. *BMC Mol Biol* 9, 44.

Modrek, B. and Lee, C. (2002). A genomic view of alternative splicing. *Nat Genet* 30, 13-9.

Moniot, B., Berta, P., Scherer, G., Sudbeck, P. and Poulat, F. (2000). Male specific expression suggests role of DMRT1 in human sex determination. *Mech Dev* 91, 323-5.

Moore, A. W., McInnes, L., Kreidberg, J., Hastie, N. D. and Schedl, A. (1999). YAC complementation shows a requirement for Wt1 in the development of epicardium, adrenal gland and throughout nephrogenesis. *Development* 126, 1845-57.

Morais da Silva, S., Hacker, A., Harley, V., Goodfellow, P., Swain, A. and Lovell-Badge, R. (1996). Sox9 expression during gonadal development implies a conserved role for the gene in testis differentiation in mammals and birds. *Nat Genet* 14, 62-8.

Muscatelli, F., Strom, T. M., Walker, A. P., Zanaria, E., Recan, D., Meindl, A., Bardoni, B., Guioli, S., Zehetner, G., Rabl, W. et al. (1994). Mutations in the DAX-1 gene give rise to both X-linked adrenal hypoplasia congenita and hypogonadotropic hypogonadism. *Nature* 372, 672-6.

Nachtigall, M. J., Kliman, H. J., Feinberg, R. F., Olive, D. L., Engin, O. and Arici, A. (1996). The effect of leukemia inhibitory factor (LIF) on trophoblast differentiation: a potential role in human implantation. *J Clin Endocrinol Metab* 81, 801-6.

Nanda, I., Kondo, M., Hornung, U., Asakawa, S., Winkler, C., Shimizu, A., Shan, Z., Haaf, T., Shimizu, N., Shima, A. et al. (2002). A duplicated copy of DMRT1 in the sex-determining region of the Y chromosome of the medaka, Oryzias latipes. *Proc Natl Acad Sci U S A* 99, 11778-83.

Nef, S., Schaad, O., Stallings, N. R., Cederroth, C. R., Pitetti, J. L., Schaer, G., Malki, S., Dubois-Dauphin, M., Boizet-Bonhoure, B., Descombes, P. et al. (2005). Gene expression during sex determination reveals a robust female genetic program at the onset of ovarian development. *Dev Biol* 287, 361-77.

Nef, S., Verma-Kurvari, S., Merenmies, J., Vassalli, J. D., Efstratiadis, A., Accili, D. and Parada, L. F. (2003). Testis determination requires insulin receptor family function in mice. *Nature* 426, 291-5.

Nichols, J., Davidson, D., Taga, T., Yoshida, K., Chambers, I. and Smith, A. (1996). Complementary tissue-specific expression of LIF and LIF-receptor mRNAs in early mouse embryogenesis. *Mech Dev* 57, 123-31.

Nichols, J., Zevnik, B., Anastassiadis, K., Niwa, H., Klewe-Nebenius, D., Chambers, I., Scholer, H. and Smith, A. (1998). Formation of pluripotent stem cells in the mammalian embryo depends on the POU transcription factor Oct4. *Cell* 95, 379-91.

Niwa, H., Burdon, T., Chambers, I. and Smith, A. (1998). Self-renewal of pluripotent embryonic stem cells is mediated via activation of STAT3. *Genes Dev* 12, 2048-60.

Niwa, H., Miyazaki, J. and Smith, A. G. (2000). Quantitative expression of Oct-3/4 defines differentiation, dedifferentiation or self-renewal of ES cells. *Nat Genet* 24, 372-6.

Noce, T., Okamoto-Ito, S. and Tsunekawa, N. (2001). Vasa homolog genes in mammalian germ cell development. *Cell Struct Funct* 26, 131-6.

Nordenskjold, A., Fricke, G. and Anvret, M. (1995). Absence of mutations in the WT1 gene in patients with XY gonadal dysgenesis. *Hum Genet* 96, 102-4.

O'Shea, J. J., Gadina, M. and Schreiber, R. D. (2002). Cytokine signaling in 2002: new surprises in the Jak/Stat pathway. *Cell* 109 Suppl, S121-31.

Olsen, L. C., Aasland, R. and Fjose, A. (1997). A vasa-like gene in zebrafish identifies putative primordial germ cells. *Mech Dev* 66, 95-105.

Oreal, E., Mazaud, S., Picard, J. Y., Magre, S. and Carre-Eusebe, D. (2002). Different patterns of anti-Mullerian hormone expression, as related to DMRT1, SF-1, WT1, GATA-4, Wnt-4, and Lhx9 expression, in the chick differentiating gonads. *Dev Dyn* 225, 221-32.

Oreal, E., Pieau, C., Mattei, M. G., Josso, N., Picard, J. Y., Carre-Eusebe, D. and Magre, S. (1998). Early expression of AMH in chicken embryonic gonads precedes testicular SOX9 expression. *Dev Dyn* 212, 522-32.

Ottolenghi, C., Uda, M., Crisponi, L., Omari, S., Cao, A., Forabosco, A. and Schlessinger, D. (2007). Determination and stability of sex. *Bioessays* 29, 15-25.

Ozisik, G., Achermann, J. C., Meeks, J. J. and Jameson, J. L. (2003). SF1 in the development of the adrenal gland and gonads. *Horm Res* 59 Suppl 1, 94-8.

Paling, N. R., Wheadon, H., Bone, H. K. and Welham, M. J. (2004). Regulation of embryonic stem cell self-renewal by phosphoinositide 3-kinase-dependent signaling. *J Biol Chem* 279, 48063-70.

Palmieri, S. L., Peter, W., Hess, H. and Scholer, H. R. (1994). Oct-4 transcription factor is differentially expressed in the mouse embryo during establishment of the first two extraembryonic cell lineages involved in implantation. *Dev Biol* 166, 259-67.

Pan, G. and Pei, D. (2005). The stem cell pluripotency factor NANOG activates transcription with two unusually potent subdomains at its C terminus. *J Biol Chem* 280, 1401-7.

Park, S. Y. and Jameson, J. L. (2005). Minireview: transcriptional regulation of gonadal development and differentiation. *Endocrinology* 146, 1035-42.

Parker, K. L., Schedl, A. and Schimmer, B. P. (1999). Gene interactions in gonadal development. *Annu Rev Physiol* 61, 417-33.

Pelegri, F., Knaut, H., Maischein, H. M., Schulte-Merker, S. and Nusslein-Volhard, C. (1999). A mutation in the zebrafish maternal-effect gene nebel affects furrow formation and vasa RNA localization. *Curr Biol* 9, 1431-40.

Piedrahita, J. A., Weaks, R., Petrescu, A., Shrode, T. W., Derr, J. N. and Womack, J. E. (1997). Genetic characterization of the bovine leukaemia inhibitory factor (LIF) gene: isolation and sequencing, chromosome assignment and microsatellite analysis. *Anim Genet* 28, 14-20.

Piera-Velazquez, S., Hawkins, D. F., Whitecavage, M. K., Colter, D. C., Stokes, D. G. and Jimenez, S. A. (2007). Regulation of the human SOX9 promoter by Sp1 and CREB. *Exp Cell Res* 313, 1069-79.

Polanco, J. C. and Koopman, P. (2007). Sry and the hesitant beginnings of male development. *Dev Biol* 302, 13-24.

Pontiggia, A., Rimini, R., Harley, V. R., Goodfellow, P. N., Lovell-Badge, R. and Bianchi, M. E. (1994). Sex-reversing mutations affect the architecture of SRY-DNA complexes. *Embo J* 13, 6115-24.

Qi, X., Li, T. G., Hao, J., Hu, J., Wang, J., Simmons, H., Miura, S., Mishina, Y. and Zhao, G. Q. (2004). BMP4 supports self-renewal of embryonic stem cells by inhibiting mitogen-activated protein kinase pathways. *Proc Natl Acad Sci U S A* 101, 6027-32.

Qin, Y. and Bishop, C. E. (2005). Sox9 is sufficient for functional testis development producing fertile male mice in the absence of Sry. *Hum Mol Genet* 14, 1221-9.

Rathjen, P. D., Toth, S., Willis, A., Heath, J. K. and Smith, A. G. (1990). Differentiation inhibiting activity is produced in matrix-associated and diffusible forms that are generated by alternate promoter usage. *Cell* 62, 1105-14.

Raymond, C. S., Kettlewell, J. R., Hirsch, B., Bardwell, V. J. and Zarkower, D. (1999). Expression of Dmrt1 in the genital ridge of mouse and chicken embryos suggests a role in vertebrate sexual development. *Dev Biol* 215, 208-20.

Raymond, C. S., Murphy, M. W., O'Sullivan, M. G., Bardwell, V. J. and Zarkower, D. (2000). Dmrt1, a gene related to worm and fly sexual regulators, is required for mammalian testis differentiation. *Genes Dev* 14, 2587-95.

Raymond, C. S., Shamu, C. E., Shen, M. M., Seifert, K. J., Hirsch, B., Hodgkin, J. and Zarkower, D. (1998). Evidence for evolutionary conservation of sex-determining genes. *Nature* 391, 691-5.

Resnick, J. L., Bixler, L. S., Cheng, L. and Donovan, P. J. (1992). Long-term proliferation of mouse primordial germ cells in culture. *Nature* 359, 550-1.

Roberts, S. G. (2005). Transcriptional regulation by WT1 in development. *Curr Opin Genet Dev* 15, 542-7.

Rongo, C., Gavis, E. R. and Lehmann, R. (1995). Localization of oskar RNA regulates oskar translation and requires Oskar protein. *Development* 121, 2737-46.

Rose-John, S. (2002). GP130 stimulation and the maintenance of stem cells. *Trends Biotechnol* 20, 417-9.

Rosner, M. H., Vigano, M. A., Ozato, K., Timmons, P. M., Poirier, F., Rigby, P. W. and Staudt, L. M. (1990). A POU-domain transcription factor in early stem cells and germ cells of the mammalian embryo. *Nature* 345, 686-92.

Roussell, D. L. and Bennett, K. L. (1993). glh-1, a germ-line putative RNA helicase from Caenorhabditis, has four zinc fingers. *Proc Natl Acad Sci U S A* 90, 9300-4.

Sadovsky, Y., Crawford, P. A., Woodson, K. G., Polish, J. A., Clements, M. A., Tourtellotte, L. M., Simburger, K. and Milbrandt, J. (1995). Mice deficient in the orphan receptor steroidogenic factor 1 lack adrenal glands and gonads but express P450 side-chain-cleavage enzyme in the placenta and have normal embryonic serum levels of corticosteroids. *Proc Natl Acad Sci U S A* 92, 10939-43.

Saffman, E. E. and Lasko, P. (1999). Germline development in vertebrates and invertebrates. *Cell Mol Life Sci* 55, 1141-63.

Schepers, G., Wilson, M., Wilhelm, D. and Koopman, P. (2003). SOX8 is expressed during testis differentiation in mice and synergizes with SF1 to activate the Amh promoter in vitro. *J Biol Chem* 278, 28101-8.

Scholer, H. R., Ciesiolka, T. and Gruss, P. (1991). A nexus between Oct-4 and E1A: implications for gene regulation in embryonic stem cells. *Cell* 66, 291-304.

Schutt, C. and Nothiger, R. (2000). Structure, function and evolution of sex-determining systems in Dipteran insects. *Development* 127, 667-77.

Sekido, R. and Lovell-Badge, R. (2008). Sex determination involves synergistic action of SRY and SF1 on a specific Sox9 enhancer. *Nature* 453, 930-34.

Shamblott, M. J., Axelman, J., Wang, S., Bugg, E. M., Littlefield, J. W., Donovan, P. J., Blumenthal, P. D., Huggins, G. R. and Gearhart, J. D. (1998). Derivation of pluripotent stem cells from cultured human primordial germ cells. *Proc Natl Acad Sci U S A* 95, 13726-31.

Shibata, N., Umesono, Y., Orii, H., Sakurai, T., Watanabe, K. and Agata, K. (1999). Expression of vasa(vas)-related genes in germline cells and totipotent somatic stem cells of planarians. *Dev Biol* 206, 73-87.

Shimamura, R., Fraizer, G. C., Trapman, J., Lau Yf, C. and Saunders, G. F. (1997). The Wilms' tumor gene WT1 can regulate genes involved in sex determination and differentiation: SRY, Mullerian-inhibiting substance, and the androgen receptor. *Clin Cancer Res* 3, 2571-80.

Silva, J., Chambers, I., Pollard, S. and Smith, A. (2006). Nanog promotes transfer of pluripotency after cell fusion. *Nature* 441, 997-1001.

Sinclair, A. H., Berta, P., Palmer, M. S., Hawkins, J. R., Griffiths, B. L., Smith, M. J., Foster, J. W., Frischauf, A. M., Lovell-Badge, R. and Goodfellow, P. N. (1990). A gene from the human sex-determining region encodes a protein with homology to a conserved DNA-binding motif. *Nature* 346, 240-4.

Slack, J. M. (2000). Stem cells in epithelial tissues. *Science* 287, 1431-3.

Smith, A. G., Heath, J. K., Donaldson, D. D., Wong, G. G., Moreau, J., Stahl, M. and Rogers, D. (1988). Inhibition of pluripotential embryonic stem cell differentiation by purified polypeptides. *Nature* 336, 688-90.

Smith, C. A., McClive, P. J., Western, P. S., Reed, K. J. and Sinclair, A. H. (1999). Conservation of a sex-determining gene. *Nature* 402, 601-2.

Song, J. H., Houde, A. and Murphy, B. D. (1998). Cloning of leukemia inhibitory factor (LIF) and its expression in the uterus during embryonic diapause and implantation in the mink (Mustela vison). *Mol Reprod Dev* 51, 13-21.

Stahl, J., Gearing, D. P., Willson, T. A., Brown, M. A., King, J. A. and Gough, N. M. (1990). Structural organization of the genes for murine and human leukemia inhibitory factor. Evolutionary conservation of coding and non-coding regions. *J Biol Chem* 265, 8833-41.

Stefani, G. and Slack, F. J. (2008). Small non-coding RNAs in animal development. *Nat Rev Mol Cell Biol* 9, 219-30.

Stewart, C. L., Kaspar, P., Brunet, L. J., Bhatt, H., Gadi, I., Kontgen, F. and Abbondanzo, S. J. (1992). Blastocyst implantation depends on maternal expression of leukaemia inhibitory factor. *Nature* 359, 76-9.

Styhler, S., Nakamura, A., Swan, A., Suter, B. and Lasko, P. (1998). vasa is required for GURKEN accumulation in the oocyte, and is involved in oocyte differentiation and germline cyst development. *Development* 125, 1569-78.

Swain, A., Zanaria, E., Hacker, A., Lovell-Badge, R. and Camerino, G. (1996). Mouse Dax1 expression is consistent with a role in sex determination as well as in adrenal and hypothalamus function. *Nat Genet* 12, 404-9.

Symes, A., Stahl, N., Reeves, S. A., Farruggella, T., Servidei, T., Gearan, T., Yancopoulos, G. and Fink, J. S. (1997). The protein tyrosine phosphatase SHP-2 negatively regulates ciliary neurotrophic factor induction of gene expression. *Curr Biol* 7, 697-700.

Takada, S., Mano, H. and Koopman, P. (2005). Regulation of Amh during sex determination in chickens: Sox gene expression in male and female gonads. *Cell Mol Life Sci* 62, 2140-6.

Takahashi-Tezuka, M., Yoshida, Y., Fukada, T., Ohtani, T., Yamanaka, Y., Nishida, K., Nakajima, K., Hibi, M. and Hirano, T. (1998). Gab1 acts as an adapter molecule linking the cytokine receptor gp130 to ERK mitogen-activated protein kinase. *Mol Cell Biol* 18, 4109-17.

Takeda, K., Noguchi, K., Shi, W., Tanaka, T., Matsumoto, M., Yoshida, N., Kishimoto, T. and Akira, S. (1997). Targeted disruption of the mouse Stat3 gene leads to early embryonic lethality. *Proc Natl Acad Sci U S A* 94, 3801-4.

Tam, P. P. and Snow, M. H. (1981). Proliferation and migration of primordial germ cells during compensatory growth in mouse embryos. *J Embryol Exp Morphol* 64, 133-47.

Tanaka, S. S., Toyooka, Y., Akasu, R., Katoh-Fukui, Y., Nakahara, Y., Suzuki, R., Yokoyama, M. and Noce, T. (2000). The mouse homolog of Drosophila Vasa is required for the development of male germ cells. *Genes Dev* 14, 841-53.

Tevosian, S. G., Albrecht, K. H., Crispino, J. D., Fujiwara, Y., Eicher, E. M. and Orkin, S. H. (2002). Gonadal differentiation, sex determination and normal Sry expression in mice require direct interaction between transcription partners GATA4 and FOG2. *Development* 129, 4627-34.

Thompson, S. W. and Majithia, A. A. (1998). Leukemia inhibitory factor induces sympathetic sprouting in intact dorsal root ganglia in the adult rat in vivo. *J Physiol* 506 (Pt 3), 809-16.

Thomson, J. A., Itskovitz-Eldor, J., Shapiro, S. S., Waknitz, M. A., Swiergiel, J. J., Marshall, V. S. and Jones, J. M. (1998). Embryonic stem cell lines derived from human blastocysts. *Science* 282, 1145-7.

Tomancak, P., Guichet, A., Zavorszky, P. and Ephrussi, A. (1998). Oocyte polarity depends on regulation of gurken by Vasa. *Development* 125, 1723-32.

Tsunekawa, N., Naito, M., Sakai, Y., Nishida, T. and Noce, T. (2000). Isolation of chicken vasa homolog gene and tracing the origin of primordial germ cells. *Development* 127, 2741-50.

Vainio, S., Heikkila, M., Kispert, A., Chin, N. and McMahon, A. P. (1999). Female development in mammals is regulated by Wnt-4 signalling. *Nature* 397, 405-9.

Van Doren, M., Williamson, A. L. and Lehmann, R. (1998). Regulation of zygotic gene expression in Drosophila primordial germ cells. *Curr Biol* 8, 243-6.

Vidal, V. P., Chaboissier, M. C., de Rooij, D. G. and Schedl, A. (2001). Sox9 induces testis development in XX transgenic mice. *Nat Genet* 28, 216-7.

Viger, R. S., Mertineit, C., Trasler, J. M. and Nemer, M. (1998). Transcription factor GATA-4 is expressed in a sexually dimorphic pattern during mouse gonadal development and is a potent activator of the Mullerian inhibiting substance promoter. *Development* 125, 2665-75.

Ware, C. B., Horowitz, M. C., Renshaw, B. R., Hunt, J. S., Liggitt, D., Koblar, S. A., Gliniak, B. C., McKenna, H. J., Papayannopoulou, T., Thoma, B. et al. (1995). Targeted disruption of the low-affinity leukemia inhibitory factor receptor gene causes placental, skeletal, neural and metabolic defects and results in perinatal death. *Development* 121, 1283-99.

Weidinger, G., Wolke, U., Koprunner, M., Klinger, M. and Raz, E. (1999). Identification of tissues and patterning events required for distinct steps in early migration of zebrafish primordial germ cells. *Development* 126, 5295-307.

Western, P. S., Harry, J. L., Graves, J. A. and Sinclair, A. H. (1999). Temperature-dependent sex determination in the American alligator: AMH precedes SOX9 expression. *Dev Dyn* 216, 411-9.

Wilhelm, D. and Englert, C. (2002). The Wilms tumor suppressor WT1 regulates early gonad development by activation of Sf1. *Genes Dev* 16, 1839-51.

Wilhelm, D., Hiramatsu, R., Mizusaki, H., Widjaja, L., Combes, A. N., Kanai, Y. and Koopman, P. (2007a). SOX9 Regulates Prostaglandin D Synthase Gene Transcription in Vivo to Ensure Testis Development. *J Biol Chem* 282, 10553-60.

Wilhelm, D., Martinson, F., Bradford, S., Wilson, M. J., Combes, A. N., Beverdam, A., Bowles, J., Mizusaki, H. and Koopman, P. (2005). Sertoli cell differentiation is induced both cell-autonomously and through prostaglandin signaling during mammalian sex determination. *Dev Biol* 287, 111-24.

Wilhelm, D., Palmer, S. and Koopman, P. (2007b). Sex determination and gonadal development in mammals. *Physiol Rev* 87, 1-28.

Williams, D. A., Rosenblatt, M. F., Beier, D. R. and Cone, R. D. (1988). Generation of murine stromal cell lines supporting hematopoietic stem cell proliferation by use of recombinant retrovirus vectors encoding simian virus 40 large T antigen. *Mol Cell Biol* 8, 3864-71.

Willson, T. A., Metcalf, D. and Gough, N. M. (1992). Cross-species comparison of the sequence of the leukaemia inhibitory factor gene and its protein. *Eur J Biochem* 204, 21-30.

Wylie, C. (1999). Germ cells. *Cell* 96, 165-74.

Wylie, C. (2000). Germ cells. *Curr Opin Genet Dev* 10, 410-3.

Xu, C., Inokuma, M. S., Denham, J., Golds, K., Kundu, P., Gold, J. D. and Carpenter, M. K. (2001). Feeder-free growth of undifferentiated human embryonic stem cells. *Nat Biotechnol* 19, 971-4.

Xu, R. H., Chen, X., Li, D. S., Li, R., Addicks, G. C., Glennon, C., Zwaka, T. P. and Thomson, J. A. (2002). BMP4 initiates human embryonic stem cell differentiation to trophoblast. *Nat Biotechnol* 20, 1261-4.

Yamamori, T., Fukada, K., Aebersold, R., Korsching, S., Fann, M. J. and Patterson, P. H. (1989). The cholinergic neuronal differentiation factor from heart cells is identical to leukemia inhibitory factor. *Science* 246, 1412-6.

Yamazaki, Y., Mann, M. R., Lee, S. S., Marh, J., McCarrey, J. R., Yanagimachi, R. and Bartolomei, M. S. (2003). Reprogramming of primordial germ cells begins before migration into the genital ridge, making these cells inadequate donors for reproductive cloning. *Proc Natl Acad Sci U S A* 100, 12207-12.

Yao, H. H., Whoriskey, W. and Capel, B. (2002). Desert Hedgehog/Patched 1 signaling specifies fetal Leydig cell fate in testis organogenesis. *Genes Dev* 16, 1433-40.

Yeom, Y. I., Fuhrmann, G., Ovitt, C. E., Brehm, A., Ohbo, K., Gross, M., Hubner, K. and Scholer, H. R. (1996). Germline regulatory element of Oct-4 specific for the totipotent cycle of embryonal cells. *Development* 122, 881-94.

Ying, Q. L., Nichols, J., Chambers, I. and Smith, A. (2003). BMP induction of Id proteins suppresses differentiation and sustains embryonic stem cell self-renewal in collaboration with STAT3. *Cell* 115, 281-92.

Yoon, C., Kawakami, K. and Hopkins, N. (1997). Zebrafish vasa homologue RNA is localized to the cleavage planes of 2- and 4-cell-stage embryos and is expressed in the primordial germ cells. *Development* 124, 3157-65.

Yoshida, K., Taga, T., Saito, M., Suematsu, S., Kumanogoh, A., Tanaka, T., Fujiwara, H., Hirata, M., Yamagami, T., Nakahata, T. et al. (1996). Targeted disruption of gp130, a common signal transducer for the interleukin 6 family of cytokines, leads to myocardial and hematological disorders. *Proc Natl Acad Sci U S A* 93, 407-11.

Yoshizaki, G., Sakatani, S., Tominaga, H. and Takeuchi, T. (2000). Cloning and characterization of a vasa-like gene in rainbow trout and its expression in the germ cell lineage. *Mol Reprod Dev* 55, 364-71.

Yu, R. N., Ito, M., Saunders, T. L., Camper, S. A. and Jameson, J. L. (1998). Role of Ahch in gonadal development and gametogenesis. *Nat Genet* 20, 353-7.

Yuan, H., Corbi, N., Basilico, C. and Dailey, L. (1995). Developmental-specific activity of the FGF-4 enhancer requires the synergistic action of Sox2 and Oct-3. *Genes Dev* 9, 2635-45.

Zhao, Y., Lu, H., Yu, H., Cheng, H. and Zhou, R. (2007). Multiple alternative splicing in gonads of chicken DMRT1. *Dev Genes Evol* 217, 119-26.

Reviewed by **Professor Rongjia Zhou,** Department of Genetics & Center for Developmental Biology, College of Life Sciences, Wuhan University, Wuhan, 430072, P. R. China

In: Developmental Gene Expression Regulation
Editor: Nathan C. Kurzfield

ISBN: 978-60692-794-6
©2009 Nova Science Publishers, Inc.

Chapter II

Regulation of Gene Expression during Aging

Eugenia Villa-Cuesta[1]

Department of Ecology and Evolutionary Biology, Box G-W,

Brown University, Providence, RI 02912, USA.

Abstract

Genetic control of proliferation, morphogenesis, and differentiation during development of multicellular organisms is crucial for the proper formation of adults. Regulation of developmental gene expression, however, goes beyond the developmental stages of an organism. The rate at which organisms age for example, is also regulated by many developmental genes.

While it is intuitive to consider aging a simple byproduct of accumulated wear and tear on the organism, it nevertheless has been shown that altering expression of single genes can extend life span significantly; demonstrating that the aging process can be genetically influenced. Many genes involved in developmental processes modulate, later in life, adult aging. For instance, altered expression of genes affecting endocrine signaling, stress responses, metabolism, and growth during developmental stages can increase the life span of model organisms. Furthermore, the study of these genes has revealed evolutionarily conserved pathways for the modulation of aging. In contrast to the precise genetic regulation that occurs during development, life span, while genetically regulated, is not so tightly controlled. For example, there are significant differences in aging phenotype even in monozygotic human twins. In this chapter, I will discuss whether aging is mainly due to an organism's post-embryonic developmental process or a haphazard process, and I will describe and compare the regulation of genetic pathways involved in both development and aging. In doing so, I will consider what these two fields of science can learn from each other to progress our understanding of the regulation of these genes.

---

[1] Eugenia Villa-Cuesta@brown.edu

Keywords: Aging, Development, Gene regulation, Evolution, Life span, Dietary Restriction, IIS, TOR, JNK, Forkhead

Introduction

The majority of pluricellular organisms are derived from a fertilized egg that by successive rounds of cell division, controlled cell growth, differentiation and morphogenesis gives rise to the formation of tissues and organs in the proper moment to form a functional reproductive adult. With rare exception, all somatic cell in an embryo have identical genetic information, the same as the zygote. The difference between cells, thus, is generated by differences in gene expression. A genetic control coordinates the differential activation of groups of genes in specific groups of cells, allowing an increase in complexity of the organism based on structures previously established, with the ultimate objective of producing an organism with appropriate organ's size and function relative to the whole body.

Regulation of developmental gene expression, however, goes beyond the developmental stages of an organism. Organ regeneration, neural plasticity and the rate at which organisms age for example, are also regulated by many developmental genes. Indeed, mutations in single genes can extend life span dramatically, allowing the animal to age normally but more slowly. Furthermore, the study of these genes has revealed evolutionarily conserved pathways for the modulation of aging [1]

Aging and Genetic Regulation
of Aging: The Debate

In an organism, with aging- the passage of time- senescence occurs. Senescence is the process of age-related deterioration. However, because few animals fail to deteriorate with the passage of time, aging and senescence are sometime interchangeable and I will consider them synonymous. Organismal senescence is characterized by the declining ability to respond to stress, and the increasing of susceptibility to disease. In laboratories, longevity is used to measure aging. Longevity is the period of time that an animal can be expected to live under ideal conditions. Two other concepts convenient to have in mind are: average longevity (life expectation) maximum longevity (life span).

The union of signaling pathways controlling development and influencing life span opens the door to a matter of intense debate [2, 3]: aging considered as a post-embryonic development process. If so, aging will be subtracted to a genetic program and therefore selected for in evolution. Development is clearly of the benefit to the organism and it is intuitive to think that natural selection has designed sequential events leading to the formation of a highly functional adult. However, is difficult to understand how evolution selected a program that actively ages the organism. Aging programs, by comparison, seem pretty useless [2].

Evolutionary Senescence Theory of Aging

The most accepted overall theories for aging is the "evolutionary senescence" theory which focuses on the lack of selection against late-life deleterious events, because such events do not decrease evolutionary "fitness". In the wild, predation, environmental degradation and infectious diseases promote the presence of younger individuals reproducing than older individuals. Because natural selection operates via reproduction, if mutations in genes are deleterious later in life but are not affecting reproduction states, can be passed on future generations. In this line of logic, Medawar (1952) proposed the "mutation accumulation" theory; the inability of natural selection to influence late-life traits produce that genes with detrimental late-life effects are continue to be passed through generations. Also, an allele improving survival in centenarians confers no evolutionary advantage either. Two outcomes on the behavior of genetic mechanism in senescence were predicted: 1- deleterious alleles affecting survival or reproduction only late in live, could accumulate in the genome over evolution because of a weak selection and 2- genes that increase the odds of successful reproduction early in life may have deleterious effects later in life [4]. This second option was posed by Williams in 1957 as "antagonistic pleiotropy" theory [5]. In the 1970s, Thomas Kirkwood added the "disposable soma theory", in which organisms have to balance between maintaining their body -soma-, and reproducing [6]. An organism invests resources into reproduction, and over time mutations and other cellular damage accumulate in the soma because the body cannot repair all of it. Following this theory, if an animal is likely to die fast because predation, natural selection would favor allocating all an animal's resources to reproduction, leaving nothing for somatic maintenance and therefore dying early. On the contrary, humans for example, that have few predators, can allocate more resources to repairing damage since they will be able to reproduce over longer period of time (reviewed in [4]).

Increased longevity has been associated with a decrement in early-life components. Reduced fecundity is the most common trade-off, although reduced larval viability, increased developmental time, and decreased body size are also been observed (reviewed in [4]). With the purpose of exemplified genes affecting early and late life, this chapter will summarized some, but not all, signaling pathways and environmental factors influencing both development and aging.

Nutrition in Development and Aging

One key environmental factor shown to influence development and aging is nutrition. In different animals, poor nutrition delays development and reduces adult body and organ size [7]. Dietary restriction (DR), a reduction in food intake while avoiding malnutrition, extends life span in diverse organism including yeast *Saccharomyces cerevisiae*, *Caenorhabditis elegans*, *Drosophila melanogaster* and mouse. DR is still the only environmental intervention shown to extend the longevity of both invertebrates and vertebrates [8].

The current theory about the response of organisms to DR is that this mechanism evolved to provide animals protection during times of famine [8]. Because DR extends life span in nearly every species tested, the ability to increase longevity when food is limited may confers some fitness advantage that caused the trait to be selected for during evolution. Hence, in periods of food scarcity when reproduction might be costly and the chances of progeny surviving low, increasing somatic maintenance at the cost of growth and reproduction would increase the evolutionary fitness of the organism [9, 10]. However, if increased life span in response to DR is an evolutionary adaptation to increase survival in the wild, it must be regulated genetically, and those pathways involved might be conserved between species. Nevertheless, is still unknown if the mechanism through DR extends life span in different organisms is conserved or if it is a case of evolutionary convergence [11].

Insulin /IGF -Insulin like Growth Factor- Signaling (IIS)

Mutations in genes that encode components of the insulin/insulin-like growth factor signaling (IIS) pathway result in a robust extension of life span that is conserved from *C. elegans* [12, 13] to Drosophila [14, 15] and mammals [16-18]. IIS pathway has also been conserved during evolution as a regulator of growth and physiology upon changes in nutrient availability [7, 19, 20].

Insulin and insulin-like growth factors (IGFs) signal via the insulin receptor (InR), and IGF receptor respectively. These receptors share common downstream signaling components, and there is some evidence of cross talk between the insulin and IGF signaling systems (reviewed in [21]). The mammalian insulin and IGF receptors are tyrosine kinases activated by ligand-induced transphosphorylation. InR, for example, consists of a dimmer and insulin binds to the α-subunit of the insulin receptor, which activates the tyrosine kinase activity present in the β-subunit. Insulin receptor substrate-1 (IRS-1) is a tyrosine-phosphorylated substrate for the insulin receptor which phosphorylated on multiple tyrosine residues lacks any enzymatic activity but acts as a scaffold to bind to Phosphoinositide-3 kinase (PI3K). Activation of PI3K results in the generation of phosphatidylinositol lipids phosphorylated at the 3'position of the inositol rings (PI3Ps), which functions as a second messenger to active downstream signaling. Dephosphorylation of PI3P lipids is catalyzed by several phosphatases, most importantly, the tumor-suppressor phosphatase and tensin homologue (PTEN). An important downstream of IIS is the serine-theronine kinase AKT. Activation of AKT requires phosphorylation by the phosphoinositide-dependent kinase-1 (PDK-1) in addition to PI3P lipids (rewied in [20]) (Figure 1). Thus, Insulin activates a cascade of events that, among others, results in the activation AKT and in turn influences gene expression through the forkhead-related FoxO family of transcriptions factors (described in FoxO secction).

In worms, the homolog of FoxO, Daf-16, has emerged as the major target of AKT; loss of *daf-16* function suppresses the majority of phenotypes associated with loss of insulin signaling in *C. elegans* [22, 23]. In contrary, in mammals, numerous phosphorylation targets of AKT have been identified besides FoxOs [24, 25]. Therefore, because many of the functions of IIS are mediated by FoxOs factors, some of the biological functions of IIS are described below and some are described in the section of biological function of FoxOs.

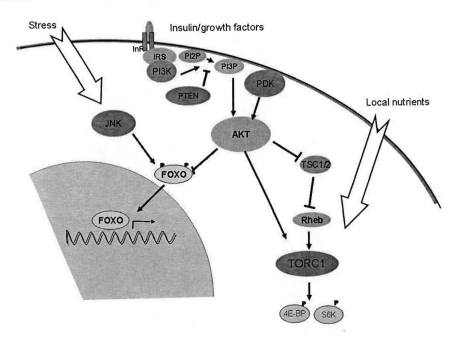

Figure 1. An integrate model for signaling pathways affecting development and aging. The cell, under different external inputs, activates IIS, TOR, and JNK pathways which will play a crucial role in transcriptional and translational regulation of gene expression. The signaling proteins shown in the figure are discussed in the text.

Biological Functions of IIS

IIS pathway, as well as the Target of rapamycin (TOR) signaling pathway (see below), plays a central role in nutrient sensing, regulating animal growth and energy balance in response to environmental conditions. In Drosophila, expression of insulin-like peptides (dILPs) is nutritionally regulated [26], as is expression of IGF1 in developing rats [27]. Genetics studies in Drosophila demonstrate that modulation IIS during development alters growth, affecting cell size and number, both in clones of cells and in the entire animal. Theses alterations in growth are sufficient to affect final body size (reviewed in [21]). Also in mice, mutations of the IGF1 receptor show delayed development and growth deficiency with a reduction in body and organ size [28, 29]. However, strong *InR* mutations in Drosophila are recessive embryonic lethal [30] [31], indicating an essential function for *InR* during normal development beyond growth control. Likewise, the phenotype of mice lacking insulin receptors is more severely altered than that of mice lacking IRS proteins. This is consistent with the position of *InR* as the first step in the signaling pathway and suggests that in both mammals and in flies, multiple substrates are required to mediate insulin action [32].

Besides growth, the primary function of insulin signaling in mammals is to maintain blood sugar homeostasis by regulation glucose uptake and metabolism [33]. The role of insulin in the regulation of food intake and energy expenditure via actions in the brain has been extensively studied. Insulin from the periphery acts in the hypothalamus to inhibit food intake and increase energy expenditure. Conversely, lower insulin levels in response to decreased food intake or energy stores act centrally on neuroendocrine pathways to suppress energy expenditure and stimulate eating (reviewed in [34]).

In flies, insulin signaling pathway also has a role in the regulation of metabolism as well as reproduction via modulation of central neuroendocrine pathways [35]. The phenotype conferred by hypomorphic *InR* or *chico* (Drosophila homolog of the insulin receptor substrate) mutations also includes female sterility [31, 36]. Interestingly, flies exhibit female sterility, increased triglyceride stores during diapause; a physiological adaptation to stressful environmental conditions (reviewed in [37]). Increased stress resistance is also manifested during diapause [37] and this, in conjunction with heightened energy stores, enhances survival and hence the probability of reproduction. Entry into diapause in flies is initiated by a decrease in the level of Juvenile Hormone (JH) [37]. The similarity between diapause and the *InR* mutant phenotype suggested that reduced InR function might be impacting on JH levels, and indeed, JH was found to be reduced by 80% as a result of *InR* mutations [38]. The increased triglyceride storage in *InR* mutant flies probably reflects a well-conserved role for insulin signaling in the central regulation of whole body energy stores in response to nutrient availability. Similarly, mice with loss of brain insulin receptors [39] or insulin receptor substrate 2 deficiency [40] have increased body fat and food intake and female infertility owing to reduced pituitary gonadotropin release exhibit. These data suggest that mutations leading to decreased InR function mimic signaling in the context of low nutrient availability and that central InR serves as a key upstream regulator of the neuroendocrine system. Thus, the level of InR activity could reflect environmental conditions (such as nutrient levels), neuroendocrine-mediated physiological program designed to enhance survival (reviewed in [35]).

IIS and Longevity

As in other biological functions, regulation of life span by the IIS pathway has been conserved during evolution. Mutations in genes components of the IIS pathway in mammals extend life span: *InR* [16], *IGF-1 receptor* [17], *Irs-2* [18] and *Irs-1* [41]. Female mice null for *Irs-1* are also healthier at older ages [41].

The role of IIS in aging was first described in *C. elegans* by a genetic screen [42] that produced a collection of mutants extending adult life span. This screen revealed for example that mutations that lower the level of *age-1*, (PI(3)K homolog) extend life span [43]. Further analysis identified that mutations in *daf-2*, which encodes an insulin/IGF-1 receptor homologue [44], cause the animal to remain active and youthful longer than normal and to live more than twice as long [13, 45]. Extension of lifespan of both *age-1* and *daf-2* require the presence of *daf-16*, implying that both normally suppress the activity of *daf-16* [13][46]. Mutations in several genes components of the IIS pathway extend lifespan in Drosophila. The first IIS mutation with this effect were in the gene *chico* [14, 38] and the *dInR* [38] and since then, it is not clear yet if dFoxO is necessary for the life span extension of both mutants. The Drosophila genome contains seven genes encoding dILPs which are thought to be the ligands for the unique dINR. dILPS are produced in IPCs (dILP producing cells). Partial ablation of the IPCs by driving expression of the pro-apoptotic gene *reaper* specifically in theses cells increase life span [47]. Overexpression of the Drosophila PTEN [48], and over-expression of dFoxO [48, 49] in fat body also extend life span of flies (see below).

Forkhead Transcription Factors

A good example of the correlation between an evolutionary diversification of protein families and an increase in complexity of cell types during development of multicellular organisms, is presented by the Forkhead DNA-binding domain class of transcription factors (Fox: Forkhead box). Since the first member identified, FoxA [50], until now, more than 100 members of this family have been described with roles in morphogenesis and differentiation of the embryo and distinct functions in the adult [51]. Based on the homology within the Fox domain, the Forkhead proteins are grouped in 19 subclasses of Fox factors, FoxA-FoxS [52]. While some are ubiquitously expressed in variety of tissue types, others are expressed in a restricted spatial-temporal manner [53]. This chapter will focus on FoxA and FoxO, the only two components of Forkhead transcription factors family involved in the regulation of both, development and aging.

FoxOs

The FoxO family of Forkhead transcriptions factors have the most different DNA sequence within their DNA-binding domains, for this reason are named "other" (O) class of Forkhead superfamily [52]. *FoxO* genes are highly conserved throughout evolution and many roles are conserved among different model organisms [54]. Mammals have four members of FoxO transcription factors: FoxO1, FoxO3, FoxO4, and FoxO6, with differential but overlapping expression patterns in development and in adulthood. Furthermore, FoxO1, FoxO3 and FoxO4 bind to the same DNA target sequence and are able to regulate the same target genes. Contrary, FoxO6 is expressed only in the brain, and it is predominately nuclear while the cellular localization of the other isoforms of FoxOs is highly regulated [52, 54]. In contrast to mammals, invertebrate models organisms have only one FoxO isoform; Daf-16 in *C. elegans* and dFoxO in Drosophila. Genetics studies in these two invertebrates demonstrate that activation of the AKT by insulin or IGF suppresses activity of Daf-16 (reviewed in [55, 56]). In mammalian cells, biochemical analyses also show that FoxO transcriptional activity is also regulated by IIS through direct phosphorilation of AKT. The AKT-phosphorylated FoxO proteins bind to 14-3-3 chaperone proteins, sequestered in the cytoplasm and therefore unable to regulate gene expression. Loss of growth factor, on the contrary, leads to AKT inactivation, FoxO dephosphorylation in its AKT sites, nuclear translocation and target gene activation [55] [57] (Figure 1).

The activity of FoxO proteins is tightly regulated by phosphorilation by a variety of kinases besides AKT, acetylatilation, and ubiquitylation. These posttranslational modifications allow FoxO factors to sensor the cell status and rapidly change their activity.

Biological Functions of FoxOs

FoxO factors have emerged as a convergence point of signaling in response to growth factor stimulation, oxidative and nutrient stress and coordinate, by modulating the expression of genes, a wide range of outputs including apoptosis, proliferation, cell differentiation, cell cycle transition, DNA damage, glucose metabolism, stress response, neuropeptide secretion, atrophy, autophagy, and longevity (reviewed in [55, 58]). Therefore, FoxO factors regulate cellular processes involved in cell-fate decisions in a cell-type- and environment-specific manner.

Cell cycle transition and DNA damage. Cell proliferation involves entry into the cell cycle form a quiescent state (G0 phase) and successful progression through G1 (gap phase), S (DNA synthesis), G2 (gap phase 2) and M (mitosis) of the cell cycle. The transitions between phases are under stringent control primarily by the cooperative activity of specific cyclin-dependent kinases (CDKs) and their regulatory subunits, the cyclins [59, 60]. Cell-cycle progression is stimulated by PI3K/AKT signaling activation by insulin that relies in the absence of FoxO induced cell-cycle inhibition. FoxO factors stabilize G1 and G2/M checkpoints by the activation of a number of cell cycle regulatory proteins, including p130, members of the Cip/Kip and INK4 CK1 families, cyclin G and Gadd45a. In addition, FoxO factors can facilitate cell cycle arrest through the inhibitions of the D-type cyclin and FoxM, a subgroup of Fox factors that positively drive G2/M phase transition (reviewed in [61]). Consequently in the absence of insulin signaling, FoxOs induce cell cycle arrest.

DNA damage affects FoxO activity via cyclin dependent kinase 2 (CDK2). CDK2 phosphorylates FoxO1 and sequester it on the cytoplasm in the absence of DNA damage. If DNA is damaged, the repressive effect of CDK2 on FoxO1 is abolished and FoxO1 translocates to the nucleus to induce apoptosis [62].

Apoptosis. Apoptosis is required for the proper development of an adult and plays a critical role in tumorigenesis. The link between activation of FoxO proteins and cell death was first established by finding that overactivation of PI3K/AKT signaling is characteristic of many human cancers. When PI3K/AKT is not activated FoxO proteins are in the nucleus triggering apoptosis by inducing the expression of death genes such as *FasL* gene [63] which encodes a protein that activates the death receptor Fas/CD95/APO-1 and promotes mitochondria-independent apoptosis. Furthermore, tumor-necrosis-factor-related apoptosis-inducing ligand (TRIAL), another death cell receptor ligand, is regulated by FoxO in prostate cancer cells [64].

In addition to death receptor ligands, genes encoding members of the pro-apoptotic BH3-only subgroup of BCL-2 family proteins (*Bim* and *bNIP3*) involved in the mitochondrial apoptotic pathway are targets of FoxO proteins (reviewed in [65]).

Skeletal muscle atrophy, autophagy and metabolism. Decreased activity of PI3K/AKT signaling pathway, nutrient deprivation and starvation leads to muscle atrophy. Skeletal muscle undergoes protein degradation induced by the activity of ubiquitin-proteosomal atrophy and lysosomal autophagy [58]. Autophagy is a protective cellular response to nutrient deprivation in which lysosomal enzymes mediate the recycling of cellular components [66]. FoxO factors are key mediators of both atrophy and autophagy in muscle in response to nutrient deprivation [58]. When cultured myotubes are undergoing atrophy, the activity of

PI3K/AKT pathways decreases, leading the activation of FoxO transcription factors and the induction of two muscle-specific E3 ubiquitin ligases, *atrogin-1/MAFbx*, *MuRF1* and other components of the ubiquitin proteasome system, resulting in skeletal muscle atrophy without apoptosis (reviewed in [58]). Energy-rich exercising muscles that are resistance to atrophy express high levels of the peroxisome-proliferator-activated receptor γ coactivator 1 α (PGC1 α) an important coregulator of many transcription factors including FoxOs. The expression of PGC1 α messenger is decreased in skeletal muscles during atrophy. Transfeccion of PGC1 α in adult fibers decreases FoxO induced atrophy and expression of *atroging-1* which could explain why exercise might combat muscle atrophy [67]. In addition, in myotubes, FoxO3 activates the transcription of autophagy-related genes including *LC3*, *Bnip3*, *Gabarapl1*, and *Atg121* (reviewed in [58]). Induction of autophagy by FoxOs factors is not restricted to mammalian muscle. In Drosophila, activation of dFoxO under starvation also induces autophagy in fat body [68].

Glucose and lipid metabolism. Besides atrophy and autophagy in skeletal muscle, FoxOs are involve in the switch form oxidation of carbohydrates to fatty acids as the major energy source in muscle metabolism [69]. Glucocorticoids, whose levels also increase during starvation, boost expression of FoxO factors (reviewed in [70]). An overexpression of FoxO1 in myocytes increases lipoprotein lipase the enzyme that hydrolyses plasma triglycerides into fatty acids and glycerol for uptake by the muscle cell [71]. In essence, an upregulation of FoxO in differentiated myotubes prepares the muscle for an increased reliance on fatty acid oxidation, similar to fasting and exercise [69, 70].

In the liver, FoxO1 mediates the expression of genes involved in both glucose and lipid metabolism. FoxO regulates glucose levels thorough transcriptional activation of various genes, including *PGC1α, phosphoennolpyruvate carboxykinase (PEPCK)* and *glucose-6-phosphatase (G6Pase)* [58]. Studies in β-cells in the pancreas suggested that the FoxO factors can act at multiple levels to increase circulating glucose by either decreasing insulin secretion or by attenuating β-cells division [58]. In addition FoxO1 regulates the expression specific apolipoprotein which has the ability to alter lipase activity and thus lipid metabolism in peripheral tissues (reviewed in [70]). Furthermore, FoxOs factors also acts in the hypothalamus regulating food intake and energy homeostasis by coordinating neuropeptide production of the hypothalamic neurons [58]. FoxO1 acts to increase food intake and bodyweight by inducing the expression of orexigenic hormones *agouti-related protein (Agrp)* and/or *neuropeptide Y (Npy)* and represses indirectly the expression of anorexigenic neuropeptides such as *proopiomelanocortin (Pomc)* [72]. Therefore, FoxO transcripts have evolved to help the organism adapts to nutrient deprivation. When food is available, FoxOs appear to increase organismal food intake; in unfavorable nutrient conditions, FoxO factors alter organismal metabolism maintaining systemic glucose and lipid levels [58].

Immune system. In the immune systems FoxO proteins act in multiple stages to limit the expansion and/or activation of the mature hematopoietic cells, however how FoxOs are activated in these cells is still unknown [58]. The ablation of FoxO3 in vivo results in T cell proliferation and hyperactivity because of the activation of NF-κB, which produces a muti-system inflammatory syndrome and suggests a role of FoxOs in suppression of inflammation in mammals (reviewed in [73]).

Stress resistance. Besides the role of FoxO factors in nutrient stress discussed above, activated FoxO proteins promote oxidative stress resistance by binding to the promoters of the genes encoding manganese superoxide dismutase (MnSOD) and catalase, two proteins that play essential roles in oxidative detoxification in mammals (reviewed in [55]). Several stress-induced pathways intermingle to FoxOs factors, including SIRT1, a member of mammalian SIR2 family, and Jun-N-terminal kinase (JNK). SIRT1 is a NAD-dependent protein deacetylase which forms a complex with FoxO in response to oxidative stress [63]. Additionally, an increase in oxidative stress also leads to activation of JNK pathway, nuclear localization of FoxO, induction of stress defense genes and increase the life span of Drosophila [74] and *C. elegnas* [75] (See below in JNK section).

Another example of cellular stress is hypoxia. FoxO3 transcription is induced by cellular hypoxia via direct binding of the hypoxia-inducible factor HIF1 to the FoxO3 promoter. Surprisingly, the increased expression for FoxO3 results in an enhance cellular survival by attenuating HIF-induced apoptosis [76].

Foxos and Longevity

Studies from invertebrates have addressed the function of FoxOs in the context of the entire organism. The postembryonic life cycle of *C. elengas* consists in four larval stages (L1-L4). When environmental conditions are not favorable, *C. elegans* can enter into an alternative larval stage, the dauer stage, rather than to progress to L3, [77]. The dauer larva is developmentally arrested and adapted to long-term survival, characterized by low metabolic activity and a long life span in response to starvation. However, when conditions improve, dauer larvae restart their life cycle at the stage of L4. Large set of genes are involved in the control of dauer formation, prolonging the lifecycle and also extending life span on adults [78]. Such is the case of Daf-16 [22, 23]. In Drosophila, the role of dFoxO in life span was evidenced by the finding that overexpression of dFoxO in adult fat body is sufficient to increase longevity [48, 49]. Also Overexpression of dFoxO prevents the age associated decline of cardiac functions [79]. Apparently, in worms and flies, FoxO activity in the cells of specific tissues is more pertinent to promote long life span than other [48, 49, 80-82] which interestingly suggests that Daf-16/dFoxO regulates the production of secondary signals or hormones that coordinate the life span of tissues throughout the organism [83]. In multiple laboratories, the role of FoxO factors in the regulation of longevity has been described as independent of dietary restriction [84-89]. Contradictory, in *C. elegans* a recent study using different method of DR, found that daf-16 was required for DR to extend life span [90] enhancing, may be, the importance of different methods of dietary restriction in the study of model organism. So far, it has not been described an implication of FoxO factors in the control of mammalian life span.

Post-Translational Modifications of Foxo Proteins

To better understand the role of FoxO factors in development, metabolism and life span is important to understand the molecular details of FoxO regulation. FoxO transcription factors are highly post-translationally modified. Most of the stimuli outcome in phosphorylation, ubiquitylation and acetylation, and these posttranslational modifications result in altered subcellular localization, protein stability, DNA binding properties and transcriptional activity (reviewed in [58]).

Phosphorylation. In response to growth and survival factors FoxO proteins become phosphorylated and localized in the cytoplasm. Contrary, FoxO proteins remain in the nucleus in response to various stress events even in the presence of growth factors [55]. In Addition to IIS pathway growth factors, the oxidative-stress-regulated mammalian sterile 20 like kinase-1 (MST1) binds and phosphorylate FoxO directly *in vitro* and *in vivo* assays [91]. Interestingly MST1 phosphorylates FoxO within the conserved DNA binding domain; therefore, phosphorylation would be expected to weaken DNA binding. MST1 also activates JNK pathway, thus MST1 might regulate the activity of FoxO proteins also through JNK-dependent pathways [91]. Other kinases such as Serum and glucocorticoide-inducible kinase (SGK), casein kinase 1, dual-specificity tyrosine-phosphorylated, regulated kinase 1A and anti-I-kappa-B kinase beta, also regulate FoxO inactivation by active nuclear export to the cytoplasm [92]. In worms, SMK1, which encodes the regulatory subunit of PPH-4.1 (protein phosphatase 4), is a co-activator of FoxO/Daf-16, suggesting that this phosphatase could also participate in the regulation the phosphorylated status of FoxO factors [93].

Acetylation. Acetylation plays a double role in FoxOs gene transcription: it can facilitate FoxO mediate transcription by acetylation of chromosomal histones, and also acetylation and regulation of FoxO proteins themselves. The proteins CBP and p300 have intrinsic histone acetyl-transferase (HAT) activity. Theses proteins promote transcription by acetylating histones and have also the ability of acetylate transcription factors [94]. Mamalian FoxOs are acetylated by CBP and p300 and are deacetylated by SIRT1 (reviewed in [57]). The precise roles of acetylation and SRIT1 in FoxO regulation are unclear. But several studies suggest that acetylation inhibits and SRIT1 activates FoxO while other suggests that acetylation activates and SIRT1 inhibits FoxO (reviewed in [57]). Interestingly, the SIR2 deacetylases extend life span in yeast, worms and flies [95-97]. The ability of SIR2 to deacetylate FoxO factors may contribute to life span extension and this will be consistent with the fact that deacetylation activates FoxO (reviewed in [98]).

Ubiquitylation. Several studies have described polyubiquitylation and the subsequent proteasomal degradation of FoxOs [57]. SKP-2 has been identified as a putative E3 ubiquitin ligase that polyubiquitylates FoxOs [99]. In *C. elegans*, RLE-1 (regulation of longevity by E3) is an E3 ubiquitin ligase of Daf-16 that regulates longevity by polyubiquitination of Daf-16 and proteosomal degradation [100]. Besides polyubiquitylation, FoxOs are also monoubiquitylated after oxidative stress [101, 102], leading to the nuclear localization of FoxO and an increase in its transcriptional activity; however the mechanism that mediates its nuclear localization is unclear [57].

FoxAs

The first identified member of the Forkhead box transcription factors family was the gene *fork head* (*fkh*), the homolog of FoxA in Drosophila. *fkh* mutants are defective in the formation of the anterior and posterior gut, and the embryos show tow spiked-headed structures [50]. In *C. elegans*, as in Drosophila, there is only one member of FoxA factors, named Pha-4. However, mammals have three members, FoxA1, FoxA2, and FoxA3 [103]. FoxA transcription factors play important roles in multiple stages of mammalian life, early development, organogenesis, metabolism and longevity of the adult [104, 105]

Biological Functions of FoxAs

In mammals FoxAs are required for early embryogenesis for the formation of specific organs and they control specialized functions of differentiated tissues. Thus, mouse embryos homozygous for a null mutation in FoxA2 presents abnormalities of the neural tube and somites, absence of notochord, and no gut tube [106]. The effects in the neural tube and notochord can be attributed to the loss of expression of *Sonic hedgehog* (*Shh*), as FoxA2 activates the expression of this gene [107]. FoxA1 and FoxA2 are required for normal development of endoderm-derived tissues, including liver, pancreas, lungs and prostate (reviewed in [103]). FoxA1 is essential for proper secretory functions of the kidneys [108] and all three FoxAs act in various tissues to regulate glucose metabolism. Furthermore, they control gluconeogenesis and fatty acid metabolism in the liver [103]. Interestingly, FoxAs cooperate with glucocorticoids to regulate the expression of hepatic genes [109] and in breast cancer, the oestrogen responses depend of estrogen receptor and the cooperation with FoxA1[110]. In Drosophila *fkh* gene is required for the proper development of the endoderm and ectoderm derived parts of the gut, the Malpighian tubules and the salivary glands [50, 111, 112]. In the salivary glands, Fkh acts as a survival factor and participates in the normal control of apoptosis. Besides, Fkh is important for secretory functions of this tissue [112, 113]. As in mammals with glucocorticoids, Fkh cooperates with a steroid hormone, 20 hydroxyecdysone (20E), to activate some of the targets genes [114]. *fkh* is also negative regulated by 20E [113, 115, 116]. In *C. elegans* Pha-4 regulates the development of the pharynx (reviewed in [117]) and knocking down *pha-4* expression by RNAi during post-embryonic development showed that Pha-4 is essential for is essential for dauer recovery, gonad and vulva development [118] .

FoxAs and Longevity

In *C. elegans*, *pha-4* is the only Forkhead transcription factors of 15 identified necessary for the increase in life span by dietary restriction (DR) [86]. Pha-4 is constitutively nuclear and *pha-4* transcripts increase two fold in response to DR [86]. Loss of Pha-4 blocks the extended longevity of animals subjected to DR, independently of IIS mediated longevity [86]. Because Pha-4 is essential early development, Ponawski and collaborators, determined that the developmental function of Pha-4 is separable from its role during adulthood in modulating the response to DR using a temperature-sensitive mutant allele of *pha-4* or conditional RNAi. Significantly, the adult-specific reduction of *pha-4* expression, after development has completed, still blocks the response to DR, nevertheless, it is not yet clear how Pha-4 activity is regulated in response to DR.

TOR: Nutrition Sensing Pathway

Target of Rapamycin (TOR) signaling pathway is one of the main pathways that integrates signals form growth factors and nutrients to control cell growth. TOR signaling influences several processes, including transcription, translation and ribosome biogenesis, that all together determine the mass of the cell and controlling that, the overall body size [119]. TOR was originally identified in yeast through mutants that confer resistance to growth inhibitory properties of the immunosuppressive macrolide rapamycin [120, 121]. In yeast TOR has two homologous genes, *TOR1* and *TOR2*, whereas other eukaryotes appear to contain only one *TOR* gene [122]; *TOR1* in fungi, *AtTOR* in plants, *CeTOR* in worms, *dTOR* in flies and *mTOR* in mammals [119]. TOR is a member of the phosphatidylinositol kinase - related –kinase (PIKK) family that structurally resemble phosphoinositide kinases but possess serine/threonine rather than lipid kinase activity [123]. TOR has many protein-protein interaction domains and consistently is part of multi-protein complexes. In cells, mTOR is found in different complexes with different proteins including, mLST8, Raptor or Rictor. TOR complex 1, TORC1, when mTOR is associated with mLST8 and Raptor and TORC2 when associated with mLST8 and Rictor (reviewed in [124]). TORC1 is characterized by its sensitivity to rapamycin, whereas, TOR2 is resistant to it [124].

The activity of mTORC1 is controlled by nutrients (amino acids) and by hormones and growth factors that activate the protein kinase AKT. Therefore IIS and TOR pathways are connected through AKT signaling (Figure 1). The signaling pathway downstream of AKT involves the protein products of the genes mutated in tuberous sclerosis, TSC1 and TSC2, and the small guanosine triphosphatase, Rheb (Ras homolog expresses in brain) [119, 124]. TSC1 and TSC2 have a negative effect on mTOR1 signaling through the repression of mTOR1 activator Rheb [119, 124] (Figure 1). By phosphorylating TSC2, AKT inhibits TSC1/TSC2 TOR inhibition towards Rheb and therefore promotes mTORC1 activation [124]. Interestingly, AKT is able to regulate directly mTORC1 independently of TSC1/TSC2 [125]. In addition to direct phosphorylation by AKT, TSC2 can be inactivated by phosphorylation by other kinases induced for examples under energy starvation and hypoxia.

In Drosophila, dTOR is regulated by IIS/AKT pathway, but it may not be only by a direct phosphorylation of TSC2 by AKT [126]. Furthermore, *C. elegans* genome lacks of TSC1/2 homologs, and have been suggested that *C. elegans* AKT (CeAKT) will be mediating its effects of CeTOR by phosphorylation and inhibition of Daf-16. Daf-16 has a potent negative effect of Daf-15 (Raptor's homolog in *C. elegans*) expression [127] and as a result, the inhibition of Daf-16 by CeAKT may allow Raptor/Daf-15 to form CeTOR1 [124].

Biological Functions of TOR

TOR complexes have different functions, although both are important in the regulation of cell cycle progression. TORC1 represses autophagy, regulates protein synthesis, couples cell size to cell-cycle progression, whereas TORC2 regulates cell cycle-dependent polarization of actin cytoskeleton (reviewed in [119, 124]). Loss-of-function studies in mice shown that mTORC1 is critical at early stages of embryogenesis, while mTORC2 is required at mid-gestation and subsequent embryogenesis [124].

Cellular growth and proliferation might be proportional to the rate of protein synthesis [128-130]. Sustained increases in the rates of protein synthesis are required during normal cellular processes such as organ regeneration, but also pathological conditions such as cardiac hypertrophy and tumour development [131, 132]. Cells can up-regulate their rates of protein synthesis either via increasing the translation efficiency of existing ribosomes and/or by increasing the capacity for translation through the production of new ribosomes (ribosome biogenesis) [133]. TORC1 regulates theses processes largely by the phosphorylation and inactivation of the repressors of mRNA translation 4E-binding proteins (4E-BPs) and by the phosphorylation and activation of the ribosomal protein S6 kinase (S6K) [134]. While 4E-BP retains cap-dependent translation in the absence of growth factors and/or nutrients, S6K acts to stimulate protein synthesis when growth factors and nutrients are available. S6K also contributes to the regulation of ribosomal RNA (rRNA) synthesis [135], a limiting step of ribosome biogenesis (reviewed in [133]). In mice and Drosophila, loss of S6K results in a small animal due to a reduced cell size. In mice, in addition to the reduction of body size, animals have type 2 diabetes mellitus phenotype, hypoinsulinaemia and glucose intolerance as result form a decreased on pancreatic β-cells [136]. S6K knockout mice are resistant to develop age and diet obesity due to an inactivation of the negative feedback loop between S6k and insulin receptor [137]. Inactivation of 4E-BP by TORC1 phosphorylation results in a release of eIF4E (eukaryotic initiation factor 4E, a component of the cap-binding complex that directs the translational machinery to the 5′end of mRNAs) and translation initiation can occurs [138]. Overexpression of eIF4E in a mouse model of B cell lymphoma accelerates tumorigenesis (reviewed in [124]).

In Drosophila, the 4E-BP has been identified as a target gene of dFoxO. Like *dfoxo* mutants, *d4EBP* mutants do not exhibit a growth phenotype in a wild-type background [139]. However, the reduced cell number phenotype of viable *dAKT* mutant combination is partially suppressed by simultaneous mutation of *d4EBP* [140]. Interestingly, reducing *dTor* expression specifically in the fat body in Drosophila acts non-cell autonomously to regulate growth in other tissues [141].

TOR Pathway and Longevity

Diminished TOR signaling can extend the life span of yeast, *C. elegans* and Drosophila. In mammals, although much of the mechanics of TOR signaling is largely conserved, it is presently unknown whether diminished mTOR activity correlates with an extension of life span. Yeast aging is usually analyzed in one of two different assays; replicative life span, which is defined as the number of times a mother yeast cell can give rise to a daughter bud, and chronological life span, defined as the length of time that a non-dividing yeast cell can remain viable in culture. *S. cerevisiae* deleted in the TOR2 gene are not viable, while deletion of TOR1 increases replicative [142] and chronological life span [143, 144].

In *C. elegans*, depleted of *CeTOR* by using RNA interference [145] and heterozygous for mutations in *daf-15* are long-lived [127]. Interestingly, *CeTOR* genetically interact with *daf-2* to influence longevity [145]. Furthermore, suppression of downstream targets of TOR that regulate translation also increases life span [146, 147]. In Drosohila, overexpression of dTSC1, dTSC2, or dominant negative forms of dTOR or dS6K all cause life span extension [148].

Reduction of amino acid intake is sufficient to extend life span in yeast [143], Drosophila [149], and rodents [150, 151]. For that reason, recently focus has turned toward the amino acid-sensing TOR pathway as another candidate mechanism through which DR may act but this remains to be proven empirically.

Stress-Responsive
Jun-N-Terminal Kinase (JNK) Pathway

The c-Jun NH_2-terminal kinases (JNKs), also known as stress-activated MAP kinases (SAPKs), are a subgroup of mitogen-activated protein kinases (MAPKs) activated by a kinase signaling cascade [152]. In addition to c-Jun, JNK also phosphorylates activation transcription factor-2 (ATF2) and other Jun family proteins involved in the AP-1 transcription factor complex (a dimeric transcription factor composed of members of c-Jun and c-Fos families) playing, therefore, a crucial role in transcriptional regulation of gene expression (reviewed in [153]). Additionally, JNKs phosphorylate non-transcription factors participating in processes independently of gene transcription [154] [155].

The mammalian JNK consists of several polypeptides encoded by three genes: *JNK1*, *JNK2*, and *JNK3*. *JNK1* and *JNK2* are ubiquitously expressed, whereas, *JNK3* is restricted to the brain, heart and testis [152]. These genes are alternatively spliced to create at least ten JNK isoforms [156]. JNK pathway is activated by exposure the cell to stress and has been implicated in opposing cell functions such as apoptosis and cell survival [157]. In one way, JNK may mediate the effects of stress on cells and, in other way, be a protection to cells from stress insult. The specific roles of JNK depend upon the cellular context, and it is activated by different physiological, pathological and environmental circumstances [74, 158].

Biological Functions of JNK

JNK signaling is involved in the regulation of cell survival, cell migration, morphogenetic movement of epithelial sheets, tissue polarity, and longevity. The diversity of the roles of JNK most likely depends on the signal that is activated or the protein expression profile of the different cell types that express it.

The role of JNK in apoptosis/survival was first studied in neuronal cell death in response to neurotrophic factor withdrawal [159]. PC-12 (rat pheochromocytoma) cells differentiate in presence of nerve growth factor (NGF), once differentiated, the withdrawal of NFG causes activation of JNK and subsequent cell death [153]. The response of PC-12 cells to JNK was similar to cells exposed to UV or heat shock [159]. Furthermore, transfection assays using dominant negative and gain-of-function of JNK pathway demonstrate that JNK contributed to the apoptotic response (reviewed in [152]).

JNK1/2 double knockout mice are embryonic lethal with severe dysregulation in the control of cell death in the hindbrain and forebrain [160, 161]. *JNK3* knockout mice (but not *JNK1* or *JNK2* knockout) have increased resistance to kainic acid-induced seizures and cell death of hippocampal neurons [162]. In Drosophila there is only one JNK gene, *basket* (*bsk*) which is also involved in cell death mechanisms. Downregulation of *bsk*, partially suppresses the massive cell death induced by overexpression of *reaper*, [163]. *reaper* appears to modulate JNK pathway through the degradation of the Drosophila inhibitor-of-apoptosis protein 1 (*DIAP1*) [163]. JNK is also involved in regulation of developmental apoptosis in flies. When the gradient of the TGF-β homolog encoded by *decapentaplegic* (*dpp*) or the fly homolog of mammalian Wnt, *wingless*, morphogen is distorted in the developing wing, JNK-mediated apoptosis is activated [164].

Also in flies, JNK is required for embryonic epithelial cell sheet movements and epithelial planar polarity. Dorsal closure in the embryo and imaginal tissue closure, are morphogenetic processes that involve epithelial cell sheet movements (reviewed in [165]). In the dorsal closure, lateral dorsal epithelia move towards the dorsal embryonic midline. The two edges of the dorsal epithelium finally meet and form a suture that seals the dorsal epidermis. JNK is required in dorsal closure, for Jun phosphorylation and expression of *dpp* in the cells that form the leading edge of the lateral epithelial cells sheet. *dpp* is required for the elongation and spread of the lateral epithelial cells that cover the dorsal surface of the developing embryo. Similar processes occur in the thorax closure (reviewed in [165]). In epithelial planar polarity, *dishevelled* is a key component of the Wnt signaling and planar polarity pathways. *dishevelled* activates JNK to arrangement cells within the plane in the eye imaginal disc (reviewed in [152]).

Mice with a single JNK2 gene and no JNK1 have open eyelids small lenses and retinal coloboma, and only 20% survive to adulthood [166]. In retina, the JNK pathway control the expression of BMP4, the analog of *dpp*, which directs the expression of transcriptional factors crucial for optic fissure closure and retinal development [166]. In addition to tissue morphogenesis in development, the mammalian JNKs are implicated in cell migration in tissue remodeling, angiogenesis, tumor invasive phenotype, and wound healing (reviewed [165]).

JNK Pathway and Longevity

Expression profiling experiments determined that JNK signaling in flies coordinates the induction of protective genes in response to oxidative stress [74]. Furthermore, flies with decreased JNK signaling are more sensitive to oxidative stress while flies with an increased

JNK signaling are more resistance [74]. According to the "free radical theory of aging", the sensitivity to oxidative stress is a determinant for life span. This happens to be the case for JNK signaling. Activation of JNK signaling can extend life span in Drosophila [74] and in *C. elegans* [75]. Interestingly, in both organisms, JNK extended life span requires FoxO. Furthermore, in Drosophila, JNK promotes nuclear translocation of dFoxO and induces the expression of dFoxO-dependent stress response genes [167]. In *C. elegans*, JNK has been shown to directly interact with and phosphorylates Daf-16 and, in response to heat stress, promotes the translocation of Daf-16 into the nucleus [75]. Theses data suggest an antagonistic relationship between JNK signaling and IIS and a mechanism by which JNK regulates longevity and stress resistance.

Conclusion

The evolutionary theories of aging predict that potentially many genes influence aging rate. Indeed, it is important the fact that a range of gene mutations have big impacts on life span in model organism. This chapter summarized the biological functions, early and late in life, of some genes known to influence aging. Doing so, one conclusion can be reached: genes influencing aging are highly pleiotropic [168]. Mutations that increase life span also regulate other developmental or physiological processes earlier in life [2, 3]. In essence, there are no genes only for aging; aging re-uses genes and genetics pathways.

However, developmental genetic programs influence aging. They do so because allow the organism to respond to change by sensing environment: nutrition, stress etc [2]. Nevertheless, in contrast to the precise genetic regulation that occurs during development, life span, while genetically regulated, is not tightly controlled. Case-in-point, there are significant differences in aging phenotype even in monozygotic human twins. Nonetheless, this chapter also focuses in the importance of flexibility during development to generate an organism in which the whole body is appropriate relative to environmental conditions.

Aging research is hardly a mature discipline, but the increased interest in aging research and the subsequent increasing number of genes modulating aging may help to discover new biological functions of developmental genes and new ways to regulate gene expression during development under different environment insults.

References

[1]   Kenyon, C., *The plasticity of aging: insights from long-lived mutants.* Cell, 2005. 120(4): p. 449-60.

[2]   Lithgow, G.J., *Why aging isn't regulated: a lamentation on the use of language in aging literature.* Exp Gerontol, 2006. 41(10): p. 890-3.

[3]   Antebi, A., *Physiology. The tick-tock of aging?* Science, 2005. 310(5756): p. 1911-3.

[4]   Austad, S.N. and T.B.L. Kirkwood, *Evolutionary Theroy in Aging Research*, in *Molecular Biology of Aging*, L. Guarente, L. Partridge, and D.C. Wallace, Editors. 2008, Cold Spring Harbor Laboratory Press: New York. p. 95-111.

[5] Williams, G.C., *Pleiotropy, natural selection, and the evolution of senescence.* Evolution, 1957. 11: p. 398-411.

[6] Kirkwood, T.B., *Evolution of ageing.* Nature, 1977. 270(5635): p. 301-4.

[7] Shingleton, A.W., et al., *The temporal requirements for insulin signaling during development in Drosophila.* PLoS Biol, 2005. 3(9): p. e289.

[8] Mair, W. and A. Dillin, *Aging and survival: the genetics of life span extension by dietary restriction.* Annu Rev Biochem, 2008. 77: p. 727-54.

[9] Harrison, D.E. and J.R. Archer, *Natural selection for extended longevity from food restriction.* Growth Dev Aging, 1988. 52: p. 65.

[10] Holliday, R., *Food, reproduction and longevity: is the extended lifespan of calorie-restricted animals an evolutionary adaptation?* Bioessays, 1989. 10(4): p. 125-7.

[11] Kennedy, B.K., K.K. Steffen, and M. Kaeberlein, *Ruminations on dietary restriction and aging.* Cell Mol Life Sci, 2007. 64(11): p. 1323-8.

[12] Johnson, T.E., *Increased life-span of age-1 mutants in Caenorhabditis elegans and lower Gompertz rate of aging.* Science, 1990. 249(4971): p. 908-12.

[13] Kenyon, C., et al., *A C. elegans mutant that lives twice as long as wild type.* Nature, 1993. 366(6454): p. 461-4.

[14] Clancy, D.J., et al., *Extension of life-span by loss of CHICO, a Drosophila insulin receptor substrate protein.* Science, 2001. 292(5514): p. 104-6.

[15] Tatar, M., A. Bartke, and A. Antebi, *The endocrine regulation of aging by insulin-like signals.* Science, 2003. 299(5611): p. 1346-51.

[16] Bluher, M., B.B. Kahn, and C.R. Kahn, *Extended longevity in mice lacking the insulin receptor in adipose tissue.* Science, 2003. 299(5606): p. 572-4.

[17] Holzenberger, M., et al., *IGF-1 receptor regulates lifespan and resistance to oxidative stress in mice.* Nature, 2003. 421(6919): p. 182-7.

[18] Taguchi, A., L.M. Wartschow, and M.F. White, *Brain IRS2 signaling coordinates life span and nutrient homeostasis.* Science, 2007. 317(5836): p. 369-72.

[19] Puig, O. and R. Tjian, *Transcriptional feedback control of insulin receptor by dFOXO/FOXO1.* Genes Dev, 2005. 19(20): p. 2435-46.

[20] Burgering, B.M., *A brief introduction to FOXOlogy.* Oncogene, 2008. 27(16): p. 2258-62.

[21] Leevers, S.J. and E. Hafen, *Growth Regulation by Insulin and Tor Signaling in Drosophila*, in *Cell Growth: Control of Cell Size*, M.N. Hall, M. Raff, and G. Thomas, Editors. 2004, Cold Spring Harbor Laboratory Press: New York. p. 167-192.

[22] Lin, K., et al., *daf-16: An HNF-3/forkhead family member that can function to double the life-span of Caenorhabditis elegans.* Science, 1997. 278(5341): p. 1319-22.

[23] Ogg, S., et al., *The Forkhead transcription factor DAF-16 transduces insulin-like metabolic and longevity signals in C. elegans.* Nature, 1997. 389(6654): p. 994-9.

[24] Brazil, D.P. and B.A. Hemmings, *Ten years of protein kinase B signalling: a hard Akt to follow.* Trends Biochem Sci, 2001. 26(11): p. 657-64.

[25] Scheid, M.P. and J.R. Woodgett, *PKB/AKT: functional insights from genetic models.* Nat Rev Mol Cell Biol, 2001. 2(10): p. 760-8.

[26] Ikeya, T., et al., *Nutrient-dependent expression of insulin-like peptides from neuroendocrine cells in the CNS contributes to growth regulation in Drosophila.* Curr Biol, 2002. 12(15): p. 1293-300.

[27] Shambaugh, G.E., 3rd, et al., *Insulin-like growth factors and binding proteins in the fetal rat: alterations during maternal starvation and effects in fetal brain cell culture.* Neurochem Res, 1993. 18(6): p. 695-703.

[28] Baker, J., et al., *Role of insulin-like growth factors in embryonic and postnatal growth.* Cell, 1993. 75(1): p. 73-82.

[29] Liu, J.P., et al., *Mice carrying null mutations of the genes encoding insulin-like growth factor I (Igf-1) and type 1 IGF receptor (Igf1r).* Cell, 1993. 75(1): p. 59-72.

[30] Fernandez, R., et al., *The Drosophila insulin receptor homolog: a gene essential for embryonic development encodes two receptor isoforms with different signaling potential.* EMBO J., 1995. 14: p. 3373-3384.

[31] Chen, C., J. Jack, and R.S. Garofalo, *The Drosophila insulin receptor is required for normal growth.* Endocrinology, 1996. 137(3): p. 846-56.

[32] Kido, Y., J. Nakae, and D. Accili, *Clinical review 125: The insulin receptor and its cellular targets.* J Clin Endocrinol Metab, 2001. 86(3): p. 972-9.

[33] Saltiel, A.R. and C.R. Kahn, *Insulin signalling and the regulation of glucose and lipid metabolism.* Nature, 2001. 414(6865): p. 799-806.

[34] Schwartz, M.W., et al., *Central nervous system control of food intake.* Nature, 2000. 404(6778): p. 661-71.

[35] Garofalo, R.S., *Genetic analysis of insulin signaling in Drosophila.* Trends Endocrinol Metab, 2002. 13(4): p. 156-62.

[36] Böhni, R., et al., *Autonomous control of cell and organ size by CHICO, a Drosophila homolog of vertebrate IRS1-4.* Cell, 1999. 97: p. 865-875.

[37] Tatar, M. and C. Yin, *Slow aging during insect reproductive diapause: why butterflies, grasshoppers and flies are like worms.* Exp Gerontol, 2001. 36(4-6): p. 723-38.

[38] Tatar, M., et al., *A mutant Drosophila insulin receptor homolog that extends life-span and impairs neuroendocrine function.* Science, 2001. 292(5514): p. 107-10.

[39] Bruning, J.C., et al., *Role of brain insulin receptor in control of body weight and reproduction.* Science, 2000. 289(5487): p. 2122-5.

[40] Burks, D.J., et al., *IRS-2 pathways integrate female reproduction and energy homeostasis.* Nature, 2000. 407(6802): p. 377-82.

[41] Selman, C., et al., *Evidence for lifespan extension and delayed age-related biomarkers in insulin receptor substrate 1 null mice.* Faseb J, 2008. 22(3): p. 807-18.

[42] Klass, M.R., *A method for the isolation of longevity mutants in the nematode Caenorhabditis elegans and initial results.* Mech Ageing Dev, 1983. 22(3-4): p. 279-86.

[43] Morris, J.Z., H.A. Tissenbaum, and G. Ruvkun, *A phosphatidylinositol-3-OH kinase family member regulating longevity and diapause in Caenorhabditis elegans.* Nature, 1996. 382(6591): p. 536-9.

[44] Kimura, K.D., et al., *daf-2, an insulin receptor-like gene that regulates longevity and diapause in Caenorhabditis elegans.* Science, 1997. 277(5328): p. 942-6.

[45] Larsen, P.L., P.S. Albert, and D.L. Riddle, *Genes that regulate both development and longevity in Caenorhabditis elegans.* Genetics, 1995. 139(4): p. 1567-83.

[46] Dorman, J.B., et al., *The age-1 and daf-2 genes function in a common pathway to control the lifespan of Caenorhabditis elegans.* Genetics, 1995. 141(4): p. 1399-406.

[47] Broughton, S.J., et al., *Longer lifespan, altered metabolism, and stress resistance in Drosophila from ablation of cells making insulin-like ligands.* Proc Natl Acad Sci U S A, 2005. 102(8): p. 3105-10.

[48] Hwangbo, D.S., et al., *Drosophila dFOXO controls lifespan and regulates insulin signalling in brain and fat body.* Nature, 2004. 429(6991): p. 562-6.

[49] Giannakou, M.E., et al., *Long-lived Drosophila with overexpressed dFOXO in adult fat body.* Science, 2004. 305(5682): p. 361.

[50] Weigel, D.J., G Kuttner,F Seifert,E Jackle,H, *The homeotic gene fork head encodes a nuclear protein and is expressed in the terminal regions of the Drosophila embryo.* Cell, 1989. 57: p. 645-658.

[51] Carlsson, P. and M. Mahlapuu, *Forkhead transcription factors: key players in development and metabolism.* Dev Biol, 2002. 250(1): p. 1-23.

[52] Kaestner, K.H., W. Knochel, and D.E. Martinez, *Unified nomenclature for the winged helix/forkhead transcription factors.* Genes Dev, 2000. 14(2): p. 142-6.

[53] Clark, K.L., et al., *Co-crystal structure of the HNF-3/fork head DNA-recognition motif resembles histone H5.* Nature, 1993. 364(6436): p. 412-20.

[54] Arden, K.C., *FOXO animal models reveal a variety of diverse roles for FOXO transcription factors.* Oncogene, 2008. 27(16): p. 2345-50.

[55] Huang, H. and D.J. Tindall, *Dynamic FoxO transcription factors.* J Cell Sci, 2007. 120(Pt 15): p. 2479-87.

[56] Burgering, B.M. and G.J. Kops, *Cell cycle and death control: long live Forkheads.* Trends Biochem Sci, 2002. 27(7): p. 352-60.

[57] van der Horst, A. and B.M. Burgering, *Stressing the role of FoxO proteins in lifespan and disease.* Nat Rev Mol Cell Biol, 2007. 8(6): p. 440-50.

[58] Salih, D.A. and A. Brunet, *FoxO transcription factors in the maintenance of cellular homeostasis during aging.* Curr Opin Cell Biol, 2008. 20(2): p. 126-36.

[59] Sherr, C.J., *Cancer cell cycles.* Science, 1996. 274(5293): p. 1672-7.

[60] Sherr, C.J. and J.M. Roberts, *CDK inhibitors: positive and negative regulators of G1-phase progression.* Genes Dev, 1999. 13(12): p. 1501-12.

[61] Ho, K.K., S.S. Myatt, and E.W. Lam, *Many forks in the path: cycling with FoxO.* Oncogene, 2008. 27(16): p. 2300-11.

[62] Huang, H., et al., *CDK2-dependent phosphorylation of FOXO1 as an apoptotic response to DNA damage.* Science, 2006. 314(5797): p. 294-7.

[63] Brunet, A., et al., *Stress-dependent regulation of FOXO transcription factors by the SIRT1 deacetylase.* Science, 2004. 303(5666): p. 2011-5.

[64] Modur, V., et al., *FOXO proteins regulate tumor necrosis factor-related apoptosis inducing ligand expression. Implications for PTEN mutation in prostate cancer.* J Biol Chem, 2002. 277(49): p. 47928-37.

[65] Fu, Z. and D.J. Tindall, *FOXOs, cancer and regulation of apoptosis.* Oncogene, 2008. 27(16): p. 2312-9.

[66] Massey, A.C., C. Zhang, and A.M. Cuervo, *Chaperone-mediated autophagy in aging and disease.* Curr Top Dev Biol, 2006. 73: p. 205-35.

[67] Sandri, M., et al., *PGC-1alpha protects skeletal muscle from atrophy by suppressing FoxO3 action and atrophy-specific gene transcription.* Proc Natl Acad Sci U S A, 2006. 103(44): p. 16260-5.

[68] Juhasz, G., et al., *Gene expression profiling identifies FKBP39 as an inhibitor of autophagy in larval Drosophila fat body.* Cell Death Differ, 2007. 14(6): p. 1181-90.

[69] Bastie, C.C., et al., *FoxO1 stimulates fatty acid uptake and oxidation in muscle cells through CD36-dependent and -independent mechanisms.* J Biol Chem, 2005. 280(14): p. 14222-9.

[70] Gross, D.N., A.P. van den Heuvel, and M.J. Birnbaum, *The role of FoxO in the regulation of metabolism.* Oncogene, 2008. 27(16): p. 2320-36.

[71] Kamei, Y., et al., *A forkhead transcription factor FKHR up-regulates lipoprotein lipase expression in skeletal muscle.* FEBS Lett, 2003. 536(1-3): p. 232-6.

[72] Kim, M.S., et al., *Role of hypothalamic Foxo1 in the regulation of food intake and energy homeostasis.* Nat Neurosci, 2006. 9(7): p. 901-6.

[73] Peng, S.L., *Foxo in the immune system.* Oncogene, 2008. 27(16): p. 2337-44.

[74] Wang, M.C., D. Bohmann, and H. Jasper, *JNK signaling confers tolerance to oxidative stress and extends lifespan in Drosophila.* Dev. Cell, 2003. 5(5): p. 811-816.

[75] Oh, S.W., et al., *JNK regulates lifespan in Caenorhabditis elegans by modulating nuclear translocation of forkhead transcription factor/DAF-16.* Proc Natl Acad Sci U S A, 2005. 102(12): p. 4494-9.

[76] Shaw, W.M., et al., *The C. elegans TGF-beta Dauer pathway regulates longevity via insulin signaling.* Curr Biol, 2007. 17(19): p. 1635-45.

[77] Cassada, R.C. and R.L. Russell, *The dauerlarva, a post-embryonic developmental variant of the nematode Caenorhabditis elegans.* Dev Biol, 1975. 46(2): p. 326-42.

[78] Riddle, D.L., M.M. Swanson, and P.S. Albert, *Interacting genes in nematode dauer larva formation.* Nature, 1981. 290(5808): p. 668-71.

[79] Wessells, R.J., et al., *Insulin regulation of heart function in aging fruit flies.* Nat Genet, 2004. 36(12): p. 1275-81.

[80] Libina, N., J.R. Berman, and C. Kenyon, *Tissue-specific activities of C. elegans DAF-16 in the regulation of lifespan.* Cell, 2003. 115(4): p. 489-502.

[81] Wolkow, C.A., et al., *Regulation of C. elegans life-span by insulinlike signaling in the nervous system.* Science, 2000. 290(5489): p. 147-50.

[82] Iser, W.B., M.S. Gami, and C.A. Wolkow, *Insulin signaling in Caenorhabditis elegans regulates both endocrine-like and cell-autonomous outputs.* Dev Biol, 2007. 303(2): p. 434-47.

[83] Apfeld, J. and C. Kenyon, *Cell nonautonomy of C. elegans daf-2 function in the regulation of diapause and life span.* Cell, 1998. 95(2): p. 199-210.

[84] Houthoofd, K., et al., *Life extension via dietary restriction is independent of the Ins/IGF-1 signalling pathway in Caenorhabditis elegans.* Exp Gerontol, 2003. 38(9): p. 947-54.

[85] Bishop, N.A. and L. Guarente, *Two neurons mediate diet-restriction-induced longevity in C. elegans.* Nature, 2007. 447(7144): p. 545-9.

[86] Panowski, S.H., et al., *PHA-4/Foxa mediates diet-restriction-induced longevity of C. elegans.* Nature, 2007. 447(7144): p. 550-5.

[87] Lakowski, B. and S. Hekimi, *The genetics of caloric restriction in Caenorhabditis elegans.* Proc Natl Acad Sci U S A, 1998. 95(22): p. 13091-6.

[88] Min, K.J., et al., *Drosophila lifespan control by dietary restriction independent of insulin-like signaling.* Aging Cell, 2008. 7(2): p. 199-206.

[89] Giannakou, M.E., M. Goss, and L. Partridge, *Role of dFOXO in lifespan extension by dietary restriction in Drosophila melanogaster: not required, but its activity modulates the response.* Aging Cell, 2008. 7(2): p. 187-98.

[90] Greer, E.L., et al., *An AMPK-FOXO pathway mediates longevity induced by a novel method of dietary restriction in C. elegans.* Curr Biol, 2007. 17(19): p. 1646-56.

[91] Lehtinen, M.K., et al., *A conserved MST-FOXO signaling pathway mediates oxidative-stress responses and extends life span.* Cell, 2006. 125(5): p. 987-1001.

[92] Greer, E.L. and A. Brunet, *FOXO transcription factors at the interface between longevity and tumor suppression.* Oncogene, 2005. 24(50): p. 7410-25.

[93] Wolff, S., et al., *SMK-1, an essential regulator of DAF-16-mediated longevity.* Cell, 2006. 124(5): p. 1039-53.

[94] Li, M., et al., *Acetylation of p53 inhibits its ubiquitination by Mdm2.* J Biol Chem, 2002. 277(52): p. 50607-11.

[95] Kaeberlein, M., M. McVey, and L. Guarente, *The SIR2/3/4 complex and SIR2 alone promote longevity in Saccharomyces cerevisiae by two different mechanisms.* Genes Dev, 1999. 13(19): p. 2570-80.

[96] Tissenbaum, H.A. and L. Guarente, *Increased dosage of a sir-2 gene extends lifespan in Caenorhabditis elegans.* Nature, 2001. 410(6825): p. 227-30.

[97] Rogina, B. and S.L. Helfand, *Sir2 mediates longevity in the fly through a pathway related to calorie restriction.* Proc Natl Acad Sci U S A, 2004. 101(45): p. 15998-6003.

[98] Calnan, D.R. and A. Brunet, *The FoxO code.* Oncogene, 2008. 27(16): p. 2276-88.

[99] Huang, H., et al., *Skp2 inhibits FOXO1 in tumor suppression through ubiquitin-mediated degradation.* Proc Natl Acad Sci U S A, 2005. 102(5): p. 1649-54.

[100] Li, W., et al., *RLE-1, an E3 ubiquitin ligase, regulates C. elegans aging by catalyzing DAF-16 polyubiquitination.* Dev Cell, 2007. 12(2): p. 235-46.

[101] van der Horst, A., et al., *FOXO4 transcriptional activity is regulated by monoubiquitination and USP7/HAUSP.* Nat Cell Biol, 2006. 8(10): p. 1064-73.

[102] Brenkman, A.B., et al., *Mdm2 induces mono-ubiquitination of FOXO4.* PLoS ONE, 2008. 3(7): p. e2819.

[103] Friedman, J.R. and K.H. Kaestner, *The Foxa family of transcription factors in development and metabolism.* Cell Mol Life Sci, 2006. 63(19-20): p. 2317-28.

[104] Tuteja, G. and K.H. Kaestner, *Forkhead transcription factors II.* Cell, 2007. 131(1): p. 192.

[105] Tuteja, G. and K.H. Kaestner, *SnapShot: forkhead transcription factors I.* Cell, 2007. 130(6): p. 1160.

[106] Weinstein, D.C., et al., *The winged-helix transcription factor HNF-3 beta is required for notochord development in the mouse embryo.* Cell, 1994. 78(4): p. 575-88.

[107] Filosa, S., et al., *Goosecoid and HNF-3beta genetically interact to regulate neural tube patterning during mouse embryogenesis.* Development, 1997. 124(14): p. 2843-54.

[108] Behr, R., et al., *Mild nephrogenic diabetes insipidus caused by Foxa1 deficiency.* J Biol Chem, 2004. 279(40): p. 41936-41.

[109] Kaestner, K.H., *The hepatocyte nuclear factor 3 (HNF3 or FOXA) family in metabolism.* Trends Endocrinol Metab, 2000. 11(7): p. 281-5.

[110] Carroll, J.S., et al., *Chromosome-wide mapping of estrogen receptor binding reveals long-range regulation requiring the forkhead protein FoxA1.* Cell, 2005. 122(1): p. 33-43.

[111] Weigel, D., et al., *Primordium specific requirement of the homeotic gene fork head in the developing gut of the Drosophila embryo.* Arch Dev Biol, 1989. 198: p. 201-210.

[112] Myat, M.M. and D.J. Andrew, *Fork head prevents apoptosis and promotes cell shape change during formation of the Drosophila salivary glands.* Development, 2000. 127: p. 4217-4226.

[113] Cao, C., Y. Liu, and M. Lehmann, *Fork head controls the timing and tissue selectivity of steroid-induced developmental cell death.* J Cell Biol, 2007. 176(6): p. 843-52.

[114] Lehmann, M. and G. Korge, *The forkhead product directly specifies the tissue-specific hormone responsiveness of the Drosophila Sgs-4 gene.* EMBO J., 1996. 15: p. 4825-4834.

[115] Renault, N., K. King-Jones, and M. Lehman, *Downregulation of the tissue-specific transcription Fork head by Broad-Complex mediates a stage-specfic hormone response.* Development, 2001. 128: p. 3729-3737.

[116] Liu, Y. and M. Lehmann, *Genes and biological processes controlled by the Drosophila FOXA orthologue Fork head.* Insect Mol Biol, 2008. 17(2): p. 91-101.

[117] Mango, S.E., *The C. elegans pharynx: a model for organogenesis.* WormBook, 2007: p. 1-26.

[118] Chen, D. and D.L. Riddle, *Function of the PHA-4/FOXA transcription factor during C. elegans post-embryonic development.* BMC Dev Biol, 2008. 8: p. 26.

[119] Jacinto, E. and M.N. Hall, *Tor signalling in bugs, brain and brawn.* Nat Rev Mol Cell Biol, 2003. 4(2): p. 117-26.

[120] Kunz, J., et al., *Target of rapamycin in yeast, TOR2, is an essential phosphatidylinositol kinase homolog required for G1 progression.* Cell, 1993. 73(3): p. 585-96.

[121] Heitman, J., N.R. Movva, and M.N. Hall, *Targets for cell cycle arrest by the immunosuppressant rapamycin in yeast.* Science, 1991. 253(5022): p. 905-9.

[122] Martin, D.E. and M.N. Hall, *The expanding TOR signaling network.* Curr Opin Cell Biol, 2005. 17(2): p. 158-66.

[123] Keith, C.T. and S.L. Schreiber, *PIK-related kinases: DNA repair, recombination, and cell cycle checkpoints.* Science, 1995. 270(5233): p. 50-1.

[124] Bhaskar, P.T. and N. Hay, *The two TORCs and Akt.* Dev Cell, 2007. 12(4): p. 487-502.

[125] Vander Haar, E., et al., *Insulin signalling to mTOR mediated by the Akt/PKB substrate PRAS40.* Nat Cell Biol, 2007. 9(3): p. 316-23.

[126] Dong, J. and D. Pan, *Tsc2 is not a critical target of Akt during normal Drosophila development.* Genes Dev, 2004. 18(20): p. 2479-84.

[127] Jia, K., D. Chen, and D.L. Riddle, *The TOR pathway interacts with the insulin signaling pathway to regulate C. elegans larval development, metabolism and life span.* Development, 2004. 131(16): p. 3897-906.

[128] Brooks, R.F., *Continuous protein synthesis is required to maintain the probability of entry into S phase.* Cell, 1977. 12(1): p. 311-7.

[129] Baxter, G.C. and C.P. Stanners, *The effect of protein degradation on cellular growth characteristics.* J Cell Physiol, 1978. 96(2): p. 139-45.

[130] Kief, D.R. and J.R. Warner, *Coordinate control of syntheses of ribosomal ribonucleic acid and ribosomal proteins during nutritional shift-up in Saccharomyces cerevisiae.* Mol Cell Biol, 1981. 1(11): p. 1007-15.

[131] Hannan, R.D., et al., *Cardiac hypertrophy: a matter of translation.* Clin Exp Pharmacol Physiol, 2003. 30(8): p. 517-27.

[132] Ruggero, D. and P.P. Pandolfi, *Does the ribosome translate cancer?* Nat Rev Cancer, 2003. 3(3): p. 179-92.

[133] Jastrzebski, K., et al., *Coordinate regulation of ribosome biogenesis and function by the ribosomal protein S6 kinase, a key mediator of mTOR function.* Growth Factors, 2007. 25(4): p. 209-26.

[134] Hay, N. and N. Sonenberg, *Upstream and downstream of mTOR.* Genes Dev, 2004. 18(16): p. 1926-45.

[135] Hannan, K.M., G. Thomas, and R.B. Pearson, *Activation of S6K1 (p70 ribosomal protein S6 kinase 1) requires an initial calcium-dependent priming event involving formation of a high-molecular-mass signalling complex.* Biochem J, 2003. 370(Pt 2): p. 469-77.

[136] Pende, M., et al., *Hypoinsulinaemia, glucose intolerance and diminished beta-cell size in S6K1-deficient mice.* Nature, 2000. 408(6815): p. 994-7.

[137] Um, S.H., et al., *Absence of S6K1 protects against age- and diet-induced obesity while enhancing insulin sensitivity.* Nature, 2004. 431(7005): p. 200-5.

[138] Gingras, A.C., B. Raught, and N. Sonenberg, *Regulation of translation initiation by FRAP/mTOR.* Genes Dev, 2001. 15(7): p. 807-26.

[139] Miron, M., et al., *The translational inhibitor 4E-BP is an effector of PI(3)K/Akt signalling and cell growth in Drosophila.* Nat Cell Biol, 2001. 3(6): p. 596-601.

[140] Junger, M.A., et al., *The Drosophila forkhead transcription factor FOXO mediates the reduction in cell number associated with reduced insulin signaling.* J Biol, 2003. 2(3): p. 20.

[141] Colombani, J., et al., *A nutrient sensor mechanism controls Drosophila growth.* Cell, 2003. 114(6): p. 739-49.

[142] Kaeberlein, M., et al., *Regulation of yeast replicative life span by TOR and Sch9 in response to nutrients.* Science, 2005. 310(5751): p. 1193-6.

[143] Powers, R.W., 3rd, et al., *Extension of chronological life span in yeast by decreased TOR pathway signaling.* Genes Dev, 2006. 20(2): p. 174-84.

[144] Bonawitz, N.D., et al., *Reduced TOR signaling extends chronological life span via increased respiration and upregulation of mitochondrial gene expression.* Cell Metab, 2007. 5(4): p. 265-77.

[145] Vellai, T., et al., *Genetics: influence of TOR kinase on lifespan in C. elegans.* Nature, 2003. 426(6967): p. 620.

[146] Hansen, M., et al., *Lifespan extension by conditions that inhibit translation in Caenorhabditis elegans.* Aging Cell, 2007. 6(1): p. 95-110.

[147] Pan, K.Z., et al., *Inhibition of mRNA translation extends lifespan in Caenorhabditis elegans.* Aging Cell, 2007. 6(1): p. 111-9.

[148] Kapahi, P. and B. Zid, *TOR pathway: linking nutrient sensing to life span.* Sci Aging Knowledge Environ, 2004. 2004(36): p. PE34.

[149] Min, K.J. and M. Tatar, *Restriction of amino acids extends lifespan in Drosophila melanogaster.* Mech Ageing Dev, 2006. 127(7): p. 643-6.

[150] Miller, R.A., et al., *Methionine-deficient diet extends mouse lifespan, slows immune and lens aging, alters glucose, T4, IGF-I and insulin levels, and increases hepatocyte MIF levels and stress resistance.* Aging Cell, 2005. 4(3): p. 119-25.

[151] Zimmerman, J.A., et al., *Nutritional control of aging.* Exp Gerontol, 2003. 38(1-2): p. 47-52.

[152] Davis, R.J., *Signal transduction by the JNK group of MAP kinases.* Cell, 2000. 103(2): p. 239-52.

[153] Kanda, H. and M. Miura, *Regulatory roles of JNK in programmed cell death.* J Biochem, 2004. 136(1): p. 1-6.

[154] Huang, C., et al., *JNK phosphorylates paxillin and regulates cell migration.* Nature, 2003. 424(6945): p. 219-23.

[155] Chang, L., et al., *JNK1 is required for maintenance of neuronal microtubules and controls phosphorylation of microtubule-associated proteins.* Dev Cell, 2003. 4(4): p. 521-33.

[156] Gupta, S., et al., *Selective interaction of JNK protein kinase isoforms with transcription factors.* Embo J, 1996. 15(11): p. 2760-70.

[157] Lin, A. and B. Dibling, *The true face of JNK activation in apoptosis.* Aging Cell, 2002. 1(2): p. 112-6.

[158] Zeitlinger, J. and D. Bohmann, *Thorax closure in Drosophila: involvement of Fos and the JNK pathway.* Development, 1999. 126: p. 3947-3956.

[159] Xia, Z., et al., *Opposing effects of ERK and JNK-p38 MAP kinases on apoptosis.* Science, 1995. 270(5240): p. 1326-31.

[160] Kuan, C.Y., et al., *The Jnk1 and Jnk2 protein kinases are required for regional specific apoptosis during early brain development.* Neuron, 1999. 22(4): p. 667-76.

[161] Sabapathy, K., et al., *Defective neural tube morphogenesis and altered apoptosis in the absence of both JNK1 and JNK2.* Mech Dev, 1999. 89(1-2): p. 115-24.

[162] Yang, D.D., et al., *Absence of excitotoxicity-induced apoptosis in the hippocampus of mice lacking the Jnk3 gene.* Nature, 1997. 389(6653): p. 865-70.

[163] Kuranaga, E., et al., *Reaper-mediated inhibition of DIAP1-induced DTRAF1 degradation results in activation of JNK in Drosophila.* Nat Cell Biol, 2002. 4(9): p. 705-10.

[164] Adachi-Yamada, T., et al., *Distortion of proximodistal information causes JNK-dependent apoptosis in Drosophila wing.* Nature, 1999. 400: p. 166-169.

[165] Xia, Y. and M. Karin, *The control of cell motility and epithelial morphogenesis by Jun kinases.* Trends Cell Biol, 2004. 14(2): p. 94-101.

[166] Weston, C.R., et al., *JNK initiates a cytokine cascade that causes Pax2 expression and closure of the optic fissure.* Genes Dev, 2003. 17(10): p. 1271-80.

[167] Wang, M.C., D. Bohmann, and H. Jasper, *JNK extends life span and limits growth by antagonizing cellular and organism-wide responses to insulin signaling.* Cell, 2005. 121(1): p. 115-25.

[168] Leroi, A.M., et al., *What evidence is there for the existence of individual genes with antagonistic pleiotropic effects?* Mech Ageing Dev, 2005. 126(3): p. 421-9.

In: Developmental Gene Expression Regulation
Editor: Nathan C. Kurzfield

ISBN: 978-60692-794-6
©2009 Nova Science Publishers, Inc.

Chapter III

Developmental Regulation of Sensory Receptor Gene Expression

Simon G. Sprecher

Department of Biology, Center for Developmental
Genetics, New York University, New York, NY, USA

Abstract

Establishment of specific cell fates requires orchestrated interaction of an array of transcription factors and signaling pathways which incorporate major developmental roles. The transition of proliferating precursor or immature cells into a certain cell type and terminal differentiation has been studied in great detail on various neuronal cell types, due to the large degree of diversity, complex functional roles, and intrinsic properties of cells in the nervous system. In the peripheral nervous system sensory specificity is established by the choice of an immature, but committed, postmitotic progenitor to express a specific sensory receptor gene. Findings of sensory receptor gene regulation stemming form the olfactory system and visual system and in mouse and fruit fly provide insight in how this highly complex process is regulated and genetically controlled. After the initial decision to become a sensory neuron the cell then decides which receptor gene to express and subsequently maintain the expression of this given receptor gene. Genetic mechanisms for this choice depend upon transcriptional regulators which are expressed in subtype of sensory neurons, thus may provide a combinatorial code to orchestrate the expression of a specific receptor gene: For instance in the fly visual system, where an array of six *rhodopsin* genes can be expressed a combinatorial code of transcription factors is required for the sensory receptor gene regulation. Interestingly during larval and adult stages the regulation of *rhodopsins* makes use of distinct developmental genetic program. Regulatory regions of *rhodopsins* display a bipartite architecture with a proximal domain required for general PR expression and a distal domain encoding subtype specificity. In the human retina cone cells can express L and M opsin genes, which are located in close proximity on the chromosome. Regulation of L and M opsin depends upon "locus control regions" (LCR) a common long range *cis*-acting element regulating both genes. Interestingly, in the mouse retina *Opsins* are not clustered and co-expression of two *Opsins* genes occurs. In the mouse olfactory system, an array of over 1200 Odorant receptor (OR) genes can be expressed. OR genes are often arranged in clusters along the chromosome and seem to depend on distant and local *cis*-acting elements. Interestingly only one of the two parental copies of an OR gene is expressed, resulting in monoallelic expression of the gene. Taken together the choice of a sensory neuron to adopt a specific sensory specificity depends upon the complex interaction of *cis*- and *trans* acting factors. Even though generally only one form of sensory receptor gene is expressed, various mechanisms may be acting to achieve a similar outcome of sensory receptor gene regulation. Moreover recent findings reveal that sensory receptor genes can be co-expressed and that this co-expression is genetically controlled, thus adding an additional layer of complexity in the regulatory properties of sensory receptor genes.

Introduction

The development of a multicellular organism from a single cell (oocyte) relies on the correct decisions of cells to proliferate, exit the cell cycle and undertaking the correct terminal cell fates. The organ system incorporating the largest variety of different cell types is the nervous system. The central nervous system consists of an array of neurons and glia cells, which for proper development require highly complex pattering mechanisms (Sprecher and Reichert, 2003; Sprecher et al., 2007b). Any kind of information processed in the central nervous system however requires appropriate sensory input from the periphery, which is conducted by the peripheral nervous system. Sensory organs are able to transform information from the environment into neuronal signals, which are then transmitted to the central nervous system and eventually higher brain centers. The expression of a certain sensory receptor gene defines to which stimuli a sensory neuron is sensitive to, thereby providing functional identity to sensory neurons. Thus the terminal cell fate decision of a sensory neuron depends upon a developmental program regulating sensory receptor gene expression.

Regulation of sensory receptors has been extensively studied in genetic model organisms including the fruit fly (*Drosophila melanogaster*) and mouse (*mus musculus*). Lessons from the visual system and olfactory system of these animal species provide essential insight in how sensory specificity is established on the regulation of receptor gene expression (Keller and Vosshall, 2003; Mazzoni et al., 2004; Mombaerts, 1999a; Mombaerts, 1999b; Mombaerts, 2004b; Morante et al., 2007; Vosshall, 2000; Vosshall, 2001; Wernet and Desplan, 2004).

In the visual system photoreceptor cells (PRs) express a specific *rhodopsin* (fly) or *Opsin* (vertebrate) gene which is sensitive to a certain range of wavelength of light, thereby defining the spectral sensitivity of a distinct PR type. In the olfactory system Odorant receptor (OR) genes define to which chemical compounds an olfactory sensory neuron (OSN) is sensitive to, therefore comparable to the visual system defining functional identity. Both *Opsin* and OR genes code for G-protein coupled receptors (GPCRs, also called seven transmembrane domain receptors). These GPCRs are activated upon a conformational change caused either by ligand binding (Odorants) or by a photoisomerization reaction (conversion of *11-cis-*

retinal to *all-trans*-retinal) in case of the Opsins/Rhodopsins, and the subsequent activation of a G-Protein signaling cascade. Upon sensory stimulation a sensory receptor neuron ultimately transmits the information to second order neurons, which acts as a first station for sensory processing.

Apart from obvious functional differences between different sensory receptor gene families interesting similarities and differences of how receptor gene expression is genetically controlled have emerged. For instance the model of "one sensory neuron, one sensory receptor gene" has until recently largely been assumed to be a key feature for a sensory neuron to achieve its proper function. However insights from both mouse and fruit fly indicate that this rule does not have to be strictly maintained for the function of sensory neurons (Mazzoni et al., 2008; Mazzoni et al., 2004; Mombaerts, 2004b; Serizawa et al., 2004). For instance, in the visual system in both animal models co-expression of sensory receptor genes occurs (Mazzoni et al., 2008). In flies the key players required for co-expression have been identified. Moreover in the fly olfactory system one specific OR gene is co-expressed with many other OR genes, which is essential for the function of the OSN. In this chapter I will discuss several examples of how correct sensory receptor gene expression is controlled on both the molecular and genetic level and the consequences for generalized models and principles of sensory receptor gene expression and regulation.

1. Regulation of Sensory Receptor Genes in the Olfactory System

The wide functional range of chemosensory neurons is simplified the recognition of an array of chemical compounds and the translation into electrical neuronal signals. Studying how a large array of chemical compounds can be properly recognized and encoded into neuronal information lead to the identification of one of the largest gene family in mammals, the *Odorant receptor* (OR) genes. In 2004 the Nobel price in Physiology was awarded to Linda Buck and Richard Axel for their discoveries of ORs and the organization of the olfactory system (Buck and Axel, 1991; Mombaerts, 2004a). Even though the anatomy and complexity of chemosensory systems seem to vary quite extensively between invertebrates and vertebrates the functions are quite similar. In mouse any given olfactory receptor neuron has to choose to express one *OR* gene out of over 1200 genes (Mombaerts, 2004b; Mombaerts, 2006; Rodriguez, 2007). What are the mechanisms controlling this process and how does one cell choose one gene out of a repertoire of over 1200 genes? Interestingly, in the fruit fly OR expression regulation seems to differ from mouse. In the fly many OSNs express two functioning OR genes. Thus findings from mouse and the fruit fly revealed surprising similarities and differences in the regulation of sensory receptor genes.

1.1. Mouse Olfactory System

The olfactory system of the mouse consists of two parts, the main olfactory system and the accessory olfactory system, which acts mainly for pheromone perception. Sensory neurons of the accessory olfactory system are located in the so called "vomeronasal organ" (Rodriguez, 2004; Rodriguez, 2007; Rodriguez and Mombaerts, 2002). OSNs project to the main olfactory bulb, whereas the vomeronasal sensory neurons project to the accessory olfactory bulb, a dorsal-posterior portion of the olfactory bulb. Sensory receptor genes expressed in both systems are distinct. In the main olfactory epithelium each OSN is believed to express one single gene out of more than 1200, in a mutually exclusive, monoallelic manner.

1.1.1. Maintained Neurogenesis in the Olfactory System

One of the most amazing features of the mouse olfactory system is that the olfactory epithelium supports continuous neurogenesis throughout life quite in contrast to most neuronal organs. Continuous generation of OSNs may be required since these neurons are constantly exposed to external substances, many of which are toxic or noxious. To not loose the sense of smell by the loss of OSNs, new neurons are born to keep up with the turnover. Development of the OSNs from the epithelium follows a stereotyped lineage: from the apical progenitor cell to basal progenitor cell to intermediate neuronal precursor to the post-mitotic OSN. The proneural and neurogenic genes Mash1, Math4C, neurogenein1/2 and NeuroD act sequentially during the development and are required for the generation of OSNs (Cau et al., 1997). Two populations of cells have been suggested to act as stem cells in the olfactory epithelium mainly based on localization of maintained neurogenesis: the horizontal basal cells and the globose basal cells, which can be identified by morphology and expression of molecular markers (Duggan and Ngai, 2007). Interestingly the horizontal basal cells appear to incorporate indeed features of multipotent progenitor cells, which give rise to OSNs. For instance horizontal basal cells display maintained quiescence a typical feature of other adult tissue stem cells (Leung et al., 2007). In tissue culture horizontal basal cells can give rise to neuronal and non-neuronal cells. How this integration of new born sensory neurons into a function sensory system occurs however remains largely elusive.

1.1.2. The Mouse OR Genes

With the discovery of one of the largest gene families in vertebrates it was shown that OR genes code for seven transmembrane gene families, which generally are intron-less and have a size of about 1 kilobase (kb) (Buck and Axel, 1991). The coding sequence of ORs is distinct from other GPCRs by the presence of several conserved motives (in the intracellular domain 1, 2 and the transmembrane domains 3, 5, 6, 7). The highly variably regions, which have been implicated for the differential response to distinct chemical compounds and may act as a "binding pocket" are within transmembrane domains 3, 4, 5 (Mombaerts, 1999b). Olfactory receptor genes are scattered on most chromosomes in a few dozen of genomic clusters in the mouse genome. Intergenic regions may vary between a few kb up to 50kb (Mombaerts, 1999b). Interestingly in the human genome the frequency of pseudogenes lead to speculate that between 40% and 80% of the identified OR genes may actually be pseudogenes, based on frame shifts and nonsense mutations. In the mouse genome pseudogenes are less frequent and do not seem to contribute to the high number of OR genes.

1.1.3. Evidence for Mutually Exclusive Monoallelic Expression of OR Genes

The large number of olfactory receptor genes in the mouse genome makes it hard to study the entire repertoire as entity. Many finding are based on specific genes, cluster or loci and may not reflect entirely the complexity of the system. Thus the model that a given OSNs only one out of all possible OR genes is not easy to proof as compared for instance to the visual system where only a few genes are expressed and the hypothesis can be properly addressed. However evidence for a "one sensory receptor gene per sensory neuron" comes from a large array of experimental studies.

Expression of OR genes initially analyzed by *in situ* hybridization (visualizing the mRNA of a given OR) revealed several interesting characteristics. Expression of a given OR is restricted to one of four zones within the olfactory epithelium (Rodriguez, 2007; Shykind, 2005). A few OR genes display atypical expression patterns in "patches" overlapping two zones (Strotmann et al., 2000). For a few genes the absolute number cells expressing a specific OR gene has been elaborated and indicate that numbers may be variable ranging from 800-1800 for some of the genes expressed in patches, 4000-8000 for genes expressed in zones, 13700 for the *P2* gene and 30000 for the *Mol2.3* gene (Conzelmann et al., 2000; Kubick et al., 1997; Ressler et al., 1993; Royal and Key, 1999).

Within each zone individual receptor are expressed sparsely in an appealingly stochastic pattern (Ngai et al., 1993; Ressler et al., 1993; Vassar et al., 1993). Quantitative analysis of *in situ* experiments suggested that each OSN only expresses one OR gene. Supporting evidence comes from studies performing double *in situ* hybridization showing for three olfactory receptor genes that there is no co-expression (Tsuboi et al., 1999). Interestingly in rats co-expression of several OR genes mRNA was found, whoever this does not necessarily show that indeed two functioning olfactory receptor genes are present in these cells, since there may be several mechanisms acting to prevent co-existence of two functioning proteins (Rawson et al., 2000).

OR genes are expressed in a monoallelic manner. For most "normal" genes in the genome both copies are expressed, the paternal and maternal copy, with some well described exceptions (see below). Single cell RT-PCR (reverse transcription polymerase chain reaction) on polymorphic alleles indicates that only one allele of both is expressed at a time. Similarly in mice containing a "targeted" gene the number of labeled cells in homozygous mice was double as compared to heterozygous ones (Chess et al., 1994; Malnic et al., 1999; Mombaerts et al., 1996). Moreover the analysis of OR gene expression in single cell using RT-PCR indicates that only one receptor gene is expressed in a given sensory receptor neuron (Chess et al., 1994; Malnic et al., 1999).

1.1.4. ORs and Axonal Targeting

In the main olfactory epithelium about 2000 populations of OSNs gather information which will be transmitted to target neurons in the olfactory bulb. OSNs expressing a specific OR receptor gene all project their axons converging to glomeruli in the olfactory bulb. Thereby the odor representation in the olfactory bulb forms a topographic map, which is maintained from the OSNs. Each OSN sends one unbranched axon to one single glomerulus, from an array of more than 1500 glomeruli in the main olfactory bulb. Within a given glomerulus the OSN axon synapses with the dendrites of second-order neurons and interneurons. Thus an additional challenge is how the proper connectivity of a given OSN type is established. Several experiments indicate that the OR gene itself is required to connect to the correct target neuron. Targeting OR loci with a reporter gene (such as LacZ or GFP, combined with an IRES- internal ribosome entry site) it was observed that OSNs expressing a given OR gene send convergent axonal projections to specific glomeruli (Mombaerts et al., 1996). Replacing the coding sequence of OR genes by the coding sequence of another OR gene lead to the observation that the axons of these neurons now project in to new ectopic glomeruli. Interestingly replacement experiments with the *M71* and *M72* genes showed that the coding region of each locus is sufficient to specify the correct axonal projection patterns (Feinstein and Mombaerts, 2004).

1.1.5. Gene Choice and Regulation of OR Gene Expression

The regulation of OR genes is a highly complex process which seems to depend upon several distinct mechanisms. First any given OSN must decide to express a given OR gene, a process which is not well understood yet. Moreover it has to be taken into account that only one copy of each OR gene is selected to be expressed in a given OSN, a process termed monoallelic expression or allelic exclusion (Figure 1A). Two models which may explain how one specific neuron expressed a single OR gene propose either a deterministic or a stochastic process (Kratz et al., 2002; Shykind, 2005). In the deterministic model the unique combination of transactivators results in the expression of a given gene which is under the control of the appropriate *cis*-acting elements (Figure 1B). This model however does not explain monoallelic expression, since both alleles would be activated equally, thus allelic exclusion would depend on an additional mechanism similar to X-inactivation (Shykind, 2005). In the stochastic model the expression of an OR gene would be chosen randomly (Figure 2C). This might be controlled by a machinery which is only able to transcribe one OR gene locus at a time, therefore would also explain monoallelic expression. This process would depend on a *cis*-acting element common to all OR genes. Support for this model comes from experiments studying the regulatory regions of OR genes. Animals containing odorant transgenes (thus additional copies of the regulatory region) recapitulate the expression pattern of the OR. Comparably small elements of only a few hundred base pairs (bp) seem to sufficient to completely recapitulate the expression pattern. Interestingly expression is observed in an exclusive manner so that transgene and endogenous allele are not co-expressed, supporting the model of a stochastic choice (Qasba and Reed, 1998; Serizawa et al., 2000; Vassalli et al., 2002). Alternatively, the presence of so called locus control regions (LCRs) may act to choose one specific gene out of a repertoire (Figure 2D), which may also function in long range expression control (this model is not necessarily excluded by the stochastic model). A genomic region of 2.1 kb adjacent to a cluster of seven OR genes, termed *H* element (named after "homology" between human and mouse), has been proposed to act as major *cis* and *trans* acting element. A series of transgenic constructs derived from large yeast artificial chromosomes (YACs) showed that the expression of several ORs depends on the *H* element. Interestingly transgenic animals containing an additional *H* element express a second functional OR gene, suggesting that the *H* element may bare some *trans*-acting enhancer element may allow the stochastic activation of only one

OR allele in an OSN (Lomvardas et al., 2006). However recent findings deleting the *H* element revealed that the *H* element acts as *cis* acting region for some OR genes in the adjacent cluster and does not seem to act as general activator in *trans* (Fuss et al., 2007).

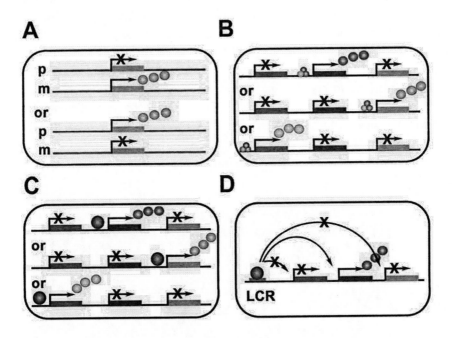

Figure 1. Model for genetic regulation of OR gene expression in the mouse olfactory system (A) Monoallelic expression of OR genes: only one of the two copies of a OR gene is expressed in a given OSN, either the paternal copy (red) or the maternal copy (blue). (B-D) Different models for the genetic control of OR gene expression; for simplicity an array of only three OR genes are pictures (red, green and blue). (B) Mutually exclusive expression explained by a model for a combinatorial code regulating OR gene expression. Depending on the presence of a specific combination of several transcription factors one specific OR gene gets activated, whereas the enhancers of other ORs do not incorporate the right combination of regulatory sequences. (C) Stochastic regulation of OR gene expression by the competition of a limited regulator. In this model the limited amount of the same factor is sufficient to only express one OR gene in one given OSN. (D) The presence of a locus control region (LCR) acts in long range to selectively only active one OR gene in any given OSN.

1.2. Fly Olfactory Systems

The olfactory system of the fly is bipartite and consists of sensory sensilla located in the antenna and the maxillary palp in the head (Figure 2A). Olfactory receptor neurons project their axons to the antennal lobe of the fly brain where they connect to their target neurons (Figure 2B). Comparable to the vertebrate olfactory bulb connectivity of olfactory receptor neurons also form a stereotypically organized array of glomeruli in both the larva and the adult fly (Stocker, 2006; Wong et al., 2002). Second order neurons, also termed projection neurons, then connect to the mushroom body and lateral horn, where sensory information will be further processed (Davis, 2005; McGuire et al., 2001). Interestingly, as the compound eye also the antenna (and maxillary pulp) develop from imaginal discs. However in contrast to the eye, which forms from an array of multipotent cells being recruited into the ommatidial cluster, the olfactory sensilla develop from a sensory organ precursors (SOPs) which give rise to a stereotype lineage. The sensilla in the olfactory system each normally contain two functioning olfactory sensory receptor neurons. The antenna incorporates about 600 olfactory sensilla and the maxillary palp about 60 sensilla. In contrast to the high number of OR genes in the mammalian genome the fly "only" contains about 60 OR genes, each expressed in a specific subset of OSNs (de Bruyne and Warr, 2006). One set of OR genes are expressed in the maxillary palp and another subset in the antenna. The different types of sensilla have been characterized according to response properties and the OR genes expressed in the OSNs. Functionality of OR genes and response properties to certain chemical compounds have been investigated by using "empty neurons". In an elegant approach using the fly genetic Gal4/UAS system functional properties of a given OR gene can be tested electrophysiological *in vivo* (de Bruyne and Warr, 2006).

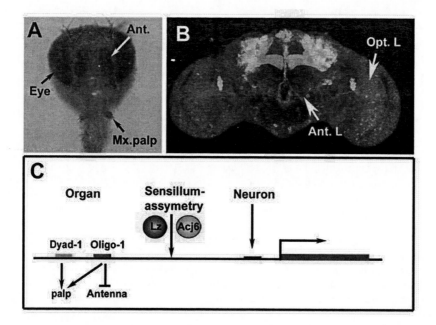

Figure 2. The fly olfactory system and OR gene choice (A) The olfactory system of the fly consists of the antenna (Ant.; arrow) and maxillary palp (Mx.palp; arrow). (B) First order processing of olfactory information occurs in the antennal lobe (Ant. L; arrow), whereas visual information is processed in the optic lobe (Opt.L; arrow). The mushroom bodies (shown in green) are a center for higher order olfactory information processing, memory and learning. (C) Model of OR gene choice largely depending on a combinatorial code of transcription factors (Modified after Ray et al., 2008). The elements acting include the regulation of palp versus antenna (Dyad-1 and Oligo-1); the transcription factors Lz and Acj6, which are required for asymmetric expression in a sensillum and a general neuron specific element.

1.2.1. Co-Expression of Or83b with other OR Genes

A large amount of evidence in the vertebrate olfactory system argues for a stringent control of mutually exclusive, monoallelic expression of OR genes. Initial mapping of OR genes in the fly antenna and maxillary palp suggested that OR genes in the fly olfactory system are also expressed in a comparable manner. The identification of a specific OR gene however lead to a quite distinct model. The *Or83b* gene is expressed in a large number of OSNs. Surprisingly in these OSNs is co-expressed with other OR genes (Elmore et al., 2003; Krieger et al., 2003; Pitts et al., 2004; Vosshall et al., 1999). This lead to the hypothesis that Or83b may either bind ligands independently of the other OR or that it could act in conjunction with the canonical OR to recognize a variety of compounds. Interestingly in Or83b mutants the localization of OR proteins to the distal (or dendritic) part of the neuron is abolished. Moreover in electrophysiological and behavioral experiments these OSNs were not able to correctly respond. Thus Or83b seems to code for an atypical OR gene with a distinct function, which has not been studied molecularly in detail.

The maxillary palp contains only three types of sensilla (with two OSN types each). Five of these OSN type express one OR gene, while one OSN type co-expresses two OR genes. Interestingly in the entire olfactory system contains 36 OSN classes which have been shown to express OR genes 9 co-express to receptors (Couto et al., 2005; Fishilevich and Vosshall, 2005; Goldman et al., 2005). Thus quite in contrast to the mouse olfactory system co-expression of ORs seems to occur. However the co-expression is not random and suggests a stringently controlled genetic program underlying the regulation of OR gene co-expression.

1.2.2. Choice of OR Gene Expression

The choice of a given OSN to express a given OR gene in the fly olfactory system is likely to be distinct from the ones acting in the mouse olfactory system. First the array of genes to choose from is significantly smaller. Second the combination of two OSNs in each sensilla expressing a specific combination of OR genes must be orchestrated. A recent study provides evidence that the required regulatory sequences for several OR genes lie within a few hundred bp upstream or downstream of the gene (Ray et al., 2007). Within these sequences several conserved sites mediate essential functions for proper expression (Figure 2C). One domain (termed Dyad-1) acts to promote expression in the maxillary palp, whereas another domain (termed Oligo-1) is also required for repression in the antenna. The transcription factor *lozenge* (*lz*) is required for the activation of a subset of OR genes, others depend upon another transcription factor *abnormal chemosensory jump6* (*acj6*). The regulation and expression of Lz and Acj6 seems to establish the asymmetric expression of ORs within a sensillum. Thus the regulation of OR genes in the fly seems largely regulated by combinatorial transcription factor action and the required regulatory elements (Ray et al., 2007).

2. Regulation of Sensory Receptor Genes in the Visual System

Comparable to the olfactory and chemosensory systems the eyes in different animal phyla differ largely in their anatomy and morphology. Moreover the eyes in several animal phyla seem to have evolved independently, even though the molecular players specifying the eyes (the retinal determination network) are evolutionary conserved. Insights in the regulation of photoreceptor gene expression mainly from the fruit fly provides insight in how certain cell types gets specified to be sensitive to a certain range of light by expressing a specific *rhodopsin* gene. Interestingly the nocturnal mouse shows distinct features in sensory receptor gene expression, possibly to increasing sensory sensitivity. In both cases, the amount of genes to choose from is small, when compared to the extremely large *OR* gene family.

2.1 The Mouse Retina

Light input in the mammalian visual system comes from two types of PRs: the cones and the rods. The mouse retina is rod dominated and only about 3% of the PRs are cones, which are evenly spaced out in the retina. While the rods express only one gene the murine *rod opsin*, cones express the blue-sensitive *S-opsin* and/or the green sensitive *M-opsin*. Rods and cones are located in the "outer retina" and are contacted by cells of the "inner retina", the first station of visual processing. Interestingly, there are also cells in the inner retina retinal ganglion cells which express the blue-light photoreceptor gene *melanopsin*. Visual input via Melanopsin is mainly required for circadian rhythm control (Foster, 2005; Panda et al., 2005; Qiu et al., 2005). Even though the molecular mechanisms underlying *Opsin* expression in the mouse retina remain largely elusive the analysis of the expression provides insight in basic principles of gene regulation. The distribution of S-opsin expressing cells and M-opsin expressing cells throughout the retina seems largely stochastic. Interestingly M-opsin and S-opsin are co-expressed in a specific type cone cells, thus the "one neurone, one receptor gene"-rule does not hold true in the mouse visual system (Applebury et al., 2000).

2.2. Human Visual System

As in mice the human retina consist of cones and rods, however in contrast to nocturnal animals the human retina is specialized for daylight and color vision. Thrichromacy in humans (seeing blue, green and red) has evolved from dichromacy (seeing green and blue) through the duplication of the genes coding for the green photopigments. Thus the two genes the green photopigment (*OPNL1MW*) and the red photopigment (*OPNL1LW*) are arranged on the chromosome in a head-to-tail orientation (Deeb, 2005). One single LCR lies significantly upstream and controls the expression of both genes, ensuring that only one gene is being transcribed (Smallwood et al., 2002; Wang et al., 1999). This process is thought to be regulated by interaction of the LCR with the promoters of the genes, activating the transcription of the selected gene.

2.3. Fly Visual System

The compound eye of the fruit fly has extensively been studied in various biological contexts. Already about century ago the fly eye was made famous with the discovery of the white mutation in 1910 by T.H. Morgan, which was part of the demonstration of hereditary transmission in *Drosophila* for which Morgan was awarded the Nobel Prize in Physiology and Medicine in 1933. During the last decade the fruit fly eye provided major genetic insights from how correct *rhodopsin* expression is regulated.

2.3.1 Anatomy of the Fly Eye

The compound eye of the fly is composed of about 800 subunits (Figure 2A), termed ommatidia. Each ommatidium consists of the eight PRs, surrounded by accessory cells. The eight PRs can be further subdivided into outers (R1-R6) and inners (R7 and R8), which also reflect their functional role. The outer PRs, R1 to R6, are the fly equivalent of the vertebrate rods and have been implicated in motion detection, dim light vision and image formation. Outer PRs express the broad spectrum photopigment *rhodopsin 1* gene and display an ommatidium of large diameter that spans the entire thickness of the retina, thus containing an increased volume of membranous structures in their rhabdomere. The increased volume of which enables them to capture photons with high efficiency (Wernet and Desplan, 2004).

The inner PRs (R7 and R8) function like vertebrate cones in color vision. In contrast to outer PRs which span the whole length of the ommatidium, inners PRs are located on top of each other and only span half the length: R7 is located on top of R8. This particular top to bottom orientation has as consequence that R7 and R8 in the ommatidium share the same light path, which allows them to compare their sensory outputs (Wernet and Desplan, 2004). Sensory integration of visual input in the fly occurs in the optic lobes(Morante et al., 2007).. The Outer PRs project into the lamina neuropile where they are primarily contacted by neurons localized in the lamina cortex. Inner PRs extend their axonal projections to the medulla neuropile where they are contacted by neurons located in the medulla cortex.

2.3.2 Characteristics of Rhodopsin Regulatory Regions

In general the regulatory regions of *rhodopsin* genes display a bipartite architecture. The relatively small proximal domain includes a region covering about 100bp and required for general PR expression. In this region lie the so called Rhodopsin Control Sequence (RCS), which are shared between all fly *rhodopsins*, even though minor differences are observed (Papatsenko et al., 2001). Further upstream lies a distal domain encoding subtype specificity of a given *rhodopsin*. The RCS incorporates a palindromic paired-class Q50 homeodomain binding site and has been thought to be activated by Pax6 genes (Cook et al., 2003; Tahayato et al., 2003). Interestingly the RCS seems to be largely interchangeable between different *rhodopsin* genes. The distal domain contains an array of transcription factors binding sites, which if mutated either leads to the loss or the ectopic expression of a specific *rhodopsin* gene.

2.3.3. Different Ommatidia Subtypes

Even though externally all ommatidia in the *Drosophila* eye appear the same, they can be subdivided into four distinct classes. First, ommatidia of the dorsal rim area (DRA), second, yellow (y) ommatidia, third the pale (p) ommatidia and fourth the dorsal-yellow ommatidia (Figure 3A). Even though p- and y-ommatidia are randomly distributed throughout the eye the ratio of p- versus y-ommatidia is always 70% yellow and 30% pale. The differences in their spectral absorbance lead to speculation that they might be involved in discrimination of different wavelengths of light. Indeed, inner PRs of p- and y-ommatidia express a different set of *rhodopsins* (Mikeladze-Dvali et al., 2005a; Wernet and Desplan, 2004). In yellow ommatidia R7 expresses the UV-sensitive Rh4 and R8 expresses the green-sensitive Rh6, whereas in pale ommatidia R7 expresses the UV-sensitive Rh3 and R8 expresses the blue-sensitive Rh5. The coupling between Rh3 and Rh5 or Rh4 and Rh6, respectively, is kept in a stringent manner. The combination of a UV-*rhodopsin* and a green/blue-*rhodopsin* has been proposed to be required for color discrimination. p-ommatidia discriminate between shorter wavelength light (UV and blue), whereas y-ommatidia discriminate between longer wavelength light (UV and green). Inner PRs of DRA ommatidia express Rh3 in both R7 and R8. The function of the DRA is quite different from the rest of the retina, which is specifically required for the perception of polarized light (Tomlinson, 2003; Wernet et al., 2003a).

2.3.4. Specification Inner PRs

The recruitment of PRs into the ommatidial clusters occurs in a highly stereotyped manner. The first cell to be specified is the R8 cell, which subsequently recruits the R2/R5-pair, the R3/R4-pair, the R1/R6-pair and finally R7 making use of Notch-, EGFR- and Sevenless signaling cascades. Even though the specification of outer PRs occurs in a highly orchestrated spatial and temporal manner after terminal differentiation they will all express the same *rhodopsin* (*rh1*). The two inner PRs however express an array of 4 different rhodopsins (*rh3*, *rh4*, *rh5* and *rh6*) and therefore have to adopt the correct subtype cell fate to express the appropriate *rhodopsin*.

In a fist step the *spalt* gene complex distinguishing inner from outer PRs (Figure 3B). The *spalt* genes are expressed specifically in R7 and R8. In *spalt* mutants the two inner PRs do develop to display morphological characteristics of outer PRs and most strikingly express Rhodopsin1 (Mollerau et al. 2001). In addition they lose inner PRs specific expression of Rh3, Rh4, Rh5 and Rh6. However, axonal projections, which are established earlier in development still project to the medulla neuropile, therefore initial specification of R7 and R8 does not seem to be affected but only later in terminal differentiation adopt an outer PR-identity. Thus the genetic program specifying PR subtypes occurs in two phases: during an early phase the establishment of neural and general PR identity and later the terminal differentiation into distinct PR types and subtypes. The *spalt* gene complex acts to maintain inner PR terminal differentiation in otherwise an "outer-PR" ground state.

The homeodomain transcription factor Prospero (Pros) and the zinc finger transcription factor Sensless (Sens) subsequently specify R7 versus R8 identity (Figure 3C). Pros is expressed specifically in R7-PR in response to EGFR, Notch and Sevenless signaling, and absent in R8 or outer PRs (Cook et al., 2003). Pros has been proposed to act directly by binding to enhancer regions of R8 specific *rhodopsins* (*rh5* and *rh6*) and thereby repressed them in R7. Loss of *pros* leads to the de-repression of *rh5* and *rh6* in R7-PRs, as well as nuclear mislocation therefore resulting in a second R8-like cell in each ommatidium. Conversely, misexpression of Pros leads to the repression of *rh5* and *rh6* in R8-PRs (Cook et al., 2003). Sens is expressed in R8-PRs where it initially acts to specify R8-PRs by promoting neuronal and specifically R8 identity (Frankfort et al., 2001; Nolo et al., 2000). During terminal differentiation Sens acts opposing to Pros, by promoting R8 features while suppressing R7-like features. In late Loss-of Sens the "R8"-PRs do not express R8 specific *rhodopsins* (*rh5* and *rh6*), but express R7-specific *rhodopsins* (*rh3* and *rh4*). Conversely, misexpression of Sens leads to the repression of *rh3* and *rh4* in R7 and the ectopic activation or *rh5* and *rh6*. Thus Sens and Pros act in an opposing manner to specify R7 versus R8 features including sensory receptor gene expression.

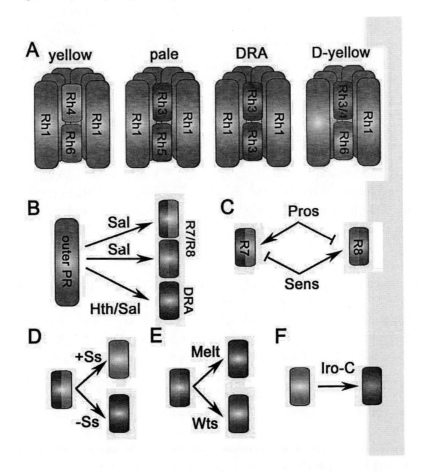

Figure 3. The fly ommatidia subtypes and regulation of *rhodopsins*(A) The adult retina consist of four different ommatidia subtypes which express a distinct combination of rhodopsins: yellow ommatidia express Rh4 (in R7) and Rh6 (in R8), pale ommatidia express Rh3 (in R7) and Rh5 (in R8), DRA ommatidia express Rh3 (in R7 and R8), dorsal yellow ommatidia co-express Rh3 and Rh4 (in R7) and express Rh6 (in R8). (B) The transcription factor Sal specifies inner PRs versus outer PRs, Hth specifies DRA inner PRs. (C) The transcription factors Pros and Sens specify R8 versus R7 (Pros for R7 features and Sens for R8 features). (D) The transcription factor Ss is necessary and sufficient to specify the p-R7 and subsequently the pale versus yellow ommatidia: Ss leads to the expression of Rh4 whereas the absence of Ss results in Rh3 expression. (E) The genetic bistable loop of Wts and Melt cross regulate each other and specify Rh5 or Rh6 expression in the R8-PR: Wts is required for Rh6 expression; Melt acts to promote Rh5 expression. (F) The Iro-C genes act to establish co-expression of Rh3 and Rh4 in the dorsal yellow ommatidia.

The dorsal rim of the eye contains a set of highly specialized ommatidia which are involved in the detection of polarized light (Figure 3B). Inner PRs of the DRA ommatidia are highly adapted in their morphology and configuration to act as polarized light sensors, which depends on strict alignment of microvilli which form the rhabdomere. Moreover the inner PR rhabdomere diameter is significantly enlarged. The Wingless (Wg) secreted morphogen is involved in the development and specification of DRA ommatidia. Wg is expressed in the head cuticle surrounding the eye. Ectopic activation of the Wg pathway in the eye transforms ommatidia in the dorsal half into DRA ommatidia, suggesting that these ommatidia are competent to respond to Wg signaling. Genes of the Iroquois-complex (IRO-C) are expressed in the dorsal half of the eye (Tomlinson, 2003; Wernet et al., 2003a). Ectopic expression of any of the three IRO-C genes *auracan*, *caupolican* and *mirror* in the rest of the eye leads to the expansion of DRA to the ventral margin. Thus the combinatorial action of Wg and IRO-C is crucial for the specification of DRA ommatidia (Tomlinson, 2003; Wernet et al., 2003a).

The homeodomain transcription factors Hth is the major factor for the development of inner DRA PRs. Hth is specifically expressed in R8 and R7 cells of DRA ommatidia. In *hth* mutants DRA ommatidia do not display the DRA specific morphology such as the increased rhabdomere diameter and the expression of Rh3 in R7 and R8. Interestingly, R7 and R8 express the untypical combination of Rh3/Rh6 (Wernet et al., 2003a). Ectopic expression of Hth is sufficient to transform all ommatidia into DRA ommatidia, with an increased inner PR diameter, expression of Rh3 and lack of normal coupling of Rh3/Rh5 or Rh4/Rh6. *hth* is both necessary and sufficient for the specification of polarized light sensors by coordinating R7 and R8 terminal differentiation (Wernet et al., 2003b).

2.3.5. Specification Inner PRs Subtypes

An intriguing and currently largely enigmatic question is how the apparent stochastic distribution of p- versus y-ommatidia in the retina is controlled (Figure 3D). The bHLH-PAS (basic helix-loop-helix-Period-Arnt-Single-minded) transcription factor *spineless* (*ss*) is the major determinant in for the specification of p- versus y-ommatidia (Wernet et al., 2006). Ss is expressed in a subset of R7 cells in a stochastic manner. About 60-80% of all R7 express Ss, suggesting that *ss* acting in y-ommatidia specification. Ss is both necessary and sufficient for the yR7 fate. In *spineless* mutants all R7 PRs express Rh3, therefore become pR7. Conversely ectopic expression of Ss in all R7 results in the transformation of all R7 into yR7 (expressing Rh4). Interestingly, Ss misexpression is also sufficient to induce R4 expression in outer PRs. Thus Ss seems to act as major activator for Rh4 expression. The coupling of Rh3 with Rh5 and Rh4 with Rh6 is stringently coordinated. Interestingly, in Ss mutants most R8 PRs express Rh5, whereas the ectopic expression of *ss* makes R8 PRs express Rh6. Therefore spineless is acting at several aspects in y-ommatidium specification: First, in R7 where it promotes Rh4 expression and represses Rh3, and secondly in the underlying R8-PR where Ss is non-autonomously required for the expression of Rh6. Thus in yR7 Ss may also control a signal to R8, which is required in this cell to adopt the yR8 fate (Wernet et al., 2006).

Even though the choice of a given ommatidium to adopt a pale or yellow fate is made by Ss in R7, the underlying R8 cell must adopt the same (p or y) fate as R7, thereby ensuring the strict coupling of inner *rhodopsins* (Mikeladze-Dvali et al., 2005b). The genetic bistable loop of the growth regulator *warts* (*wts*) and the tumor suppressor *melted* (*melt*) acts to promote an unambiguous decision in R8 to adopt either a yR8-fate (expressing Rh6) or a pR8-fate (expressing Rh5). Wts is expressed in yR8 where it promotes Rh6 expression and represses Rh5, whereas Melt is expressed in pR8 and promotes Rh5 expression and represses Rh6 (Figure 3E). Wts represses *melt* and vice versa. In *wts* mutants all R8 PRs express Melt and therefore Rh5. In *melt* mutants all R8 PRs express *wts* and therefore Rh6. Ectopic activation of *wts* or *melt* leads to the repression of the opponent (Mikeladze-Dvali et al., 2005b). These novel roles of a growth regulator and a tumor suppressor in the specification of postmitotic neurons are rather unexpected.

2.3.6. Alternative Mechanisms of Specifying the Same Set of Rhodopsin Fates

Interestingly the same set of *rhodopsins* expressed in the adult R8-PR (*rh5* and *rh6*) is also expressed in the larval eye. As the life cycle of all holometabolous insects, such as *Drosophila*, is bipartite, the animal uses of a set of sensory organs which develop during embryonic stages, as compared to the sensory organs of the adult fly which only get terminally specified during pupation. The comparably simple larval eye (also termed Bolwig organ) consisting only of about 12 PRs. Interestingly, larval PRs express *rh5* and *rh6* in a similar ratio as retinal PRs: about four PRs express Rh5 and about eight PRs express Rh6 (Sprecher et al., 2007a). No other *rhodopsins* are expressed in the larval eye. Development of larval PRs occurs in a two step. First, primary precursor cells get specified which requires the proneural gene *ato* and the RDN genes *sine oculis* and *eyes absent* in combination with Hh signaling (Daniel et al., 1999; Green et al., 1993; Suzuki and Saigo, 2000). Subsequently, primary precursors signal to the surrounding cells via EGFR signaling to develop as secondary precursors. Primary precursors will develop into the Rh5-subtype, whereas secondary precursors develop into the Rh6-subtype.

Even though both larval PR-subtypes express R8 specific *rhodopsins* the genetic programs determine correct *rhodopsin* regulation does not depend on *wts* and *melt* as compared to the adult retina. In the larva the combinatorial action of three transcription factors orchestrate subtype specification. The transcription factor Sal is expressed in the Rh5-subtype, where it is required for expression of *rh5*. In *sal* mutants the presumptive Rh5-subtype is devoid of rhodopsin expression, therefore displaying an intriguing "empty" state (Sprecher et al., 2007a). Interestingly, the role of Sal in this case is quite different from the function in adult ommatidia development where Sal specifies inner PR cell fate (see above). The orphan nuclear receptor Seven-up (Svp) acts in the opposite manner. Svp is specifically expressed in the Rh6-subtype, where it is required to repress Sal and to promote Rh6-expression. In *svp* mutants all PRs express Rh6. The homeodomain transcription factor Orthodenticle (Otd) also acts in larval PR-subtype specification. The expression of *otd* is not restricted to a specific subtype, but is only required in the Rh5-PRs. In *otd* mutants all PRs express Rh6, thus in the Rh5-subtype Otd promote the expression of Rh5 and the repression of Rh6. Thus the same sensory sensitivity and *rhodopsin* expression can be achieved using distinct genetic mechanisms (Sprecher et al., 2007a).

2.3.7. A Genetic Program Altering Sensory Specificity by Switching Rhodopsin Expression

During development of an organism multipotent proliferating stem cells will give rise to or precursor cells, which often still have the potential to develop into an array of different cell types. These cells subsequently will adopt a specific cell fate, which in most cases is orchestrated by intrinsic and extrinsically acting factors. Generally this process of terminal differentiation is unidirectional and once the choice is made to become a specific cell type the cell is stringently committed to differentiate. This basic principal is also believed to be true for sensory neurons. An interesting exception for this rule stems from the so called fruit fly eyelet (also termed Hofbauer-Buchner eyelet), a small photosensory organ located between the retina and the optic ganglia in the adult fly. The eyelet develops from the larval eye and represents what remains from the larval eye in the adult fly. The eyelet consists of about four PRs all of which express *rh6* (Sprecher and Desplan, 2008). Interestingly during metamorphosis the eight Rh6-subtype of the larval eye degenerate, while the four Rh5-subtype remain present. However the Rh5-subtype switches *rhodopsin* expression from Rh5 (in the larva) to Rh6 (in the eyelet). The steroid hormone Ecdysone acts as trigger for this sensory receptor switch in the Rh5-subtype, as well as for the degeneration of the Rh6-subtype. Inhibiting the function of the *Ecdysone Receptor* gene product Interestingly this process of sensory plasticity neither utilizes the mechanisms which act in the adult R8 cell, the wts/melted pathway, nor the one acting in larval PR-subtype specification (Sal, Svp and Otd interaction). Thus three largely independent programs depending on EcR acts for the genetic control of sensory receptor gene expression for *rh5* and *rh6* (Sprecher and Desplan, 2008). The genetic mechanism of how EcR regulates rhodopsin expression however remains elusive (Figure 4).

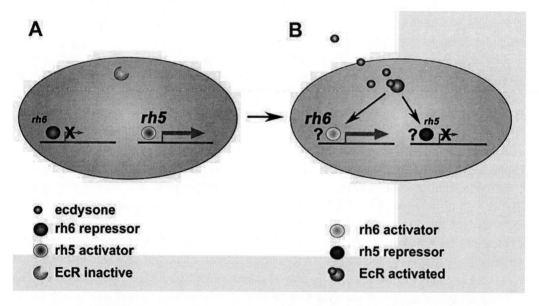

Figure 4. The genetic regulation of *rhodopsins* switch from the larval eye to the adult eyelet(A) During larval stages four larval PRs express Rh5, depending on the combinatorial action of Sal and Otd, which probably act through a combination of activators for Rh5 and repressors for Rh6. During metamorphosis the steroid hormone ecdysone triggers the switch from Rh5 to Rh6. (B) in the adult these PRs now express Rh6: Ecdysone receptor acts through a combination of activators for Rh6 and repressors for Rh5 to establish this genetic respecification of terminally differentiated PR neurons.

2.3.8. Genetic Regulation of Sensory Receptor Gene Co-Expression in the Fly Eye

Many observations of sensory receptor gene expression described suggested that a given sensory neuron can only express a single sensory receptor gene. Even though this "one neuron one receptor" rule holds true for many cases, recent findings start to challenge this paradigm. In the fruit fly retina R7 PRs typically either express *rh3* (pale subtype) or *rh4* (yellow subtype). Interestingly in the dorsal part of the eye *rh3* and *rh4* are co-expressed in R7, while the underlying R8 cell still expresses either *rh5* or *rh6*. The genetic regulation of the co-expression of *rh3* and *rh4* is under the control of the genes of the Iroquois-complex (*auracan*, *caupolican* and *mirror*; see above) (Mazzoni et al., 2008). Loss of the IRO-C in the dorsal ommatidia inhibits the co-expression, either Rh4 or Rh3 are expressed in R7 (Figure 3F). Conversely, the ectopic expression of either *auracan* or *caupolican* induced co-expression of Rh3 and Rh4 (Mazzoni et al., 2008). Thus genes of the IRO-C are necessary and sufficient for co-expression of Rh3/Rh4 breaking the "one neuron one receptor" rule.

3. Conclusion

A terminal step in cell fate determination in sensory organs is the establishment of sensory receptor gene expression. This choice of any sensory neuron to express a specific sensory receptor gene incorporates several challenges for a differentiating cell. Insights from different animal model systems lead to the description and discovery of an array shared and not shared features and mechanisms. The mouse olfactory system has provided substantial and elemental insights in the monoallelic, mutually exclusive expression of a specific OR gene, even though the mechanisms of how this mutually exclusive expression is genetically and mechanistically regulated remains elusive. A generally proposed rule that a given sensory neuron can only express one sensory receptor gene has been shown to be context dependent and examples form the visual system in mice and flies (as well as the olfactory system in flies) shows that exceptions exist. In the fly visual system the co-expression of two sensory receptor genes is stringently genetically controlled in a developmental manner. A surprising phenomenon it the sensory receptor gene switch in functional terminally differentiated neurons, in the fly larval eye. Interestingly, even though the key player for this re-specification has been identified, the genetic and molecular mechanisms remain unknown. Thus the comparison of mechanisms acting in vertebrates and invertebrates shows quite distinct characteristics in the regulation of sensory receptor gene expression. Many mechanistic aspects remain elusive and fundamental questions remain to be answered, thus provide a wealth of possibilities to study both the development of sensory organ systems as well as to give insights in the genetic and molecular mechanisms of developmental gene expression.

Acknowledgments

I would like to thank my colleagues at New York University for fruitful discussion and comments on the manuscript. This work has been supported by the Swiss National Science Foundation, the Novartis Foundation and the Janggen-Pöhn Stiftung.

References

Applebury, M. L., Antoch, M. P., Baxter, L. C., Chun, L. L., Falk, J. D., Farhangfar, F., Kage, K., Krzystolik, M. G., Lyass, L. A. and Robbins, J. T. (2000). The murine cone photoreceptor: a single cone type expresses both S and M opsins with retinal spatial patterning. *Neuron* 27, 513-23.

Buck, L. and Axel, R. (1991). A novel multigene family may encode odorant receptors: a molecular basis for odor recognition. *Cell* 65, 175-87.

Cau, E., Gradwohl, G., Fode, C. and Guillemot, F. (1997). Mash1 activates a cascade of bHLH regulators in olfactory neuron progenitors. *Development* 124, 1611-21.

Chess, A., Simon, I., Cedar, H. and Axel, R. (1994). Allelic inactivation regulates olfactory receptor gene expression. *Cell* 78, 823-34.

Conzelmann, S., Levai, O., Bode, B., Eisel, U., Raming, K., Breer, H. and Strotmann, J. (2000). A novel brain receptor is expressed in a distinct population of olfactory sensory neurons. *Eur J Neurosci* 12, 3926-34.

Cook, T., Pichaud, F., Sonneville, R., Papatsenko, D. and Desplan, C. (2003). Distinction between color photoreceptor cell fates is controlled by Prospero in Drosophila. *Dev Cell* 4, 853-64.

Couto, A., Alenius, M. and Dickson, B. J. (2005). Molecular, anatomical, and functional organization of the Drosophila olfactory system. *Curr Biol* 15, 1535-47.

Daniel, A., Dumstrei, K., Lengyel, J. A. and Hartenstein, V. (1999). The control of cell fate in the embryonic visual system by atonal, tailless and EGFR signaling. *Development* 126, 2945-54.

Davis, R. L. (2005). Olfactory memory formation in Drosophila: from molecular to systems neuroscience. *Annu Rev Neurosci* 28, 275-302.

de Bruyne, M. and Warr, C. G. (2006). Molecular and cellular organization of insect chemosensory neurons. *Bioessays* 28, 23-34.

Deeb, S. S. (2005). The molecular basis of variation in human color vision. *Clin Genet* 67, 369-77.

Duggan, C. D. and Ngai, J. (2007). Scent of a stem cell. *Nat Neurosci* 10, 673-4.

Elmore, T., Ignell, R., Carlson, J. R. and Smith, D. P. (2003). Targeted mutation of a Drosophila odor receptor defines receptor requirement in a novel class of sensillum. *J Neurosci* 23, 9906-12.

Feinstein, P. and Mombaerts, P. (2004). A contextual model for axonal sorting into glomeruli in the mouse olfactory system. *Cell* 117, 817-31.

Fishilevich, E. and Vosshall, L. B. (2005). Genetic and functional subdivision of the Drosophila antennal lobe. *Curr Biol* 15, 1548-53.

Foster, R. G. (2005). Neurobiology: bright blue times. *Nature* 433, 698-9.

Frankfort, B. J., Nolo, R., Zhang, Z., Bellen, H. and Mardon, G. (2001). senseless repression of rough is required for R8 photoreceptor differentiation in the developing Drosophila eye. *Neuron* 32, 403-14.

Fuss, S. H., Omura, M. and Mombaerts, P. (2007). Local and cis effects of the H element on expression of odorant receptor genes in mouse. *Cell* 130, 373-84.

Goldman, A. L., Van der Goes van Naters, W., Lessing, D., Warr, C. G. and Carlson, J. R. (2005). Coexpression of two functional odor receptors in one neuron. *Neuron* 45, 661-6.

Green, P., Hartenstein, A. Y. and Hartenstein, V. (1993). The embryonic development of the Drosophila visual system. *Cell Tissue Res* 273, 583-98.

Keller, A. and Vosshall, L. B. (2003). Decoding olfaction in Drosophila. *Curr Opin Neurobiol* 13, 103-10.

Kratz, E., Dugas, J. C. and Ngai, J. (2002). Odorant receptor gene regulation: implications from genomic organization. *Trends Genet* 18, 29-34.

Krieger, J., Klink, O., Mohl, C., Raming, K. and Breer, H. (2003). A candidate olfactory receptor subtype highly conserved across different insect orders. *J Comp Physiol A Neuroethol Sens Neural Behav Physiol* 189, 519-26.

Kubick, S., Strotmann, J., Andreini, I. and Breer, H. (1997). Subfamily of olfactory receptors characterized by unique structural features and expression patterns. *J Neurochem* 69, 465-75.

Leung, C. T., Coulombe, P. A. and Reed, R. R. (2007). Contribution of olfactory neural stem cells to tissue maintenance and regeneration. *Nat Neurosci* 10, 720-6.

Lomvardas, S., Barnea, G., Pisapia, D. J., Mendelsohn, M., Kirkland, J. and Axel, R. (2006). Interchromosomal interactions and olfactory receptor choice. *Cell* 126, 403-13.

Malnic, B., Hirono, J., Sato, T. and Buck, L. B. (1999). Combinatorial receptor codes for odors. *Cell* 96, 713-23.

Mazzoni, E. O., Celik, A., Wernet, M. F., Vasiliauskas, D., Johnston, R. J., Cook, T. A., Pichaud, F. and Desplan, C. (2008). Iroquois complex genes induce co-expression of rhodopsins in Drosophila. *PLoS Biol* 6, e97.

Mazzoni, E. O., Desplan, C. and Celik, A. (2004). 'One receptor' rules in sensory neurons. *Dev Neurosci* 26, 388-95.

McGuire, S. E., Le, P. T. and Davis, R. L. (2001). The role of Drosophila mushroom body signaling in olfactory memory. *Science* 293, 1330-3.

Mikeladze-Dvali, T., Desplan, C. and Pistillo, D. (2005a). Flipping coins in the fly retina. *Curr Top Dev Biol* 69, 1-15.

Mikeladze-Dvali, T., Wernet, M. F., Pistillo, D., Mazzoni, E. O., Teleman, A. A., Chen, Y. W., Cohen, S. and Desplan, C. (2005b). The Growth Regulators warts/lats and melted Interact in a Bistable Loop to Specify Opposite Fates in Drosophila R8 Photoreceptors. *Cell* 122, 775-87.

Mombaerts, P. (1999a). Odorant receptor genes in humans. *Curr Opin Genet Dev* 9, 315-20.

Mombaerts, P. (1999b). Seven-transmembrane proteins as odorant and chemosensory receptors. *Science* 286, 707-11.

Mombaerts, P. (2004a). Love at first smell--the 2004 Nobel Prize in Physiology or Medicine. *N Engl J Med* 351, 2579-80.

Mombaerts, P. (2004b). Odorant receptor gene choice in olfactory sensory neurons: the one receptor-one neuron hypothesis revisited. *Curr Opin Neurobiol* 14, 31-6.

Mombaerts, P. (2006). Axonal wiring in the mouse olfactory system. *Annu Rev Cell Dev Biol* 22, 713-37.

Mombaerts, P., Wang, F., Dulac, C., Chao, S. K., Nemes, A., Mendelsohn, M., Edmondson, J. and Axel, R. (1996). Visualizing an olfactory sensory map. *Cell* 87, 675-86.

Morante, J., Desplan, C. and Celik, A. (2007). Generating patterned arrays of photoreceptors. *Curr Opin Genet Dev* 17, 314-9.

Ngai, J., Chess, A., Dowling, M. M., Necles, N., Macagno, E. R. and Axel, R. (1993). Coding of olfactory information: topography of odorant receptor expression in the catfish olfactory epithelium. *Cell* 72, 667-80.

Nolo, R., Abbott, L. A. and Bellen, H. J. (2000). Senseless, a Zn finger transcription factor, is necessary and sufficient for sensory organ development in Drosophila. *Cell* 102, 349-62.

Panda, S., Nayak, S. K., Campo, B., Walker, J. R., Hogenesch, J. B. and Jegla, T. (2005). Illumination of the melanopsin signaling pathway. *Science* 307, 600-4.

Papatsenko, D., Nazina, A. and Desplan, C. (2001). A conserved regulatory element present in all Drosophila rhodopsin genes mediates Pax6 functions and participates in the fine-tuning of cell-specific expression. *Mech Dev* 101, 143-53.

Pitts, R. J., Fox, A. N. and Zwiebel, L. J. (2004). A highly conserved candidate chemoreceptor expressed in both olfactory and gustatory tissues in the malaria vector Anopheles gambiae. *Proc Natl Acad Sci U S A* 101, 5058-63.

Qasba, P. and Reed, R. R. (1998). Tissue and zonal-specific expression of an olfactory receptor transgene. *J Neurosci* 18, 227-36.

Qiu, X., Kumbalasiri, T., Carlson, S. M., Wong, K. Y., Krishna, V., Provencio, I. and Berson, D. M. (2005). Induction of photosensitivity by heterologous expression of melanopsin. *Nature* 433, 745-9.

Rawson, N. E., Eberwine, J., Dotson, R., Jackson, J., Ulrich, P. and Restrepo, D. (2000). Expression of mRNAs encoding for two different olfactory receptors in a subset of olfactory receptor neurons. *J Neurochem* 75, 185-95.

Ray, A., van Naters, W. G., Shiraiwa, T. and Carlson, J. R. (2007). Mechanisms of odor receptor gene choice in Drosophila. *Neuron* 53, 353-69.

Ressler, K. J., Sullivan, S. L. and Buck, L. B. (1993). A zonal organization of odorant receptor gene expression in the olfactory epithelium. *Cell* 73, 597-609.

Rodriguez, I. (2004). Pheromone receptors in mammals. *Horm Behav* 46, 219-30.

Rodriguez, I. (2007). Odorant and pheromone receptor gene regulation in vertebrates. *Curr Opin Genet Dev* 17, 465-70.

Rodriguez, I. and Mombaerts, P. (2002). Novel human vomeronasal receptor-like genes reveal species-specific families. *Curr Biol* 12, R409-11.

Royal, S. J. and Key, B. (1999). Development of P2 olfactory glomeruli in P2-internal ribosome entry site-tau-LacZ transgenic mice. *J Neurosci* 19, 9856-64.

Serizawa, S., Ishii, T., Nakatani, H., Tsuboi, A., Nagawa, F., Asano, M., Sudo, K., Sakagami, J., Sakano, H., Ijiri, T. et al. (2000). Mutually exclusive expression of odorant receptor transgenes. *Nat Neurosci* 3, 687-93.

Serizawa, S., Miyamichi, K. and Sakano, H. (2004). One neuron-one receptor rule in the mouse olfactory system. *Trends Genet* 20, 648-53.

Shykind, B. M. (2005). Regulation of odorant receptors: one allele at a time. *Hum Mol Genet* 14 Spec No 1, R33-9.

Smallwood, P. M., Wang, Y. and Nathans, J. (2002). Role of a locus control region in the mutually exclusive expression of human red and green cone pigment genes. *Proc Natl Acad Sci U S A* 99, 1008-11.

Sprecher, S. G. and Desplan, C. (2008). Switch of rhodopsin expression in terminally differentiated Drosophila sensory neurons. *Nature* in Press.

Sprecher, S. G., Pichaud, F. and Desplan, C. (2007a). Adult and larval photoreceptors use different mechanisms to specify the same Rhodopsin fates. *Genes Dev* 21, 2182-95.

Sprecher, S. G. and Reichert, H. (2003). The urbilaterian brain: developmental insights into the evolutionary origin of the brain in insects and vertebrates. *Arthropod Struct Dev* 32, 141-56.

Sprecher, S. G., Reichert, H. and Hartenstein, V. (2007b). Gene expression patterns in primary neuronal clusters of the Drosophila embryonic brain. *Gene Expr Patterns* 7, 584-95.

Stocker, R. F. (2006). Olfactory coding: connecting odorant receptor expression and behavior in the Drosophila larva. *Curr Biol* 16, R16-8.

Strotmann, J., Conzelmann, S., Beck, A., Feinstein, P., Breer, H. and Mombaerts, P. (2000). Local permutations in the glomerular array of the mouse olfactory bulb. *J Neurosci* 20, 6927-38.

Suzuki, T. and Saigo, K. (2000). Transcriptional regulation of atonal required for Drosophila larval eye development by concerted action of eyes absent, sine oculis and hedgehog signaling independent of fused kinase and cubitus interruptus. *Development* 127, 1531-40.

Tahayato, A., Sonneville, R., Pichaud, F., Wernet, M. F., Papatsenko, D., Beaufils, P., Cook, T. and Desplan, C. (2003). Otd/Crx, a dual regulator for the specification of ommatidia subtypes in the Drosophila retina. *Dev Cell* 5, 391-402.

Tomlinson, A. (2003). Patterning the peripheral retina of the fly: decoding a gradient. *Dev Cell* 5, 799-809.

Tsuboi, A., Yoshihara, S., Yamazaki, N., Kasai, H., Asai-Tsuboi, H., Komatsu, M., Serizawa, S., Ishii, T., Matsuda, Y., Nagawa, F. et al. (1999). Olfactory neurons expressing closely linked and homologous odorant receptor genes tend to project their axons to neighboring glomeruli on the olfactory bulb. *J Neurosci* 19, 8409-18.

Vassalli, A., Rothman, A., Feinstein, P., Zapotocky, M. and Mombaerts, P. (2002). Minigenes impart odorant receptor-specific axon guidance in the olfactory bulb. *Neuron* 35, 681-96.

Vassar, R., Ngai, J. and Axel, R. (1993). Spatial segregation of odorant receptor expression in the mammalian olfactory epithelium. *Cell* 74, 309-18.

Vosshall, L. B. (2000). Olfaction in Drosophila. *Curr Opin Neurobiol* 10, 498-503.

Vosshall, L. B. (2001). The molecular logic of olfaction in Drosophila. *Chem Senses* 26, 207-13.

Vosshall, L. B., Amrein, H., Morozov, P. S., Rzhetsky, A. and Axel, R. (1999). A spatial map of olfactory receptor expression in the Drosophila antenna. *Cell* 96, 725-36.

Wang, Y., Smallwood, P. M., Cowan, M., Blesh, D., Lawler, A. and Nathans, J. (1999). Mutually exclusive expression of human red and green visual pigment-reporter transgenes occurs at high frequency in murine cone photoreceptors. *Proc Natl Acad Sci U S A* 96, 5251-6.

Wernet, M. F. and Desplan, C. (2004). Building a retinal mosaic: cell-fate decision in the fly eye. *Trends Cell Biol* 14, 576-84.

Wernet, M. F., Labhart, T., Baumann, F., Mazzoni, E. O., Pichaud, F. and Desplan, C. (2003a). Homothorax Switches Function of Drosophila Photoreceptors from Color to Polarized Light Sensors. *Cell* 115, 267-279.

Wernet, M. F., Labhart, T., Baumann, F., Mazzoni, E. O., Pichaud, F. and Desplan, C. (2003b). Homothorax switches function of Drosophila photoreceptors from color to polarized light sensors. *Cell* 115, 267-79.

Wernet, M. F., Mazzoni, E. O., Celik, A., Duncan, D. M., Duncan, I. and Desplan, C. (2006). Stochastic spineless expression creates the retinal mosaic for colour vision. *Nature* 440, 174-80.

Wong, A. M., Wang, J. W. and Axel, R. (2002). Spatial representation of the glomerular map in the Drosophila protocerebrum. *Cell* 109, 229-41.

In: Developmental Gene Expression Regulation
Editor: Nathan C. Kurzfield

ISBN: 978-60692-794-6
©2009 Nova Science Publishers, Inc.

Chapter IV

Normal and Injury-Induced Gene Expression in the Developing Postnatal Rat Inner Ear

Gross, Johann[a], Ralf-Jürgen Kuban[b], Ute Ungethüm[b] and Birgit Mazurek[a]

[a]Department Othorhinolaryngology, Charité-Universitätsmedizin Berlin, Molecular Biology Research Laboratory, [b]Charité-Laboratory of Functional Genomics, Charitéplatz 1, 10117-Berlin, Germany

Abstract

This chapter deals with an organotypic culture system to examine transcriptional events contributing to cell survival of the organ of Corti (OC), the modiolus (MOD) and the stria vascularis (SV) of newborn rats. mRNA profiling using Affymetrix gene chips was carried out in tissue obtained immediately after preparation and after 24 h in culture. The probe sets of 45 genes were subjected to a cluster analysis. A number of identified genes represented three major processes associated with the preparation of the cultures: mechanical injury and inflammation, hypoxia and excitotoxicity. The inflammatory response ontology was represented by inflammatory cytokines Interleukin-1beta (Il-1b), Interleukin 6 (Il-6), TNF-alpha converting enzyme (Tace) and Intercellular adhesion molecule (Icam). The hypoxia response was represented by the increase of Hypoxia-inducible factor-1 alpha (Hif-1a),

Glucose transporter 1 (Glut1) and Glucose transporter 3 (Glut3). The excitotoxic damaging process included changes in the glutamate transporters and NMDAR receptors. We identified Tace expression as a novel gene in the inner ear with a potentially important role in inner ear injury. The MOD region belongs to the most vulnerable regions of the ear characterized by a particularly high increase of Il-1b, Il-6 and Hif-1alpha mRNA expression.

Cell survival in culture is maintained by a complex regulation of pro-death genes on the one hand and protective genes on the other. In general, genes encoding proteins involved in triggering or executing cell death are downregulated and genes encoding protective acting proteins are upregulated. We found caspase 2, caspase 6 and calpain downregulated and the mitochondrial superoxide dismutase Sod2, the heat shock proteins Hsp27 and Hsp 70 and the insulin like growth factor binding proteins Igfbp3 and Igfbp5 upregulated. For two genes (Tace, Bax) we observed a differential response of coding and non-coding sequences. These data provide new insights into the role of the various members of the pro-death and pro-survival genes in protecting inner ear cells from injury-induced damage during the developing period.

Introduction

The inner ear is composed of the vestibular system that controls balance and the cochlea that mediates sound perception. The cochlea consists of three main complex structures, each serving a specific function: the organ of Corti (OC), the modiolus (MOD) and the stria vascularis (SV). The OC with its auditory receptors, the inner and outer hair cells, is located in the scala media compartment of the cochlea. This scala contains the endolymph, a very specific fluid with a high concentration of potassium and a low concentration of sodium determining the transduction of the sound-induced stimuli into electrical impulses. The ion composition of the endolymph is maintained by the SV. The hair cells have both afferent and efferent innervations. The connection between hair cells and the cochlear nucleus in the brainstem is provided by bipolar spiral ganglion neurons (SGN) through central and peripheral processes. The SGNs are located in the MOD, the conical shaped central axis in the cochlea (Slepecky, 05).

Damage to the cochlear structures results in two functional impairments: hearing loss and/or tinnitus. Both are among the most frequent disorders of the neurosensory system with a prevalence of about 10 % of the population (Gates and Mills, 05). In general, functional impairments of the inner ear result not only from damage to the neurosensory cells, but also from changes in the surrounding tissue. Important causes for hearing loss are noise, presbyacusis, ototoxic substances and hypoxia/ischemia. Temporary or permanent noise induces damage to the hair cells as well as changes in the SV and the spiral ligament (Hirose and Liberman, 03). Susceptibility to noise together with a reduced endocochlear potential and acute cellular changes in the SV and spiral ligament are inherited in mice in an autosomal dominant manner (Ohlemiller and Gagnon, 07). Presbyacusis, another frequent cause of hearing loss and tinnitus, can have its basis in independent changes occurring in the OC, the MOD or the SV (Ohlemiller, 04). Further, there is clear evidence that cisplatin ototoxicity includes histological lesions and effects the OC, the spiral ganglion and the SV (van Ruijven et al., 05). Inner ear pathology of severe neonatal asphyxia may include the degeneration of spiral ganglia, hair cells and edematous changes in the SV (Koyama et al., 05). Therefore, we focused our studies on the differential response of OC, MOD and SV to culture-induced injury (Gross et al., 07; Gross et al., 08)

Causes producing hearing loss or tinnitus are associated with various pathogenetic mechanisms. For example, noise acts via mechanical injuries, but also via reduced blood flow producing hypoxia/ischemia. Mechanical energy of noise destroys not only the organ of Corti, but also the lateral wall of the cochlear duct (Ulehlova, 83). Damage of the inner ear induced by ototoxic substances and hypoxia/ischemia are mediated by oxidant stress. Presbyacusis seems to be closely associated with hypoxia/ischemia (Pujol et al., 90). A specific mechanism of damage to inner hair cells by hypoxia/ischemia is excitotoxicity, the degeneration of hair cells and SGNs as a result of excessive stimulation of the glutamatergic receptors.

Organotypic cultures of the organ of Corti from neonatal rats (3-5 day old) were introduced as a useful experimental model to study the structure and function of one of the most vulnerable cells of the cochlea, the hair cells (Sobkowicz et al., 93). Similarly, cultures from two other main parts of the cochlea, the SV (Achouche et al., 91; Melichar and Gitter, 91) and the SGNs (Cheng et al., 99; Dazert et al., 97) were successfully introduced. The cells can be kept in a living status for several days or even weeks. These cultures are valuable tools to study changes in gene expression induced by injuries of different types. Obviously, preparing the tissue and growing in culture is associated with mechanical and metabolic stress which strongly interrupt the homeostasis of the gene expression. Because most of the cells survive in the culture for several days or weeks we assume that the changes induced in culture reflect effective endogenous mechanisms of repair and regeneration and may thus provide new insights into the role of various members of the pro-death and pro-survival genes in protecting inner ear cells from injury-induced damage.

We hypothesize that three groups of genes are involved in the expressional changes associated with preparation and growth of the cultures, (1) transcripts of genes associated with cell damaging processes; (2) transcripts of genes associated with triggering or executing cell death and (3) transcripts of genes associated with the protection from cell death. (1) Inner ear tissue allowed to grow in culture includes at least three types of damaging mechanisms: mechanical injuries, hypoxia/ischemia and excitotoxicity. The pathophysiology of mechanic injury includes several secondary events, for example oxidative stress and inflammatory response (Liu et al., 07). In this chapter we used the pro-inflammatory criteria Interleukin-1beta (Il-1b), Interleukin 6 (Il-6), TNF-alpha converting enzyme (Tace) and Intercellular adhesion molecule 1 (Icam1) as indicators of preparatory injury. Well-known characteristic markers of hypoxia/ischemia are Hypoxia inducible factor 1alpha (HIF-1a) and Glucose transporter 1 (Glut1) and Glut3 (Airley and Mobasheri, 07). Markers of stimuli depolarizing the neuronal membrane of the SGNs and their radial fibers and causing the excitotoxic death of their synaptic sensory cells through excessive stimulation of the glutamatergic receptors are glutamate transporters (neuronal Glu-n; glial Glu-g) and receptors (Grina - glutamate binding subunit of the N-methyl-D-aspartate [NMDA] receptor; Grin2b – NMDA receptor, 2B subunit; Grm4 - metabotropic glutamate receptor, type 4). (2) Transcripts of genes closely associated with triggering or executing cell death are members of the Bax/Bcl family (Bax - Bcl2-associated X protein), members of the caspase (aspartate specific cysteine proteases) and calpain (calcium activated cysteine proteases) families. The Bax/Bcl genes are involved in triggering cell death, and caspases and calpains are enzymes involved in executing cell death. Cell-death stimuli cause Bax genes to translocate from the cytosol to the mitochondria, where it causes membrane permeabilization. This releases cytochrome c leading to caspase activation and commitment to cell death (Carvalho et al., 04; Gross et al., 99). (3) Transcripts of genes supporting cell survival are members of the superoxide dismutases (SODs), heat shock proteins (HSP) and the insulin like growth factor (IGFs). It is assumed that the balance of the activity of these genes may be the basis of cell survival under culture conditions. SODs are important for elimination of reactive oxygen species (ROS), HSPs are important factors for protecting proteins from denaturation and IGFs are important for repair, regeneration and growth.

Gene expression methods such as oligonucleotide microarrays are effective techniques for identifying differentially regulated genes that may serve as new diagnostic or therapeutic targets. In this chapter, microarray analyses were utilized for the quantification of the mRNA levels in freshly prepared OC, MOD and SV of 3-5 day old rats and in cultures after 24 hours. Analyses of the expression of the transcripts of pathogenic, pro-death and pro-survival groups of genes could be used to extend our current understanding of the differential response of these genes in the OC, MOD and SV in the developing postnatal period.

Material and Methods

Explant Cultures

Each cochlea from 3- to 5-day-old Wistar rats was dissected into the OC, the MOD and the SV (Sobkowicz et al., 93). Dissection of the cochlea was performed in buffered saline glucose solution (BSG; 116 mM NaCl, 27.2 mM Na_2HPO_4, 6.1 mM KH_2PO_4, glucose 11.4 mM) at 4 °C under a laminar flow hood using a dissecting microscope (Stemi, SV6, Zeiss Germany). Most of the spiral ganglion cell bodies remained with the modiolus. The modiolus was cut into two halves to reduce the tissue thickness. The cochlear parts were well preserved for up to 24 h in culture (Gross et al., 07). For culture, the fragments were incubated in 4-well tissue culture dishes (1.9-cm^2 culture surface per well, Nunc, Wiesbaden, Germany) in 500 µl Dulbecco's modified Eagle medium/F12 nutrient (1:1) mixtures (DMEM/F12, Gibco, Karlsruhe, Germany) supplemented with 10% fetal bovine serum (FBS, Biochrom AG, Berlin, Germany), 50 mM glucose, insulin–transferrin–Na–Selenit-Mix 2 µl/ml (Roche Diagnostics GmbH, Mannheim, Germany), penicillin 100 U/ml (Grünenthal GmbH, Aachen, Germany) at 37 °C and 5% CO_2 in a humidified tissue culture incubator (Cheng et al., 99a; Lowenheim et al., 99).

To characterize the viability of the explants we used the LDH release assay and the propidium iodide/calcein AM assay (PI; 1 µg/ml; Molecular Probes, Eugene, Oregon, USA; calcein AM 10 µM, Molecular Probes Europe, Leiden, The Netherlands; Gross et al., 07; Noraberg et al., 99). To measure cell viability by visualizing healthy and damaged nuclei, we used Hoechst dye 33342 (SIGMA) combined with PI, as described previously followed by confocal microscopy analyses (Ciancio et al., 88). All studies were performed in accordance with the German Prevention of Cruelty to Animals Act and permission was obtained from the Berlin Senate Office for Health (T0234/00).

RNA Isolation and Quantification

Total RNA was isolated using the RNeasy kit (Qiagen, Hilden, Germany) strictly according to the manufacturer's protocol. The RNA content was determined with the RiboGreen RNA quantitation kit (Molecular Probes, Göttingen, Germany). For the microarray study, the total RNA isolated from the OC, MOD and SV of six animals were pooled to obtain a sample of 7.5 µg total RNA each. The RNA samples originated from three independent series of RNA preparations within 1 year: Series 1: Samples OC1, OC2 (to test the reproducibility of the microarray approach), MOD and SV from freshly prepared tissue indicated as controls (Co). Series 2: Samples originated from culture of the OC, MOD and SV under normoxic (Cu–No) and hypoxic conditions. Series 3 was a repetition of Series 2. The data of the hypoxic cultures are reported elsewhere (Gross et al., 07; Gross et al., 08; Mazurek et al., 03). Altogether 16 independent RNA samples and 4 arrays were used for the microarray study.

Cdna Microarray Analysis

The quality of the RNA used for the microarray was analyzed by an Agilent 2100 Bioanalyzer (Agilent Technologie, Palo Alto, CA, USA). The microarray analysis was carried out as described recently (Chaitidis et al., 05). Briefly, from total RNA, double-stranded cDNA was synthesized using T7 (dT)24 primers containing a T7 RNA polymerase promoter sequence. To produce biotin-labeled cRNA from the cDNA, an in-vitro transcription was carried out. The quality and quantity of each RNA/DNA sample were assessed by gel electrophoresis and spectrophotometric analysis. Labeled and processed cRNA samples (6 µg) were hybridized to the Affymetrix Rat Neurobiology U34 Array (RN-U34; 1322 gene transcripts, Affymetrix, Santa Clara, USA). After the arrays had been washed and stained in an Affymetrix-Fluidics station, they were scanned on an Agilent scanner. Hybridization was performed by the Laboratory for Functional Genome Research, Charité–University Medicine Berlin (RJK, UU). All processes were subjected to the QC protocols as described in the Affymetrix Gene Chip Manual. Raw data were quantified by means of the Microarray Suite (MAS 5.0) software. Expression data were normalized using the Genespring settings for Affymetrix gene chip arrays. Each chip was normalized to the 50th percentile of the measurements of that chip. Each gene was normalized to the median of the measurements of the probe sets of that gene. Results from probe sets on each array were collected as data on Excel spreadsheets.

Tables 1-3 present genes analyzed in the present chapter (gene names, accession numbers of the mRNA used for the target sequence on the RN-U34 chip and of the corresponding NCBI reference sequence, probe location and length). To reduce the annotation discrepancies we referred the probe sequences to the NCBI reference sequence using the BLAST analysis.

The RN-U34 chip contains three different probes for Tace, and two for Grina, all expressed at a statistically significant level, at least in one experimental group. *No significant expression in control samples, therefore the cutoff value of the normalized signal (= 300) was used. Cds-coding sequence, pcds-partial coding sequence (cds + UTR), aff-

affinity, UTR-untranslated region. Glu-n: [a]No similiarity with NM_013032, solute carrier family 1, neuronal/epithelial high affinity glutamate transporter, system Xag, member 1, Slc1a1. Glu-g corresponds to solute carrier family 1, glial high affinity glutamate transporter, member 3, Slc1a3.

Table 1. Transcripts used as markers of pathogenic processes

Gene	Acc.-Nr. (RefSeq)	Probe location (length)	Gene name, function
Il-1b*	E01884 (NM_031512)	406-666 (261) cds	Interleukin-1beta, pro-inflammatory
Il-6*	M26744 (NM_012589)	682-1006 (325) pcds	Interleukin 6, anti-oxidant
Tace-1	AJ012603 (NM_020306)	2304-2706 (403) cds	TNF-alpha converting enzyme,
Tace-2	AJ012603 (NM_020306)	3487-3619 (133) UTR	(disintegrin and metallopeptidase
Tace-3	AJ012603 (NM_020306)	3656-4009 (354) UTR	domain 17, Adam17)
Icam	D00913 (NM_012967)	2123-2538 (416) UTR	Intercellular adhesion molecule 1
Hif-1a	Y09507 (NM_024359)	1980-2490 (511) cds	Hypoxia inducible factor 1alpha
Glut1	S68135 (NM_138827)	2465-2553 (89) UTR	Glucose transporter 1, Slc2a1
Glut3	D13962 (NM_017102)	947-1474 (528) UTR	Glucose transporter 3, Slc2a3
Glu-n[a]	D63772	3144 - 3726 (583) UTR	Glutamate transporter, neuronal, high aff.
Glu-g	S59158 (NM_019225)		
Grina-1	S61973 (NM_153308)	3401-3833 (433) UTR	Glutamate transporter, glial, high aff.
Grina-2	S61973 (NM_153308)	1102-1201 (100) cds	Glutamate receptor, ionotropic, NMDA
Grin2b	U11419 (NM_012574)	1192-1579 (388) cds	associated protein 1,glutamate binding
Grm4	M90518 (NM_022666)	4646-5240 (595) pcds	Glutamate receptor 2B, ionotropic
		3917-4463 (547) UTR	Glutamate receptor, metabotropic

Table 2. Genes associated with cell death

Gene	Acc.-Nr.	Probe location (length)	Gene name , function
Bax-U49[a]	U49729 (AF235993)	488-571 (85) pcds	Bcl2-associated X protein (Bax),
Bax-S76[b]	S76511 (AF235993)	794-927 (134) cds	splice variant k
Bax-U59[c]	U59184 (AF235993)	1024-1075 (52) pcds	
Bcl2l1-S78	S78284 (NM_001033670)	487-611 (125) cds	Bcl2-like 1, nuclear gene encoding
Bcl2l1-U34	U34963 (NM_001033670.1)	542-731 (190) cds	mitochondrial protein, transcript var.3
Casp1	S79676 (NM_012762)	659-1097 (439) cds	
Casp2	U77933 (NM_022522)	2781-3333 (553) UTR	Caspase 1, initiator of apoptosis
Casp3	U49930 (NM_012922)	1087-1259 (173) UTR	Caspase 2, initiator of apoptosis
Casp6	AF025670 (NM_031775)	762-941 (193) cds	Caspase 3, executioner of apoptosis
			Caspase 6, executioner of apoptosis
Capn1-1	U53859 (NM_017118)	848-1037 (190) pcds	
Capn1-2	U53859 (NM_017118)	1141-1284 (144) UTR	Calpain 1, small subunit 1, involved
Capn2	L09120 (NM_017116.2)	3017-3513 (523) UTR	in
			apoptosis
			Calpain 2

Abbr.: See legend Table 1. The chip contains three different probes for Bax, two different probes for Capn1. [a]NM_017059 (33-117); [b]NM_017059.1 (340-473); [c]no similarity with NM_017059. var.-variant.

Table 3. Genes associated with cell survival

Gene	Acc.-Nr	probe position (length)	Gene name , function
Sod1	M21060 (NM_017050)	126-536 (411) cds	Super oxide dismutase 1, Cu-Zn
Sod2	Y00497 (NM_017051)	924-1476 (553) UTR	
Sod3	Z24721 (NM_012880)	1146-1710 (565) UTR	Sod 2, mitochondrial, MnSOD
			Sod 3, extracellular, Cu-Zn
Hsp27	M86389 (NM_031970.3)	77-226 (150) cds	
Hsp60	X54793 (NM_022229)	1730-2156 (427) UTR	
Hsp70	L16764 (NM_212504)	4085-4377 (293) UTR	Heat shock protein, 27kDa
Hsp90	S45392	1869-2060 (193) cds	Hsp60,chaperonin, mitoch.
	(NM_001004082)	(97%)	Hsp70, 70kDa
Igf1			Hsp 90, 90kDa
Igf2	M15481 (NM_178866.4)	48-385 (567) pcds	
Igfbp2	X17012 (NM_031511)	3921-4500 (580) UTR	Insulin like growth factor 1
Igfbp3	J04486 (NM_013122)	977-1226 (250) pcds	Igf 2
Igfbp5	M31837 (NM_012588)	1841-2321 (481) UTR	Igf binding protein 2
Igfbp6	M62781 (NM_012817)	1077-1533 (457) pcds	Igf binding protein 3
Igf1r	M69055 (NM_013104)	447-872 (487) pcds	Igf binding protein 5
	M27293 (NM_052807)	497-1082 (586) cds	Igf binding protein 6
			Igf1 receptor

Abbr.: See legend Table 1.

Statistical Analysis of Microarray Data

Several approaches were chosen to analyze the reliability of the microarray data and to estimate the thresholds of relevant expression levels and fold changes. The calculations used the normalized signals calculated as the base 2 logarithm. Correlation analyses of the normalized signals between corresponding pairs of samples resulted in the following equation: $y = 1.05 \ x - 428$; $R^2 = 0.93$, n=143. The mean variation coefficient (VC) of all duplicates was calculated as VC= 18.4 $\pm$ 16.3 %, n = 143. It is known that the inter-probe variation of signal strength on a microarray is dependent not only on biological factors but also on other factors like probe-target hybridization and the subsequent dissociation which occurs during stringent washing of the microarray (Held et al., 06). All replicate samples were within fold changes of 2.0 and 0.5.We used the k-mean approach (1) to classify the homeostatic expression patterns in freshly prepared tissue and (2) to classify the culture-induced changes of gene expression. For both calculations we used the log2 of the normalized signals or of the fold changes. Clustering of differentially expressed genes can facilitate identification of genes sharing co-regulation by transcriptional regulators. All of the expressions were divided into high, middle or low co-expression patterns only.

Results

Cellular Injury Characterization

Figure 1 illustrates images of organotypic cultures of the OC, MOD and SV. Images A,B, C are from living cultures of tissue grown at 37° C for 24 h magnified x 40 using phase-contrast microscopy. Images D,E, F are stained with calcein AM, a widely used green fluorescent cell marker for living tissue. The staining with propidium iodide, an indicator of dead cells was very weak (images not shown). Cultures were in a viable condition after 24 hours in culture with only a few dead, PI stained cells. To quantify the life-death status on the cellular level, we counted the numbers of Hoechst 33342 and PI-stained nuclei (Gross et al., 08). In freshly prepared tissue, about 2-9 % of all nuclei were found to be PI-stained, indicating cell damage that occurred during preparation of the cochlear tissue. In the OC and SV, no statistically significant increase of the percentage of PI-stained nuclei was observed during normoxic culture. In the MOD, 23 % of the cells were damaged in normoxic culture. Previously, we used another two criteria to analyze the viability of cultures, the LDH release assay and the numbers of intact hair cells of the OC. Neither the LDH test nor hair cell staining and counting revealed cell loss in normoxic cultures compared to the freshly prepared tissue (Gross et al., 07). Overall, the findings have demonstrated that the culture conditions maintain nearly all cells of the OC and the SV in a viable status. MOD cells have a higher rate of damage and appear to be the most vulnerable cells under the culture conditions used.

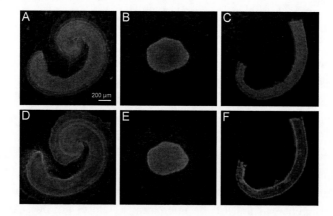

Figure 1. Images of organotypic cultures of organ of Corti, modiolus and stria vascularis. A,B, C: Images from living cultures of tissue grown at 37° C for 24 h in culture, magnified x40 using the Leica DMIL phase-contrast microscope. D,E, F: Images from cultures stained with calcein AM.

Genes Involved in Damaging Mechanisms

Figure 2 illustrates the basal expression levels of the genes associated with damaging processes in freshly prepared tissue and the fold changes after 24 h in culture. The expression of Il-1b and Il-6 was found to be absent in freshly prepared tissue, but strongly increased in culture (except Il-6 in OC). Tace and Icam fragments are statistically significantly expressed in freshly prepared tissue and further increased in culture. Remarkably, both Tace transcripts responded differently, the Tace1 fragment increased, while the Tace3 fragment decreased. The hypoxia markers Hif-1a, Glut1 and Glut 3 are significantly expressed in all regions and increased their expression by a factor of 1.8-8. Among several glutamate transporters and receptors, which could be identified by the RN-U34 chip, two glutamate transporters (Glu-g and Glu-n) and three glutamate receptors (Grina, Grin2b, Grm4) show a statistically significant expression. Culture induces a differential response of these genes (see later).

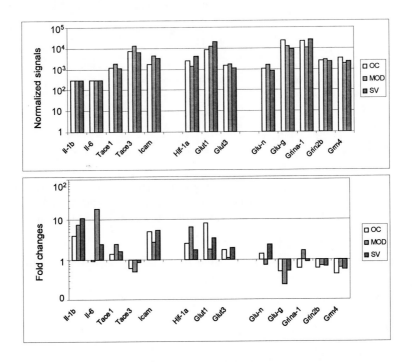

Figure 2. Expression of genes associated with damaging mechanisms. Top: expression pattern in freshly prepared tissue; bottom: fold changes after 24 h in culture. Data of Tace-2 and Grina-2 are not shown because of the similarity of the changes. Il-1beta and Il-6 expression levels in controls were set as 300 corresponding to the cut-off value in raw signal measurement.

To classify the gene expression profiles and the fold changes, we employed the k-mean approach (Table 4). The highest basal expression was observed for Tace2 (UTR), Glut1, Glu-g and Grina and the lowest for the Il-1b and Il-6. This expression pattern may reflect the important role of the substrate glucose and the glutamate systems in the homeostasis of the developing inner ear. Figure 3 illustrates the fold changes induced in culture. The highest increase (2-6 folds) occurs for Il-1beta, Il-6 and Hif-1alpha (cluster 1). Genes of cluster two (Tace-1, Icam, Glu-n, Glut1, Glut3) increased by a factor of 2-3, whereas the genes of cluster 3, (Tace-2, Glu-g, Grina, Grin2b and Grm4) decreased slightly. Remarkably, in the MOD, the increase of cluster-1 genes was found to be higher than in the OC and the SV (p< 0.05), whereas that of cluster-2 genes tended to be lower.

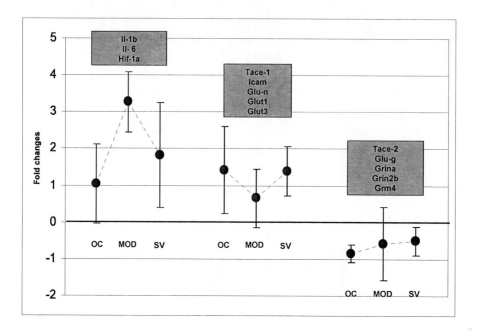

Figure 3. K-means clustering of the changes in the expression of genes associated with damaging mechanisms. Basal expression level is given in Table 4. Ordinate indicates the log2 of fold changes induced in culture as compared to the corresponding controls. Each box indicates genes belonging to the same cluster.

Table 4. K-mean analysis of genes involved in damaging mechanisms

Cluster, genes	Region
C1 Tace-2, Glut1,	OC 13.7 ± 0.9
Glu-g, Grina,	MOD 13.5 ± 0.2
	SV 13.7 ± 1.0
C2 Tace-1, Icam,	OC 10.8 ± 0.6
Hif-1alpha, Glut3, Glut-n,	MOD 11.0 ± 0.6
Grin2b, Grm4	SV 10.9 ± 0.9
C3 Il-1b, Il-6	OC 8.2
	MOD 8.2
	SV 8.2

Numbers indicate mean and SD of log2 of the normalized signals. Significance between the clusters: $p <$ 0.000.

Genes Involved in Cell Death Processes

The basal expression levels of genes involved in cell death processes are illustrated in Figure 4. Among the genes of the Bax/Bcl family, Bax-U59 appears to be highest expressed, among the genes of the caspases and calpains it is Casp2 and Capn that are highest expressed. Unexpectedly, the Bax-variants responded quite differently in culture. The highly expressed

Bax-U59 decreased its expression by about 50 %. In contrast, Bax-S76 and Bax-U49 increased by a factor of about 2-3 in culture. The two Bcl2 transcripts, Bcl2ll-S78 and Bcl2ll-U34, show a low level of expression and responded slightly different during culture. The caspase family is represented on the chip by Casp1,-2, 3 and -6. The basal expression levels as well as the culture-induced changes differ dramatically. A relatively high expression is seen for Casp2- and Casp6-fragments, whereas Casp3 is relatively low expressed. Culture induces a differential expression, with Casp2 and Casp6 decreasing and Casp3 increasing. Both Calp1 and Calp2 show a statistically significant expression level in freshly prepared tissue. After 24 h in culture, Calp1 was found to be down-regulated by a factor of 0.5, whereas Capn 2 did not change significantly.

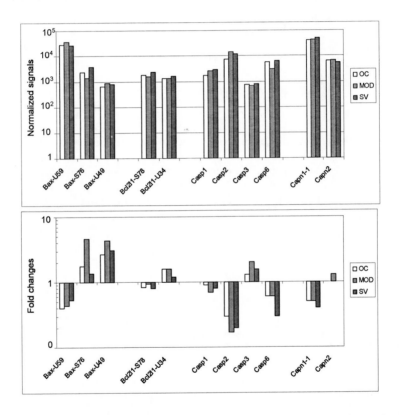

Figure 4. Expression of genes associated with cell death.Top: expression pattern in freshly prepared tissue; bottom: fold changes after 24 h in culture. The chip contained two probes for Capn1 (UTR). Data are not shown because of the similarity of the values.

The k-mean analysis of the basal expression confirms that the cluster with the highest expression contains Bax-U59, Casp2 and Capn1 and that with the lowest expression Bax-U49 and Casp3 (Table 5). A different pattern of fold changes appears in culture (Figure 5). Bax-U59, Casp2, Casp6 and Capn1 appear in one cluster which decreased by a factor of 2-4. Interestingly, an increase was observed for Bax-S76 and Bax-U49. No change was found for both Bcl2ll, Casp1, Casp3, and Calp2.

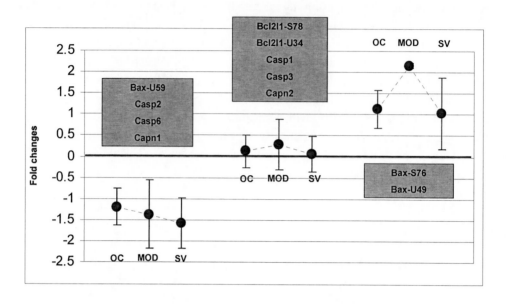

Figure 5. K-means clustering of the changes in the expression of genes associated with cell death. Basal expression level is given in Table 5. Ordinate indicates the log2 of fold changes induced in culture as compared to the corresponding controls. Each box indicates genes belonging to the same cluster.

Table 5. K-mean analysis of genes involved in cell pro-death processes

Cluster, genes	Region	
C1 Bax-U-59	OC	14.4 ± 1.3
Caspase 2	MOD	14.8 ± 0.8
Calpain 1	SV	14.6 ± 1.0
C2 Bax S76, Bcl2l1-S78, Bcl2l1-U34,	OC	11.6 ± 0.9
Caspase 1, Caspase 6,	MOD	11.4 ± 0.9
Calpain 2	SV	11.9 ± 0.6
C3 Bax-U49	OC	9.8 ± 0.6
Caspase 3	MOD	9.9 ± 0.5
	SV	10.0 ± 0.6

Significance between the clusters: $p < 0.001$.

Genes Involved in Cell Protective Processes

Figure 6 shows the expression pattern of genes involved in cell-survival processes in freshly prepared tissue. Among the SOD family, the highest expression level is observed for Sod1 and the lowest for Sod2. Whereas Sod1 did not change its expression in culture, the Sod2-mRNA level increases by a factor of 8-14 in all regions. Among the HSP transcripts, Hsp60 and Hsp90 show a high expression and Hsp27 and Hsp70 a low expression. Culture induces an increase of the expression level of only Hsp27 and Hsp70 mRNA. The IGF family consists of two ligands, Igf1 and Igf2, and several IGF binding proteins. Igf1, Igf2 and Igfbp2 belong to genes with a high expression level, whereas Igfbp3, 5, 6 belong to the low-level expressed genes. Culture induces an increase of Igfbp3 and Igfbp5 in all regions. As observed for the genes involved in damaging mechanisms and genes involved in cell death, no regional differences are present.

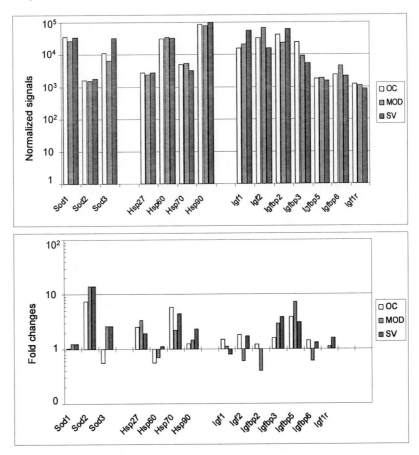

Figure 6. Expression of genes associated with cell survival. Top, expression pattern in freshly prepared tissue; bottom, fold changes after 24 h in culture. The chip contained two further Hsp27 probes with similar expression pattern.

The k-mean analysis of the basal expression shows that the highly expressed cluster contains genes of the SOD, the HSP and the IGF families (Table 6). The k-mean analysis of the fold changes of these genes shows that Sod2 appeared as a cluster of its own because of its extraordinarily high increase (Figure 7). Cluster 2 containing Hsp27, Hsp70, Igfbp3 and Igfbp5 also increases by a factor of 2-7. All other genes did not change.

The raw microarray data set (series no. GSE5446) can be accessed at the Gene Expression Omnibus (GEO) website (**http://www.ncbi.nlm.nih.gov/geo/**).

Table 6. K-mean analysis of genes involved in cell survival processes

Cluster, genes	Region
C1 Sod1,	OC 15.1 ± 0.8
Hsp60, Hsp90	MOD 15.2 ± 0.8
Igf1, Igf2, Igfbp2	SV 15.4 ± 0.9
C2 Sod3	OC 14.0 ± 0.8
Igfbp3	MOD 12.9 ± 0.3
	SV 13.7 ± 1.8
C3 Sod2	OC 11.1 ± 0.7
Hsp27, Hsp70	MOD 11.2 ± 0.9
Igfbp5, Igfbp6, Igf1r	SV 10.9 ± 0.7

Significance between the clusters: $p < 0.000$.

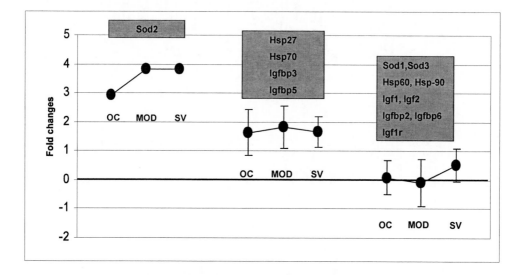

Figure 7. K-means clustering of the changes in expression of genes associated with cell survival. Basal expression level is given in Table 6. Ordinate indicate the log2 of fold changes induced in culture as compared to the corresponding controls. Each box indicates genes belonging to the same cluster.

Discussion

This chapter examines transcriptional events associated with the preparation of short-term organotypic cultures of the OC, MOD and SV of newborn rats. There are two major observations: (1) Preparation of the organotypic cultures induces changes in the expression of genes involved in inflammation, hypoxia/ischemia and glutamate toxicity. (2) Cell survival is maintained by a complex regulation of pro-death and protective genes. In general, pro-death genes are down-regulated and protective genes are up-regulated.

Damaging Processes Associated With the Preparation Of Organotypic Cultures

Preparing organotypic cultures induces regionally different responses of the genes associated with inflammatory, hypoxic and excitotoxic damage. In the OC and SV, the two regions where the death rate level is low, a parallel moderate (2-4 fold) increase of inflammatory markers (Il-1, Il-6, Tace-1, Icam), hypoxia markers (Hif-1a, Glut-1 and Glut3) and the neuronal glutamate transporter Glu-n is observed. Concurrently, the glial glutamate transporter Glu-g, the NMDA Grina-, Grin2b- (ionotropic) and the Grm4- (metabotropic) receptors are decreased. In the MOD, the region with a higher cell death rate, a clearly higher (6-fold) increase of Il-1, Il-6 and Hif-1a and only a small increase of the cluster containing Tace, Icam, Glut1, Glut3 and Glu-n were found (about 1.5 fold). It may well be that these differences contribute to the higher cell death rate in the MOD region.

Inflammatory responses are frequently observed in inner ear diseases including trauma, damage due to noise over-stimulation and infections (Fujioka et al., 06). Inflammation is characterized by the secretion of the pro-inflammatory cytokines TNF-alpha, Il-1beta and Il-6, (Chetty et al., 08; Fujioka et al., 06; Ichimiya et al., 00; Kawasaki et al., 08; Nam, 06). The activation mechanism and function of these cytokines in hearing loss are still unclear. Suppression of inflammation of the cochleae by inhibiting TNF-alpha signal has been demonstrated to reduce hearing loss in the experimental inner ear inflammation model (Fujioka et al., 06). Specifically, TNF-alpha, Il-6 and Il-1beta regulate excitatory and inhibitory neurotransmission (Kawasaki et al., 08). Another role of Il-6 is its anti-oxidative stress effect by up-regulating several anti-apoptotic genes, including those of the BCL family, or cell survival signals (Chetty et al., 08; Fujioka et al., 06). Inhibition of the Il-6 signal was suggested even as a therapeutic approach in the acute phase of noise-induced hearing loss (Fujioka et al., 06). Experimental inner ear inflammation studies have shown that in addition to these classical cytokines, intercellular adhesion molecule-1 (Icam) is expressed (Shi and Nuttall, 07). ICAM-1 is the ligand for leukocyte adhesion molecule LFA-1 (leukocyte function-associated antigen 1) and *in vivo* mediates leukocyte adhesion to endothelial cells and transmigration through the vascular wall. Intercellular adhesion molecules were shown to be important factors for a lateral wall inflammatory response from loud-sound stimulation (Shi and Nuttall, 07).

The observed changes of the tumor necrosis factor-alpha converting enzyme (Tace or Adam17) expression are of particular interest. As far as we know, this is the first study to show the expression of Tace in the inner ear. In the present model, TNF-alpha mRNA was found absent (Mazurek et al., 06a). In contrast, Tace shows a significant basal expression level and the coding sequence (cds) fragment increased in culture and the non-cds fragment decreased. Tumor necrosis factor-alpha (TNF-alpha) is a pro-inflammatory cytokine which is shed in its soluble form by TACE, a disintegrin and metalloproteinase (ADAM; Colon et al., 01). TACE has been described to be responsible for the proteolytic release or shedding of several proteins apart from TNF-alpha, such as p55 and p75 tumor necrosis factor receptor (TNFR) and interleukin 1R-II. Hurtado et al. (2002) have previously demonstrated that TACE is up-regulated in response to oxygen glucose deprivation. It may have a role to play in inhibiting apoptosis via induction of TNF-alpha release (Charbonneau et al., 07). It was demonstrated that TACE overexpression exerts a neuroprotective effect by up-regulating glutamate uptake (Romera et al., 04). Recently it was reported that low oxygen concentrations and TNF-alpha enhance Tace mRNA levels in synovial cells through direct binding of hypoxia-inducible factor-1 (HIF-1) to the 5′ promoter region (Charbonneau et al., 07). The differential response of the cds and non-cds fragments may be an expression of splice variants with different functions.

The co-expression of inflammatory and hypoxic markers is well known. The increase of Hif-1alpha and its target genes Glut-1 and Glut 3 may be an indicator of the presence of hypoxia/ischemia during tissue preparation or during culture. However, the increase of Hif-1alpha mRNA may also result from activation by cytokines and growth factors. The response of several other target HIF-1 target genes was described recently (Mazurek et al., 06b).

The decrease of Glu-g and NMDA receptor transcripts combined with the increase of Glu-n suggests that glutamate toxicity may play a part in the present experimental model. In general, glutamate transporters carry glutamate and aspartate and may thus regulate the neurotransmitter concentration in the tissue. As extracellular glutamate may act excitotoxically to neurons and glia when unregulated, low synaptic levels of this neurotransmitter must be maintained via a rapid and robust transport system (Nickell et al., 07). The differential response of the neuronal (increase of Glu-n) and glial (decrease of Glu-g) isoforms and the increase of glutamine synthetase expression (Mazurek et al., 06a) provide evidence for changes in the functioning of the glutamate-glutamine cycle. It is known that the glutamate system is not only involved in modulating synaptic transmission but may also include functions in other processes, e.g. metabolism and microvascular function (Gillard et al., 03; Ottersen et al., 98). The decrease in expression of all three NMDA receptors Grina, Grin2b and Grm4 may reflect an effective adaptation to the culture conditions by reducing the excitotoxic capacity and to limit the damaging consequences of glutamate. During the last few years, it has become increasingly clear that the expression of glutamate receptors and transporters is not limited to the neuronal tissue (Gillard et al., 03; Ottersen et al., 98). This explains why the members of the glutamate system were found to be expressed in all regions.

Endogenous Mechanisms to Maintain Cell Survival In Culture

K-mean analysis (Figures 5, 7) of the expression patterns of pro-death and pro-survival genes revealed the transcriptional mechanisms for maintaining survival under culture conditions: transcripts of genes involved in triggering or executing cell death are down-regulated and transcripts for genes involved in protection from cell death are up-regulated. This regulation includes genes encoding for proteins of Bcl-2, caspase, calpain, superoxide dismutase, heat shock protein and insulin like growth factor binding protein families.

An important molecule of the BCL-2 family that promotes cell death is BAX. Its function is countered by BCL-2 and BCL-XL. BCL-2 family proteins are key regulators of apoptosis pathways and either promote or suppress mitochondrial outer membrane permeabilization (Zhai et al., 08). The BAX gene has been shown to encode spliced variants, the function of these alternatively spliced variants is not yet known (Oltvai et al., 93). However, there exist forms of BAX that promote cell survival (Zhou et al., 98). To elucidate the heterogeneous response of the different Bax-fragments we aligned the fragments to the corresponding reference genes (Figure 8). Bax-U49 and Bax-S76 align significantly to the reference mRNA described as Bcl2-associated X protein (Bax; Acc-Nr. NM_017059). Bax-U59 does not show any similarity to this reference mRNA but produces significant alignments to the non-coding sequences of Bax protein splice variant k mRNA (AF235993). Recombinant protein from this Bax splice variant was found to induce a loss in mitochondrial membrane potential contributing to apoptosis (Carvalho et al., 04). Thus, the observed down-regulation of Bax-U59 and the up-regulation of Bax-U49/Bax-S76 may reflect different variants of the Bax transcripts exerting different roles in the cell death process. In contrast to Bax, both antiapoptic Bcl2l1 fragments slightly increase or do not change its expression significantly. The corresponding reference gene for both probe sets is defined as being Bcl2-like 1 (Bcl2l1, Bcl-xL), nuclear gene encoding mitochondrial protein, transcript variant 3, mRNA (Acc.-No. NM_001033670). The probe set Bcl2l1-S78 detects variant 3 and variant 4 (98 % identity), and the probe set Bcl2l1-U34 detects variant 1,-2 and-3 (98 % identity). Bcl-xL protects cells from apoptosis induced by a variety of stimuli. The inhibitory effects of Bcl-xL on the extrinsic apoptosis have been shown to be cell-specific (Wang et al., 08). Interestingly, Bcl-2 transcript level was found absent in our model (Mazurek et al., 06a).

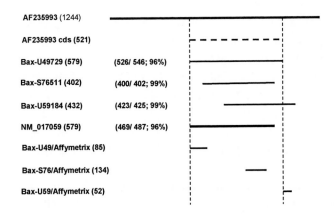

Figure 8. Nucleotide sequence alignment of the different variants of Bax compared to the reference sequences. Indicated are the accession numbers, the length of the nucleotide, the percentage of identities and the position in relation to the Bcl2-associated X protein (Bax), splice variant k (AF2359939).

Interestingly, Casp2, Casp6 and Capn1 mRNA levels were found increased together with Bax-U59 (Figure 5). Casp2 is as an initiator of apoptosis sensing changes in the mitochondrial potential (Chong et al., 05; Fan et al., 05; Fuentes-Prior and Salvesen, 04)

Casp6 belongs to the executioner of apoptosis. Calpains are also involved in apoptosis because of their ability to cleave and thus activate proteins that regulate apoptosis, including caspases. Inhibitors of calpains were found to protect the auditory sensory cells from hypoxia-induced cell damage (Cheng et al., 99).

SOD2 seem to play a special role in cell survival because it appears in the corresponding group as a cluster of its own. Mitochondria are a significant source of reactive oxygen species

(ROS) after anoxic insults (Arundine et al., 04). The generation of ROS may result in mitochondrial dysfunction and cytochrome c release (Chong et al., 05). Oxidative stress may play a particular role in this release mechanism, a key event in the apoptosis (Polster and Fiskum, 04). A wide variety of compounds induce transcription of SOD2, for example Il-1, Il-6 and TNF-alpha (Zelko et al., 02).

Hsp27, Hsp70, Igfbp-3 and -5 were among the differentially expressed genes associated with cell survival. HSP27 belongs to the family of small heat shock proteins and has been involved in several protective mechanisms (Sanz et al., 01). For example, HSP27 acts at the level of actin filaments to confer resistance to actin fragmentation by oxidative stress generated by H_2O_2 (Huot et al., 95). Further, HSP27 seems to provide neuroprotection against TNF-alpha induced oxidative stress by abolishing the burst of intracellular ROS (Mehlen et al., 96). We interpret this up-regulation as an important endogenous mechanism of cell protection.

The HSP70 family includes several constitutive as well as stress-inducible proteins (Arya et al., 07). HSP70 inhibits apoptosis acting mainly on the caspase-dependent pathway at several steps both upstream and downstream of caspase activation (Mayer and Bukau, 05). Overproduction of HSP70 leads to increased resistance against apoptosis-inducing agents such as tumor necrosis factor-alpha, while down-regulation of HSP70 levels by antisense technology leads to increased sensitivity to these agents.

The response of the members of the IGF family was recently discussed in detail (Gross et al., 08). It is important to note that regulation occurs not via the ligands itself but via the IGF-binding proteins. The present analysis shows that Igfbp3 and -5 are up-regulated, whereas all the other members of the IGF family remained unchanged. The interaction of IGFBP-3 with IGFs depends on the prevailing conditions and may result in inhibitory or stimulatory responses (Firth and Baxter, 02; Schneider et al., 00). On the one hand, IGFBP-3 may inhibit IGF-mediated effects on growth and proliferation and thus support the anti-apoptotic functions of IGF-1 mediated by IGFR1; on the other hand it is known to have pro-apoptotic functions (Firth and Baxter, 02). Knockdown of IGFBP-3 significantly decreased inner ear size and disrupted hair cell differentiation and semicircular canal formation (Li et al., 05). The concurrent increase of Igfbp3 and-5 mRNA and Hsp27 and Hsp70 let us assume that Igfbp3 and -5 have a protective role in the present model.

Consequences of the Changed Gene Expression

Activation of genes involved in cell damaging mechanisms during development of the inner ear could have two principal consequences: (1) Neurosensory cells of the OC or neurons of the spiral ganglion may die and lead to hearing loss and reorganization of the signaling pathway along the perception of sound in the inner ear and in the CNS. How inflammation could induce cell death has been, and still remains, the subject of intensive studies. One of the best characterized cytotoxic mechanisms induced by proinflammatory cytokines is the activation of inducible nitric oxide synthase (iNOS), which mediates the synthesis of high levels of nitric oxide (NO) shown to be toxic for neurons (Hunot and Hirsch, 03). During culturing, we observed a particularly high expression of inducible

Nos2/nitric oxide synthase in all regions (Mazurek et al., 06b). Different mechanisms may account for NO-induced cell toxicity. First, NO can induce nitrosylation of proteins and alter their function. Second, NO can disrupt iron homeostasis by interfering with iron regulated proteins such as ferritin and transferrin receptor by up- or down-regulating their expression. (2) Without cell death, plastic changes of the neurosensory cells and of the spiral ganglion neurons or auditory neurons in the central nervous system could appear. Plasticity refers to the neurons' ability to adapt and change over time (McClung and Nestler, 08). For example, following cochlear ablation, auditory neurons in the central nervous system (CNS) undergo alterations in morphology and function (Durham et al., 00). These consequences could result in hearing loss or the appearance of tinnitus. Tinnitus is the perception of sound in the absence of external stimulation. Tinnitus may strongly influence the quality of life of patients and have an important impact on public health systems because of its high prevalence and the suffering involved.

The molecular basis of plasticity appears to be gene expression which is controlled by a series of DNA-binding proteins known as transcription factors. Cellular stress modify the expression of transcription factors that allow nuclear entry, change protein stability, enhance DNA binding, or allow binding to essential co-factors. One of the best known transcription factor involved in brain plasticity is CREB (cAMP response element-binding protein). It is interesting to note that Creb expression is increased in OC in culture (2.2 fold) (Mazurek et al., 06a). CREB is under the control of NMDA, thus the observed changes of genes involved in glutamate metabolism may be involved in modulation of synaptic activity (McClung and Nestler, 08). Evidence has been accumulated to implicate neuroplasticity in tinnitus, including a role for the cochlear N-methyl-D-aspartate (NMDA) receptor (Guitton and Dudai, 07; Puel, 07). It was reported that the 2B subunit of the NMDA receptor (NR2B), a molecule which is implicated in the long-term potentiation and behavioral plasticity in the mammalian brain, is involved in both salicylate-induced and noise-induced tinnitus. In the present chapter the expression of 2B subunit of the NMDA receptor appears in a down-regulated cluster.

We are aware of some limitations of the present study. (1) We are lacking data about the expression at the protein levels. Changes in the mRNA levels are not necessarily associated with changes in protein expression. However, it is interesting to note that the trends in changes of the mRNA levels are frequently synergistic to changes in the protein level which indicates that the analyzed factors are also regulated at the transcriptional level (Tian et al., 04). (2) For technical reasons, we used duplicate pooled samples for the microarray analysis and six independently pooled samples for confirmatory RT-PCR (Gross et al., 07; Gross et al., 08; Mazurek et al., 06b). The fact that we received a good correlation between the microarray and the RT-PCR data and that normoxic and hypoxic cultures demonstrated almost identical gene expression patterns adds to the reliability of our microarray data. (3) Gene expression is a time-dependent process. The time window (3 and 24 hours), in which genes are activated after exposure to impulse noise had been studied (Kirkegaard et al., 06). It was found that at 3 hours after trauma, most of the regulated genes are immediate early genes. At the second point in time (24 hours post-exposure), the majority of the differentially expressed genes are involved in the inflammatory response of the tissue. In the tissues, response to oxidative stress genes was up-regulated at both time points investigated. For this study, we focused our attention on the effectors rather than on the immediate early genes.

Conclusion

Organotypic OC, MOD and SV cultures preserve cells in a living status at least for a few days and can be used as a model to identify critical steps in cell fate determination following injury. Because tissue preparation and culturing induces changes that are sub threshold for cell death, insights into the endogenous protective mechanisms can be acquired.

In this chapter, we demonstrate that the maintenance of cell survival after severe stress in the early developmental period is associated with a cascade of gene expression events comprising pro-death and pro-survival genes (Figure 9). Critical genes for the prevention of cell death following traumatic, hypoxic or excitotoxic pathways are genes from the Bax/Bcl family, Casp2 and Casp6, Sod2, Hsp27, HSP70, Igfbp3 and Igfbp5. For two genes (Tace, Bax) we observed a differential response of coding and non-coding sequences. This differential response might indicate the existence of functional elements within the mRNA exerting different roles in the regulation of gene expression as described for other genes (Zamorano et al., 06). Activation of genes involved in cell damaging mechanisms during development of the inner ear could have two principal consequences, cell death or plastic changes of the survived cells. These consequences could result in hearing loss or tinnitus. However, caution should be exercised when extrapolating the present findings to the intact inner ear. Future studies will be needed to determine whether these mechanisms are valid *in vivo*.

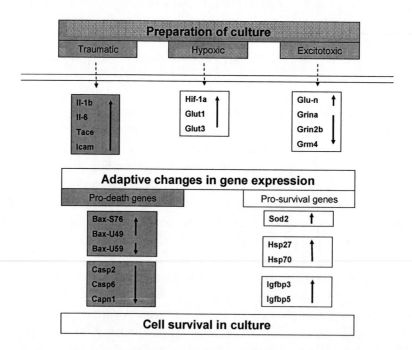

Figure 9 Schematic summary of changes in gene expression associated with the preparation of organotypic cultures of the organ of Corti, modiolus and stria vascularis. Culture includes changes in genes associated with traumatic, hypoxic and glutamate toxicity. Changes in the pro-death to pro-survival gene ratio alter the cell homeostasis towards cell survival as a function of the nature and severity of the damage induced at different regions.

References

Achouche, J; Liu, DS; Tran Ba, HP; Huy, PT. Primary culture of strial marginal cells of guinea pig cochlea: growth, morphologic features, and characterization. *Ann Otol Rhinol Laryngol,* 1991, 100, 999-1006.

Airley, RE; Mobasheri, A. Hypoxic regulation of glucose transport, anaerobic metabolism and angiogenesis in cancer: novel pathways and targets for anticancer therapeutics. *Chemotherapy,* 2007, 53, (4):233-256.

Arundine, M; Aarts, M; Lau, A; Tymianski, M. Vulnerability of central neurons to secondary insults after in vitro mechanical stretch. *J Neurosci,* 2004, 24, (37):8106-8123.

Arya, R; Mallik, M; Lakhotia, SC. Heat shock genes - integrating cell survival and death. *J Biosci,* 2007, 32, (3):595-610.

Carvalho, AC; Sharpe, J; Rosenstock, TR; Teles, AF; Youle, RJ; Smaili, SS. Bax affects intracellular Ca2+ stores and induces Ca2+ wave propagation. *Cell Death Differ,* 2004, 11, (12):1265-1276.

Chaitidis, P; O'Donnell, V; Kuban, RJ; Bermudez-Fajardo, A; Ungethuem, U; Kuhn, H. Gene expression alterations of human peripheral blood monocytes induced by medium-term treatment with the TH2-cytokines interleukin-4 and -13. *Cytokine,* 2005, 30, (6):366-377.

Charbonneau, M; Harper, K; Grondin, F; Pelmus, M; McDonald, PP; Dubois, CM. Hypoxia-inducible factor mediates hypoxic and tumor necrosis factor alpha-induced increases in tumor necrosis factor-alpha converting enzyme/ADAM17 expression by synovial cells. *J Biol Chem,* 2007a, 282, (46):33714-33724.

Cheng, AG; Huang, T; Stracher, A; Kim, A; Liu, W; Malgrange, B; Lefebvre, PP; Schulman, A; Van de Water, TR. Calpain inhibitors protect auditory sensory cells from hypoxia and neurotrophin-withdrawal induced apoptosis. *Brain Res,* 1999a, 850, (1-2):234-243.

Chetty, A; Cao, GJ; Manzo, N; Nielsen, HC; Waxman, A. The role of IL-6 and IL-11 in hyperoxic injury in developing lung. *Pediatr Pulmonol,* 2008, 43, (3):297-304.

Chong, ZZ; Li, F; Maiese, K. Oxidative stress in the brain: novel cellular targets that govern survival during neurodegenerative disease. *Prog Neurobiol,* 2005, 75, (3):207-246.

Ciancio, G; Pollack, A; Taupier, MA; Block, NL; Irvin, GL, III. Measurement of cell-cycle phase-specific cell death using Hoechst 33342 and propidium iodide: preservation by ethanol fixation. *J Histochem Cytochem,* 1988, 36, (9):1147-1152.

Colon, AL; Menchen, LA; Hurtado, O; De Cristobal, J; Lizasoain, I; Leza, JC; Lorenzo, P; Moro, MA. Implication of TNF-alpha convertase (TACE/ADAM17) in inducible nitric oxide synthase expression and inflammation in an experimental model of colitis. *Cytokine,* 2001, 16, (6):220-226.

Dazert, S; Battaglia, A; Ryan, AF. Transfection of neonatal rat cochlear cells in vitro with an adenovirus vector. *Int J Dev Neurosci,* 1997, 15, 595-600.

Durham, D; Park, DL; Girod, DA. Central nervous system plasticity during hair cell loss and regeneration. *Hear Res,* 2000, 147, (1-2):145-159.

Fan, TJ; Han, LH; Cong, RS; Liang, J. Caspase family proteases and apoptosis. *Acta Biochim Biophys Sin (Shanghai),* 2005, 37, (11):719-727.

Firth, SM; Baxter, RC. Cellular actions of the insulin-like growth factor binding proteins. *Endocr Rev,* 2002, 23, (6):824-854.

Fuentes-Prior, P; Salvesen, GS. The protein structures that shape caspase activity, specificity, activation and inhibition. *Biochem J,* 2004, 384, (Pt 2):201-232.

Fujioka, M; Kanzaki, S; Okano, HJ; Masuda, M; Ogawa, K; Okano, H. Proinflammatory cytokines expression in noise-induced damaged cochlea. *J Neurosci Res,* 2006, 83, (4):575-583.

Gates, GA; Mills, JH. Presbycusis. *Lancet,* 2005, 366, (9491):1111-1120.

Gillard, SE; Tzaferis, J; Tsui, HC; Kingston, AE. Expression of metabotropic glutamate receptors in rat meningeal and brain microvasculature and choroid plexus. *J Comp Neurol,* 2003, 461, (3):317-332.

Gross, A; McDonnell, JM; Korsmeyer, SJ. BCL-2 family members and the mitochondria in apoptosis. *Genes Dev,* 1999, 13, (15):1899-1911.

Gross, J; Machulik, A; Amarjargal, N; Moller, R; Ungethum, U; Kuban, RJ; Fuchs, FU; Andreeva, N; Fuchs, J; Henke, W; Pohl, EE; Szczepek, AJ; Haupt, H; Mazurek, B. Expression of apoptosis-related genes in the organ of Corti, modiolus and stria vascularis of newborn rats. *Brain Res,* 2007, 1162, 56-68.

Gross, J; Machulik, A; Moller, R; Fuchs, J; Amarjargal, N; Ungethuem, U; Kuban, RJ; Szczepek, AJ; Haupt, H; Mazurek, B. mRNA expression of members of the IGF system in the organ of Corti,the modiolus and the stria vascularis of newborn rats. *Growth Factors,* 2008, Jun 12:1 [Epub ahead of print].

Guitton, MJ; Dudai, Y. Blockade of cochlear NMDA receptors prevents long-term tinnitus during a brief consolidation window after acoustic trauma. *Neural Plast,* 2007, 2007:80904., 80904.

Held, GA; Grinstein, G; Tu, Y. Relationship between gene expression and observed intensities in DNA microarrays--a modeling study. *Nucleic Acids Res,* 2006, 34, (9):e70.

Hirose, K; Liberman, MC. Lateral wall histopathology and endocochlear potential in the noise-damaged mouse cochlea. *J Assoc Res Otolaryngol,* 2003, 4, (3):339-352.

Hunot, S; Hirsch, EC. Neuroinflammatory processes in Parkinson's disease. *Ann Neurol,* 2003, 53 Suppl 3:S49-58.

Huot, J; Lambert, H; Lavoie, JN; Guimond, A; Houle, F; Landry, J. Characterization of 45-kDa/54-kDa HSP27 kinase, a stress-sensitive kinase which may activate the phosphorylation-dependent protective function of mammalian 27-kDa heat-shock protein HSP27. *Eur J Biochem,* 1995, 227, (1-2):416-427.

Hurtado, O; Lizasoain, I; Fernandez-Tome, P; Alvarez-Barrientos, A; Leza, JC; Lorenzo, P; Moro, MA. TACE/ADAM17-TNF-alpha pathway in rat cortical cultures after exposure to oxygen-glucose deprivation or glutamate. *J Cereb Blood Flow Metab,* 2002, 22, (5):576-585.

Ichimiya, I; Yoshida, K; Hirano, T; Suzuki, M; Mogi, G. Significance of spiral ligament fibrocytes with cochlear inflammation. *Int J Pediatr Otorhinolaryngol,* 2000, 56, (1):45-51.

Kawasaki, Y; Zhang, L; Cheng, JK; Ji, RR. Cytokine mechanisms of central sensitization: distinct and overlapping role of interleukin-1beta, interleukin-6, and tumor necrosis factor-alpha in regulating synaptic and neuronal activity in the superficial spinal cord. *J Neurosci,* 2008, 28, (20):5189-5194.

Kirkegaard, M; Murai, N; Risling, M; Suneson, A; Jarlebark, L; Ulfendahl, M. Differential gene expression in the rat cochlea after exposure to impulse noise. *Neuroscience,* 2006, 142, (2):425-435.

Koyama, S; Kaga, K; Sakata, H; Iino, Y; Kodera, K. Pathological findings in the temporal bone of newborn infants with neonatal asphyxia. *Acta Otolaryngol,* 2005, 125, (10):1028-1032.

Li, Y; Xiang, J; Duan, C. Insulin-like growth factor-binding protein-3 plays an important role in regulating pharyngeal skeleton and inner ear formation and differentiation. *J Biol Chem,* 2005, 280, (5):3613-3620.

Liu, XY; Li, CY; Bu, H; Li, Z; Li, B; Sun, MM; Guo, YS; Zhang, L; Ren, WB; Fan, ZL; Wu, DX; Wu, SY. The Neuroprotective Potential of Phase II Enzyme Inducer on Motor Neuron Survival in Traumatic Spinal Cord Injury In vitro. *Cell Mol Neurobiol,* 2007, ..

Lowenheim, H; Kil, J; Gultig, K; Zenner, HP. Determination of hair cell degeneration and hair cell death in neomycin treated cultures of the neonatal rat cochlea. *Hear Res,* 1999, 128, (1-2):16-26.

Mayer, MP; Bukau, B. Hsp70 chaperones: cellular functions and molecular mechanism. *Cell Mol Life Sci,* 2005, 62, (6):670-684.

Mazurek, B; Machulik, A; Amarjargal, N; Kuban, RJ; Ungethuem, U; Fuchs, J; Haupt, H; Gross, J. Gene expression of organ of Corti (OC), modiolus (MOD) and stria vascularis (SV) of newborn rats. *Gene Expression Omnibus (GEO) website (http://www ncbi nlm nih gov/geo/),* 2006a, ID GSE5446.

Mazurek, B; Rheinlander, C; Fuchs, FU; Amarjargal, N; Kuban, RJ; Ungethum, U; Haupt, H; Kietzmann, T; Gross, J. [Influence of ischemia/hypoxia on the HIF-1 activity and expression of hypoxia-dependent genes in the cochlea of the newborn rat.]. *HNO,* 2006b, 54, 689-697.

Mazurek, B; Winter, E; Fuchs, J; Haupt, H; Gross, J. Susceptibility of the hair cells of the newborn rat cochlea to hypoxia and ischemia. *Hear Res,* 2003, 182, (1-2):2-8.

McClung, CA; Nestler, EJ. Neuroplasticity mediated by altered gene expression. *Neuropsychopharmacology,* 2008, 33, (1):3-17.

Mehlen, P; Kretz-Remy, C; Preville, X; Arrigo, AP. Human hsp27, Drosophila hsp27 and human alphaB-crystallin expression-mediated increase in glutathione is essential for the protective activity of these proteins against TNFalpha-induced cell death. *EMBO J,* 1996, 15, (11):2695-2706.

Melichar, I; Gitter, AH. Primary culture of vital marginal cells from cochlear explants of the stria vascularis. *Eur Arch Otorhinolaryngol,* 1991, 248, 358-365.

Nam, SI. Interleukin-1beta up-regulates inducible nitric oxide by way of phosphoinositide 3-kinase-dependent in a cochlear cell model. *Laryngoscope,* 2006, 116, (12):2166-2170.

Nickell, J; Salvatore, MF; Pomerleau, F; Apparsundaram, S; Gerhardt, GA. Reduced plasma membrane surface expression of GLAST mediates decreased glutamate regulation in the aged striatum. *Neurobiol Aging,* 2007, 28, (11):1737-1748.

Noraberg, J; Kristensen, BW; Zimmer, J. Markers for neuronal degeneration in organotypic slice cultures. *Brain Res Brain Res Protoc,* 1999, 3, (3):278-290.

Ohlemiller, KK. Age-related hearing loss: the status of Schuknecht's typology. *Curr Opin Otolaryngol Head Neck Surg,* 2004, 12, (5):439-443.

Ohlemiller, KK; Gagnon, PM. Genetic dependence of cochlear cells and structures injured by noise. *Hear Res,* 2007, 224, (1-2):34-50.

Oltvai, ZN; Milliman, CL; Korsmeyer, SJ. Bcl-2 heterodimerizes in vivo with a conserved homolog, Bax, that accelerates programmed cell death. *Cell,* 1993, 74, (4):609-619.

Ottersen, OP; Takumi, Y; Matsubara, A; Landsend, AS; Laake, JH; Usami, S. Molecular organization of a type of peripheral glutamate synapse: the afferent synapses of hair cells in the inner ear. *Prog Neurobiol,* 1998, 54, (2):127-148.

Polster, BM; Fiskum, G. Mitochondrial mechanisms of neural cell apoptosis. *J Neurochem,* 2004, 90, (6):1281-1289.

Puel, JL. Cochlear NMDA receptor blockade prevents salicylate-induced tinnitus. *B-ENT,* 2007, 3 Suppl 7:19-22.

Pujol, R; Rebillard, G; Puel, JL; Lenoir, M; Eybalin, M; Recasens, M. Glutamate neurotoxicity in the cochlea: a possible consequence of ischaemic or anoxic conditions occurring in ageing. *Acta Otolaryngol Suppl Stockh,* 1990, 476, 32-36.

Romera, C; Hurtado, O; Botella, SH; Lizasoain, I; Cardenas, A; Fernandez-Tome, P; Leza, JC; Lorenzo, P; Moro, MA. In vitro ischemic tolerance involves upregulation of glutamate transport partly mediated by the TACE/ADAM17-tumor necrosis factor-alpha pathway. *J Neurosci,* 2004, 24, (6):1350-1357.

Sanz, O; Acarin, L; Gonzalez, B; Castellano, B. Expression of 27 kDa heat shock protein (Hsp27) in immature rat brain after a cortical aspiration lesion. *Glia,* 2001, 36, (3):259-270.

Schneider, MR; Lahm, H; Wu, M; Hoeflich, A; Wolf, E. Transgenic mouse models for studying the functions of insulin-like growth factor-binding proteins. *FASEB J,* 2000, 14, (5):629-640.

Shi, X; Nuttall, AL. Expression of adhesion molecular proteins in the cochlear lateral wall of normal and PARP-1 mutant mice. *Hear Res,* 2007, 224, (1-2):1-14.

Slepecky, NB. Structure of the Mammalian Cochlea. *The Cochlea,* 2005, 44-129.

Sobkowicz, HM; Loftus, JM; Slapnick, SM. Tissue culture of the organ of Corti. *Acta Otolaryngol Suppl Stockh,* 1993, 502, 3-36.

Tian, Q; Stepaniants, SB; Mao, M; Weng, L; Feetham, MC; Doyle, MJ; Yi, EC; Dai, H; Thorsson, V; Eng, J; Goodlett, D; Berger, JP; Gunter, B; Linseley, PS; Stoughton, RB; Aebersold, R; Collins, SJ; Hanlon, WA; Hood, LE. Integrated genomic and proteomic analyses of gene expression in Mammalian cells. *Mol Cell Proteomics,* 2004, 3, (10):960-969.

Ulehlova, L. Stria vascularis in acoustic trauma. *Arch Otorhinolaryngol,* 1983, 237, (2):133-138.

van Ruijven, MW; de Groot, JC; Klis, SF; Smoorenburg, GF. The cochlear targets of cisplatin: an electrophysiological and morphological time-sequence study. *Hear Res,* 2005, 205, (1-2):241-248.

Wang, Y; Zhang, B; Peng, X; Perpetua, M; Harbrecht, BG. Bcl-xL prevents staurosporine-induced hepatocyte apoptosis by restoring protein kinase B/mitogen-activated protein kinase activity and mitochondria integrity. *J Cell Physiol,* 2008, 215, (3):676-683.

Zamorano, R; Suchindran, S; Gainer, JV. 3'-Untranslated region of the type 2 bradykinin receptor is a potent regulator of gene expression. *Am J Physiol Renal Physiol,* 2006, 290, (2):F456-F464.

Zelko, IN; Mariani, TJ; Folz, RJ. Superoxide dismutase multigene family: a comparison of the CuZn-SOD (SOD1), Mn-SOD (SOD2), and EC-SOD (SOD3) gene structures, evolution, and expression. *Free Radic Biol Med,* 2002, 33, (3):337-349.

Zhai, D; Jin, C; Huang, Z; Satterthwait, AC; Reed, JC. Differential regulation of Bax and Bak by anti-apoptotic Bcl-2 family proteins Bcl-B and Mcl-1. *J Biol Chem,* 2008, 283, (15):9580-9586.

Zhou, M; Demo, SD; McClure, TN; Crea, R; Bitler, CM. A novel splice variant of the cell death-promoting protein BAX. *J Biol Chem,* 1998, 273, (19):11930-11936.

In: Developmental Gene Expression Regulation
Editor: Nathan C. Kurzfield

ISBN: 978-60692-794-6
©2009 Nova Science Publishers, Inc.

Chapter V

The TGF-B Superfamily: A Multitask Signalling Pathway for the Animal Kingdom

Marco Patruno

Department of Experimental Veterinary Sciences, Faculty of Veterinary
Medicine, Legnaro (PD, Italy). E-mail: *marco.pat@unipd.it*

Abstract

The TGF-β superfamily consists of numerous members, including TGF-β proper, bone morphogenetic proteins (BMP) and growth differentiation factors (GDF). All TGF-β are dimeric cytokines present a biological active carboxy terminal domain of 110–140 amino acids following proteolysis.

Bone morphogenetic proteins (BMPs), first identified for their involvement in vertebrate bone formation, are now widely recognized as key factors in the regulation of many fundamental developmental processes in all deuterostomes. The active gradient established by BMP secreted ligands is one of the essential factors responsible for generating the positional information that underlies developmental patterning, including the regeneration of lost parts.

Myostatin or GDF-8, a recently discovered GDF subfamily member, acts as a negative regulator in maintaining the mammalian proper muscle mass during both embryogenesis and post-natal muscle development. Unlike most other members of the BMP/GDF superfamily, mammalian myostatin is secreted as a latent complex, usually linked to regulatory proteins, and its mature dimer produces an effect almost exclusively on muscle tissue.

The present chapter deals with the importance of the TGF-β family of growth factors in relation to regulatory spheres through the animal kingdom, focusing in particular on the expression of BMP molecules during regeneration and myostatin in "non-canonical" animal models; it also focuses on the regulative actions of myostatin during the development of vertebrates and in different experimental conditions, including *in vitro* chick co-culture and endurance training. Data available in literature indicate that there is substantial scope for future research in the area of TGF-β /myostatin linked to development.

"There is grandeur in this view of life, with its several powers, having been originally breathed by the Creator into a few forms or into one; and that, whilst this planet has gone cycling on according to the fixed law of gravity, from so simple a beginning endless forms most beautiful and most wonderful have been, and are being evolved."

Charles Darwin, The Origin of Species.

1. Introduction

The regulatory proteins involved in establishing basic body plans are often arranged in a *family*: their primary structure is similar having a common evolutionary origin. The term *super-family* was coined because some families share an almost identical domain within the same portion of the protein. Although the structure of all molecules belonging to a super-family is similar in certain domains, very often the molecules themselves are not as *close* as the members of a certain family are to each other; this concept is expressed in the terms *orthologous* (cross-species homology) and *paralogous* (cross-locus homology within a species), both of which were coined by evolutionary biologists when the phenomenon of gene duplication was discovered.

Growth factors, a comprehensive group of polypeptides widely used as multifunctional modulators of fundamental cell activities, are usually localized and activated in the extracellular matrix (ECM) of a certain tissue. A peculiar characteristic of growth factors is their ability to display multifunctional properties allowing them to act as specific stimulators or inhibitors of activities such as proliferation and differentiation depending on the type of target cells involved and their synergistic/antagonistic interactions with other factors.

The Transforming Growth Factor-Beta (TGF-β) superfamily is made up of at least 30 structurally related molecules involved in the regulation of important developmental mechanisms (Massaguè 2000). All members of this superfamily contain a carboxy-terminal region, composed of nine conserved cysteine residues (except for BMPs which have seven), representing the mature peptide. The latter is secreted from the cell to act in an autocrine or paracrine fashion, forming homodimers or heterodimers. All TGF-βs, except for GDNF (see below), bind two types of transmembrane receptors, thus activating Smad proteins and allowing the formation of a transcription factor complex (phosphorylated Smads) to enter the nucleus. This superfamily comprises a large group of structurally related genes whose secreted products act as multifunctional regulators of growth and developmental processes. Sequence analyses of this multigene family have identified remarkable amino acid conservation across species and, based upon their amino acid identity, many TGF-β subfamilies have been identified. Indeed, the superfamily is characterised by a conserved carboxy terminal feature (Kingsley 1994). Figure 1 illustrates a phylogenetic tree of transforming growth factor-β family ligands based on the conserved region of the protein. The largest known subfamily, the Dpp/BMP (*decapentaplegic*/bone morphogenetic protein) group, has members in flies, nematodes and vertebrates.

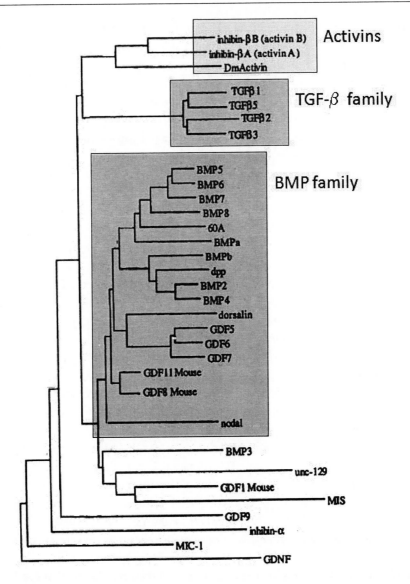

Figure 1. Phylogenetic relationship of TGF- β family ligands. Three different isoforms (TGFβ-1, TGFβ-2, TGFβ-3) can be distinguished within the transforming growth factor-β superfamily as well as several functionally proteins such as the putative products of the Vg1 gene from Xenopus laevis and the decapentaplegic complex of Drosophila, mammalian Mullerian inhibiting substance, inhibins, activins and bone morphogenetic proteins. Recently, a new member of this family has been identified, myostatin (or GDF-8), which act as a negative regulator of skeletal muscle growth.

Mouse TGF-β: transforming growth factor-β; Drosophila Dpp: decapentaplegic protein; Mouse BMP: bone morphogenetic proteins; Mouse GDF: growth differentiation factors; Mouse MIS: Muellerian inhibiting factor; Mouse MIC-1: macrophage inhibitory cytokine; X. laevis ADMP: anti-dorsalizing morphogenetic protein 1 precursor; C. elegans Daf-7: dauer larva formation TGF- β homologue; C. elegans Dbl-1: (dpp, Bmp-like gene); C. elegans Unc-129: encodes a member of the transforming growth factor-beta family; C. elegans F39G3.8: sequence familiarity to a member of the transforming growth factor-β family.

The following is a brief description of TGF-β molecules from different families:

- TGFs proper (TGF-β1, 2, 3 and 5); TGF-β1 plays an important role in the wound healing processes of both embryos and adults since it increases the amount of extracellular matrix produced by epithelial cells (Nodder and Martin 1997; O'Kane and Ferguson 1997); furthermore, this molecule, implicated in neuronal differentiation, is involved in the stimulation and/or inhibition of cell proliferation (Tomoda et al. 1996) as well as in the regulation of epithelial-mesenchymal interactions fundamental to pattern determination during development. It has been observed that TGF-β3 appears to be the pivotal molecule responsible for perfect scar embryo formation in mice and sheep (Shah et al. 1992).
- Activins and nodal: these proteins are essential in distinguishing the left and right sides of the vertebrate body axes, and have the well known function of specifying different regions of the developing mesoderm. As recently shown, MicroRNAs, a new class of non-coding regulatory RNA molecules, may control all the Nodal/TGF-β signaling during mesoderm development (Martello et al. 2007)
- BMP or bone morphogenetic proteins: a highly conserved group of secreted regulatory factors that play an important role in early embryonic patterning. BMPs, originally identified for their role in bone formation (Wozney et al. 1988), are now known to play an important role in many fundamental developmental processes, such as neural induction in vertebrates and invertebrates (Holland and Holland 1999). In *Drosophila*, the structural homologue of vertebrate BMP4, decapentaplegic (*dpp*), appears to act early in development by establishing the dorso-ventral axis and the position of the central nervous system (DeRobertis and Sasay, 1996; Holland and Holland 1999).
- GDF or growth and differentiation factors 8 and 11 (also known as myostatin and BMP11 respectively): these are TGF-β members that function as chalone molecules (the term chalone refers to a negative regulator produced by a given tissue that acts to inhibit its growth). GDF8 is a recognised negative regulator of vertebrate muscle growth (McPherron et al. 1997; McPherron and Lee 1997) while GDF11 appears to be a negative regulator of proliferation of cells other than muscle cells (McPherron et al. 1999).
- New sequenced proteins, as the Glial derived neurotrophic factor (GDNF), a growth factor important for many neuronal populations of the central, peripheral and autonomous nervous systems. Moreover, GDNFm which functions outside the nervous system as a classic morphogen during renal development, regulates spermatogonial differentiation. GDNF differs from other TGF-β since it acts through a RET (*rearranged during transfection*) receptor tyrosine kinase. The ligand-binding specificity is determined by a glycosyl phosphatidylinositol (GPI)-linked ligand-binding subunit known as GDNF family receptor α (GFRα1, GFRα2, GFRα3 and GFRα4, Sariola and Saarma 2003).

This chapter describes discoveries concerning this family of growth factors, particularly concerning the expression of TGF-β molecules in regeneration mechanisms and during fish and mammal muscle development. Moreover, this text focuses on the regulative actions of myostatin in different experimental conditions.

2. TGF-β Pathway

The Transforming Growth Factor-beta superfamily (TGF-β) is composed of secreted proteins that modulate many biochemical pathways in several cell types. These proteins are dimeric cytokines involved in a wide range of processes including cell proliferation, differentiation, apoptosis, axial patterning, wound healing and tissue remodelling (Raftery and Sutherland 1999; Schmierer and Hill, 2007). All TGF-βs are synthesized as precursor molecules which are proteolytically processed in order to remove the signal peptide necessary for targeting the protein to the secretory pathway, and in order to generate the amino-terminal propeptide and the carboxy-terminal mature protein dimmer, which is the biologically active molecule (this second cleavage, triggered by a furin-type protease, occurs at a conserved four amino acid RXRR site).

However, the mature TGF-β dimer is not biologically active since it remains associated with the amino terminal propeptide (also known as LAP or Latency Associated Peptide). The LAP propeptide dimer is associated with the TGF-β dimer by non-covalent interactions, forming a small latent complex of about 100 kDa (Gentry and Nash 1990), whereas the mature form of the TGF-β dimer and the LAP dimer are disulphide bonded (Saharinen et al. 1999). TGF-βs are only activated following the breakdown of the non-covalent interactions between the LAP and TGF-β, enabling the latter molecule to bind to a family of cell surface transmembrane type I and II serine threonine kinase receptors. The TGF-β binding induces the formation of a heterotetramer in which the type II receptors unidirectionally transphosphorylate a dimer of type I receptors. The activation of the receptors catalyses the phosphorylation of receptor substrates (a class of Mad-related proteins, the Smads) which transduce the TGF-β signal directly to the nucleus (Massagué 1998, 2000; Raftery and Sutherland 1999); another group of recently discovered Smads (inhibitory Smads or I-Smads) may block TGF-β signaling by means of an auto-inhibitory feedback loop (Figure 2). Regarding BMP signaling, two classes of receptors have been described in literature: type I (several types, named ALK) and type II (three types) both of which are indispensable for signal transduction; some of them are specific for BMPs, while others are also signaling receptors for Activins (e.g. type IIB) thus confirming the putative common ancestral role of TGF receptors. Overall, it has been shown that Activins, Nodal and TGF-β ligands trigger the phosporylation and induction of some type I (ALK4, ALK5 and ALK7) receptors, thus affecting Smad2 and Smad3 whereas BMP ligands activate another kind of type I receptors (ALK1, ALK2, ALK3 and ALK6), which specifically induce Smad1, Smad5 and Smad8 (Feng and Derynck, 2005). Smad family members were first identified in *D. melanogaster* (mad) and *C. elegans* (sma) and subsequently in all vertebrates studied (Raftery and Sutherland 1999; Massague and Wotton , 2000). In *Drosophila*, decapentaplegic (DPP), screw and a third BMP ligand (Gbb) share a common set of receptors that include the type II (punt) and the type I (thick veins and saxophone) receptors (Raftery and Sutherland 1999).

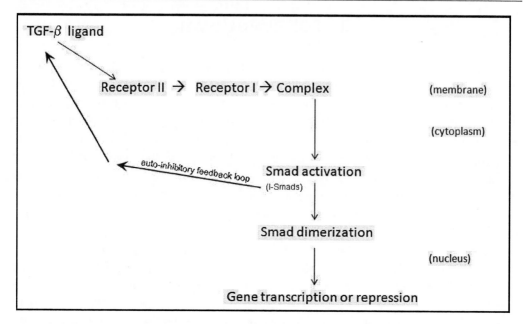

Figure 2. A schematic representation of TGF-β signaling (see text).

In *C. elegans*, five genes related to the TGF-β superfamily have been identified by sequence homology, the signaling appearing to be quite similar to that in the fruitfly. Herpin et al. (2005, 2007) tested whether BMP homologs/homologues of the lophotrochozoan mollusca *Crassostrea gigas* (Cg-BMPR1 and Cg-TGFbsfR2) can function in the context of a vertebrate TGF-β superfamily signaling pathway by overexpressing them during early zebrafish embryogenesis.

The above study showed for the first time, albeit indirectly, the most ancestral and functional mode of signal transduction for this superfamily, with an extracellular ligand binding domain TGFbsfR2 type II receptor capable of interacting with both BMP and activin ligands. Recently, Molina et al. (2007) and Orii and Watanabe (2007) showed, by means of an RNAi approach, that the BMP pathway is functionally conserved in patterning basic body axes in ecdysozoa and lophotrochozoa (Protostomia) as well as in Deuterostomes. Also the inhibitory molecules acting on the TGF-β system are conserved, the best known being follistatin and the notch pathway (Carlson et al. 2008).

It has recently been discovered that TGF-βs can be controlled by MicroRNAs (miRNAs), thus increasing the complexity of the regulation of growth factor cascades (Martello et al. 2007; Padgett and Reiss 2007).

3. TGF-β and Regeneration

3.1. Overview

Regeneration is the replacement by a fully developed organism of lost parts through growth or somatic tissue remodelling (Wolpert 1998). The capacity to regenerate varies greatly among different phyla and several/numerous studies contain remarkable examples of the regeneration of entire animals or body parts from amputated structures. This phenomenon has constantly given rise questions concerning the origin of cells that produce the regenerated structures and, for example, the part played by the nervous system in this process. The more plausible hypotheses put forward to explain the generation of progenitor cells after amputation are: 1) that cells close to the amputated surface may re-enter the cell cycle and dedifferentiate, or 2) amputation may trigger the division of reserve or stem cells. One example of cell dedifferentiation is limb regeneration in urodele amphibians (Brockes 1997), while an important contribution of stem cells during regeneration has been described in planarians (Baguña et al. 1994). The role of the nervous system in regeneration has been extensively studied in urodele amphibians, and Singer and Craven (1948) were the first to demonstrate that the early phases of newt limb regeneration are nerve-dependent, while the later phases of morphogenesis are nerve-independent. Subsequent studies have focused on the characterisation of highly conserved neurotrophic factors or molecules involved in regeneration, such as the *Hydra* head activator (HA), transferrin and members of the fibroblast growth factors (FGF) or transforming growth factors (TGF-β) superfamily. Although it is known that they are similar in many ways to molecular mechanisms governing normal development, the molecular mechanisms controlling the regeneration process have not yet been well characterised (Muneoka and Sassoon 1992). However, especially in terms of positional information, it is evident that the regenerating tissue is located in a cellular environment totally different from that of the developing embryo. Moreover, limbs from developing vertebrate embryos can form normally in the absence of nerves, whereas regeneration is arrested if the limb is denervated at the early stages of regeneration (Wallace 1981). Besides having a high regenerative capability, unlike adults, vertebrate embryos show an outstanding capacity for wound healing. In embryos, post-wound inflammation is minimal and wound closure is achieved by a complex mechanism of cellular movement and gene expression unlike that present in adults. Myofibroblasts, for example, have not developed in the embryo and therefore no scar is produced following wound closure (Nodder and Martin 1997). However, (wound) healing, which is essentially a repair process, shares a common molecular pathway with regeneration only up to early post-amputation phases, after which they diverge. Yet regeneration must incur the same problems of cellular identity and positioning as those encountered in developing invertebrate organisms in which the genetic regulation of cell fate, position and function is well established (Davidson et al. 1998). One major difference, however, is that of the tissue environment in which the process takes place. Unlike the regenerating processes in embryos, in adults this mechanism involves the development of new cells in established environment and, more importantly, an inflammatory response occurs. This fact appears to be of fundamental importance for the perfect and quick wound healing, which takes place in embryos, since inflammation is minimal or absent (Martin et al. 2003).

Technically, regeneration occurs in different phases: 1) rapid repair (or wound healing) to ensure survival; 2) active proliferation during which increased cell cycle activity produces new tissues; 3) morphogenesis, involving tissue reconstruction, regulatory proteins playing pivotal roles. A distinction can be made between two modalities of regeneration: morphallaxis where cells differentiate from existing tissues with little new growth (e.g. in coelenterates, endodermal or epithelio-muscular cells may dedifferentiate and participate in the regenerative process) (Burnett 1966; Schmidt and Adler 1984); epimorphosis, where proliferation results in the formation of a blastema comprising an accumulating mass of undifferentiated stem cells (Goss 1969). Most cases of limb regeneration in vertebrates, especially in urodele amphibians, are epimorphic (Goss 1987). In invertebrates both mechanisms (morphallaxis and epimorphosis) contribute to an equal extent in regeneration events (Goss 1969). Regeneration in invertebrates may involve the replacement of substantial amounts of the body and, in some groups (e.g. sponges and cnidarians), regeneration underpins asexual reproduction since whole organisms can be reformed from a small fragment of an adult body. Echinoderms are pentaradial deuterostomes that are able to undergo extensive regeneration (Hyman 1955). Some of the many species can autotomize limbs undergo asexual reproduction by fission (Hyman 1955). A remarkable example of regenerative capability comes from the asteroid *Linckia,* which is potentially able to regenerate an entire individual from one "single" arm (Cuenot 1948). Regeneration by both epimorphosis and morphallaxis occurs in echinoderms; epimorphosis with blastema formation is found in groups in which rapid regeneration takes place. This may correlate with the more vulnerable species in which predation is an important factor in the adaptive value of regeneration, as occurs in with crinoids and ophiuroids, in particular. The blastema is a discrete centre of proliferative activity providing a pool of "stem" cells which can give rise to all the new structures. Morphallaxis, instead, involves a more widespread proliferation, usually interesting cells derived from existing tissues by dedifferentiation, transdifferentiation and/or migration. Morphallaxis, on the other hand, is a slower process respect epimorphosis and is a feature in animals (e.g. asteroids) that have a more flexible, less fragile dermal skeleton and are at a smaller risk of predation. In crinoids, a blastema with substantial cell cycle activity appears in 24 to 48 hours of wounding or amputation, whereas in the asteroids studied so far there is little indication of cell division until 12 to 14 days post ablation (Candia Carnevali et al. 1995, 1997; Moss et al. 1998).

In vertebrates the regenerative capacity is more limited than that seen in invertebrates. However, the replacement of organs and reforming of appendages such as limbs, tails and antlers, are not uncommon features and lower vertebrates are capable of regenerating many areas of the CNS. Fish can regenerate fins, taste barbels and parts of the peripheral and central nervous system. Newts and salamanders (urodele amphibians) can regenerate many of their body-parts, even after metamorphosis, although this ability differs among species. In frogs and toads (anura amphibians), the regenerative ability is extensive right up to the pro-metamorphic larval stage. For example, limb regeneration in *Xenopus* may occur at all levels up to developmental stage 51 (pro-metamorphic larval stage) but it is lost at developmental stage 57 (post-metamorphic) (Cannata et al. 1992). In *Xenopus* this process involves the formation of a blastema, thus constuting typically epimorphic regeneration. It is still uncertain why epimophorsis is extensive in urodeles but not in other vertebrates (Goss 1987). Indeed, the factors limiting the regenerative potential of birds, reptiles and mammals have yet to be identified. However, some growth factor superfamilies such as Fibroblast growth factors (FGF),Transforming growth factors (TGF-β) and *HOX* genes (which control the fundamental architectural body plan of the embryo) have been well investigated in vertebrates, and their importance in the arm regeneration process as well as during development has been confirmed in some organisms (Muneoka and Sassoon 1992).

3.2. TGF-B Homologs in Regenerating Invertebrates

Regeneration is common in, and may form an important and central part in the life cycle of, many invertebrates. This ability undoubtedly depends upon a remarkable morphogenetic plasticity that allows the expression of developmental programmes at each, including the adult, life stage. As invertebrates frequently lose tissues following predation or during unfavourable environmental conditions, they have the potential for specific reorganisation and expression of a new developmental programme at almost every stage of life (Thorndyke et al. 2001a).

The role of TGF-β homologs in repair responses following injury to peripheral tissues is well known (Newfeld et al. 1999). During regeneration TGF-βs show a broad spectrum of activities: they are involved not only in the mediation of the inflammatory response but also in the regulation of ECM organization, specifically controlling the expression of its components, principally collagen and fibronectin (Noble et al. 1992; Sporn and Roberts 1992) while inhibiting the proteases (metalloproteases) responsible for their digestion. TGF-βs are also directly involved in phenomena of cell migration and adhesion by regulating the expression of integrins and other cell adhesion molecules during tissue repair. Moreover they can directly or indirectly regulate the expression of other growth factors and modulate their effects by acting synergistically or antagonistically with them (Lindholm et al. 1992).

In invertebrates TGF-β homologs have been found in all species studied (mainly of the BMP subfamily, probably the most ancient TGF-β ligands); they are always involved in regenerating mechanisms.

Among marine animals, sponges and coelenterates show the highest regenerative capabilities. During regeneration, sponge cells (archeocytes) display many features characteristic of totipotent cells, although they may also be derived from other cell types by dedifferentation (Connes 1974). These cells gather at the wound site, form a blastema and then differentiate into other cell types.

In the most widely studied class of Coelenterata, the hydrozoans, it seems that the replacement of lost parts occurs via undifferentiated "reserve" cells, and it is known that even differentiated adult cells may transdifferentiate to form a complex regenerate (Burnett 1966; Schmidt and Adler 1984). In this context, BMP homologs are surely involved in hydrozoan jellyfish development and transdifferentiation (Reber-Müller et al. 2006). Recently, the characterization of a Hydra BMP homolog named *hyBMP5-8b*, important in the specification of the aboral region of the polyp and the discovery of a chordin-like protein that inhibits BMP activity, has confirmed the ancient role of BMP signalling as a metazoan "invention" in establishing axial polarity (Reinhardt et al. 2004; Rentzsch et al. 2007).

Among the Platyhelminthes, planarians are the most well-known regenerating animals. The origin and role of totipotent stem cells (neoblasts) is still under debate and new insights into planarian *HOX* genes have made this animal an excellent model for regeneration studies at the molecular level (Bayascas et al. 1997). Recently, Molina MD et al. (2007) found that during planarian regeneration the basic dorsoventral axis is re-established and maintained by the BMP pathway.

The ribbonworm, *Lineus sanguineus* (nemertine), expresses a *Pax-6* homologue in the nervous system during regeneration (Loosli et al. 1996). This gene belongs to the class of homeobox-containing genes, which are highly conserved through evolution, thus confirming once again that in the attempt to understand 'ancient' molecules and mechanisms, considerable benefit can be gained by investigating of lower species. In annelids, regeneration can be striking and, as occurs in nemertines, there is the formation of a blastema without any apparent contribution from true neoblastic-totipotent cells.

Regeneration in molluscs has received even less attention although these animals can replace many parts of their body, including cephalic organs, after amputation or autotomy (Hyman 1965). Epimorphic regeneration remains the predominant mechanism and a blastema is formed from cells that have dedifferentiated from the underlying tissues (Ferretti and Géraudie 1998).

In arthropods, as in molluscs, no asexual reproduction occurs, but autotomy is a very common feature, even in the absence of regeneration. Moreover, arthropods can regenerate any appendage provided they continue to moult. Indeed, in adults, regeneration is strictly dependent upon the moulting cycle (Truby 1985). Extensive research into the regeneration of imaginal discs in *Drosophila melanogaster* larvae has contributed to a better understanding of the positional properties of cells and the formation of the proximodistal axis (French et al. 1976; Couso et al. 1993) The dpp protein is involved in the patterning of leg imaginal discs both during development and regeneration (Theisen et al. 2007). Myoglianin, another Drosophila protein sharing a significant sequence homology with a TGF-β member (with myostatin in particular), is expressed in muscle embryonically although its role during regeneration is not known (Lo and Frasch, 1999).

Interest in the regeneration of protochordates has increased in recent years, particularly because they share chordate characteristics, such as the notochord and the neural tube, at least in the larval stages. In a recent study on regeneration in the neural complex of adult ascidians, it was discovered that two types of migratory cells may be involved in this process: one a typically undifferentiated or totipotent element, the other derived from the proliferation of epithelial cells at the dorsal strand of the neural gland (Bollner et al. 1995). TGF-β homologs (nodal in particular) are essential in the regulation of neural tube formation of *Ciona intestinalis* embryo (*Mita* and *Fujiwara*, 2007), although any direct involvement of this family in regenerative mechanisms has not yet been demonstrated.

In echinoderms, TGF-β homologs have been identified in sea urchin embryos where they play a critical role in both early developmental and regulatory events. *Univin* was the first gene member of the TGF-β family to be characterised in echinoderms (Stenzel et al. 1994; Thorndyke et al. 2001b).

Another gene belonging to the BMP subfamily was sequenced in *S. purpuratus* (Ponce et al. 1999) and shown to be related to the BMP5-7 subfamily; in the same work, only two genes were isolated from a cDNA library constructed from a 14-h embryonic stage: one, a BMP5-7 homologue and the other, the previously-described/above-described/already known *Univin*. The authors suggested that other TGF-β subfamily members, which might be present in adult echinoderms, could possibly utilised for the development of the adult body plan. Postembryonic growth phases are produced by different developmental mechanisms in echinoderms; therefore, new genes other than those used for embryogenesis might be employed. In this context, a potent biological tool for confirming this idea is the arm regeneration process. Sequence data corroborate the presence of more than two (*univin* and

BMP 5-7) TGF-β ligands expressed at different stages of development (e.g. *TgBMP2/4* and *SpBMP2/4*). *TgBMP2/4* is abundantly expressed throughout all embryonic phases and, in particular, at an advanced stage of development in the sea urchin *T. gratilla*. In the later pluteus stage of the sea urchin, *T. gratilla* (Hwang et al. 1997) *Tg*BMP2/4 is thought to play an important regulatory role in gut patterning and spicule formation. Furthermore, a detailed study on sea urchin animal-vegetal embryonic axis has confirmed that BMP2/4 plays a role in influencing the regulation of cell fate within the ectoderm (Angerer et al. 2000). In view of the importance of BMPs in a wide range of regulatory functions, it seems likely that these genes may also be crucial to growth in adult echinoderms, and regulate the expression of other related genes in particular physiological conditions, such as during regeneration.

Results obtained in *Antedon mediterranea*, a crinoid echinoderm (Candia Carnevali et al. 1998a; 1998b; 2000), show that this factor, or at least an antigen which cross reacts positively with a specific antiserum against mammalian TGF-β1, is normally present in various components of the nervous system in non-regenerating arms, and is specifically localized in cell bodies and processes at the level of the brachial nerve and the basi-epithelial plexuses of both the epidermis and the coelomic epithelium. The same research group showed that during regeneration TGF-βs had a wider tissue distribution and seemed to be involved to a larger extent outside the nervous components (Candia Carnevali et al. 1998b, 2000; Patruno et al. 2001, 2002). The apical blastema also maintained its strong reactivity during the advanced stages. At the cellular level, a specific reaction for TGF-β was found all around the plasma membrane of the blastemal cells. Recent experiments allowed the identification and cloning of a native TGF-β homolog from crinoids, thus confirming these findings at the molecular level: a gene fragment (*An*BMP2/4) was found to include a deduced protein sequence containing the seven conserved c-terminal cysteine residues characteristic of the TGF-β family (Patruno et al. 2003). Using a Real Time PCR approach, the same research team is now making a quantitative evaluation of *An*BMP2/4 mRNA expression, which seems to be up-regulated during regeneration (Barbaglio et al. unpublished observations). Moreover, in the ophiuroid *A. filiformis,* two members of the BMP family have been sequenced; both are involved in regeneration (Bannister et al. 2005, 2008).

Findings from research carried out on crinoids and ophiuroids suggest that TGF-β homologs play a basic role as constitutive factors involved in both developmental processes and the maintenance/regulation of regenerative functions. Indeed, the presence of BMPs in the growing blastema may be critical for regulating cell proliferation and the synthesis and organization of the ECM components. During regeneration, the widespread localization of these factors close to the amputation site and at the blastema level, is consistent with the idea that TGF-β homologs are essential factors for repair and regenerative events also in echinoderms. However, the recent sequencing of the echinoderm genome (using the sea urchin *Strongylocentrotus purpuratus*) has revealed the presence of at least 14 TGF-β homologs, indicating that our understanding of all the roles played by these factors is, as yet, very limited.

It is therefore reasonable to suggest that during both regeneration and development, the specification of the dorsal ventral axis and the commitment of blastemal cells depends on a dual mechanism, with a TGF-β ortholog acting on presumptive neuronal cells and BMP orthologs determining coelomic epithelium proliferation; other factors, such as FGF10 and

HOX genes, important in vertebrate regeneration but not yet identified in crinoids, are surely actively involved, together with the TGF-β family, in the reparative/regenerative process of these animals (Figure 3).

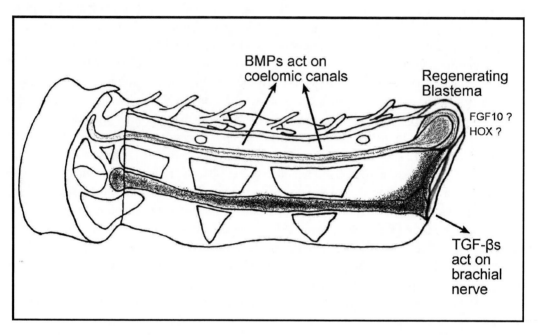

Figure 3. The main studied factors involved in the regenerative epimorphic process of crinoid echinoderms.

4. Myostatin (GDF8) and Muscle Development

4.1. Overview

The growth and differentiation factor 8, a molecule better known as myostatin (mstn), plays a negative role in the regulation of skeletal muscle mass; the protein belongs to the TGF-β superfamily (McPherron et al. 1997). As stated above, the predicted TGF-β aminoacid sequences of many vertebrate and invertebrate species in the active carboxy-terminal region are identical. In order to bind its receptor and thus elicit its biological function, the active receptor binding region of myostatin must, with the assistance of the latency associated peptide (LAP), detach from the precursor protein (Joulia-Ekaza and Cabello, 2006). Myostatin has been characterized as a potent negative skeletal muscle growth factor (McPherron et al. 1997; McPherron and Lee 1997) that inhibits satellite cell activation, myoblast proliferation (Thomas et al. 2000) and myogenic differentiation in both pre-natal and post-natal muscles (Rios et al. 2002). The increase of muscle mass observed in knockout mice is due to both hyperplasia (i.e. *de novo* muscle fibre formation) and hypertrophy (i.e. an increase in muscle fibre diameter). Mstn natural mutations produce a double-muscled phenotype of cattle (Marchigiana, Belgian Blue and Piedmontese) as well as gross muscle hypertrophy in humans (Schuelke et al. 2004). It has also been suggested that mstn may play a role in myogenesis, in addition to the control of muscle homeostasis in post-natal life

(Kambadur et al. 2004), an hypothesis confirmed by findings made in experiments in which mstn knockout was found to ameliorate the dystrophic phenotype in a mouse model of muscular dystrophy (Bogdanovich et al. 2002).

The development of skeletal muscles depends on myogenic mechanisms that lead to the establishment of primary and secondary myotubes during prenatal life (Draeger et al. 1987; Buckingham et al. 2003). In vertebrates, myogenesis is controlled by a complex network of intra- and extracellular signals classified as determination and differentiation factors (see Buckingham et al. 2003; Parker et al. 2003; Buckingam, 2006). Trunk and limb skeletal muscles originate from paraxial mesoderm which segments into somites on either side of the neural tube and the notochord. Muscle progenitor cells migrate from the somite to form primary myotubes via the activation of a number of myogenic regulatory factors (MRFs, Figure 5) together with coactivators, the myocyte enhancer factor 2 family (MEF2) (Parker et al. 2003). The development of secondary myotubes is controlled by the same MRFs (Buckingam, 2006) whereas the molecules involved in the establishment of tertiary myotubes are still under investigation. The latter are a third generation of myotubes observed during foetal life in some mammals (Picard et al. 2002). Among mammals studied, in only the pig have tertiary myotubes been found in post-natal skeletal muscles (Mascarello et al. 1992). A post-natal hyperplastic phase has also been found in the rat, new muscle fibres being present in the interstitial space of skeletal muscle (Tamaki et al. 2002, 2003). It seems that post-natal waves of myogenesis are under the influence of satellite cells (Mauro 1961) which, during embryogenesis, originate from the central area of the dermamyotome (Gros et al. 2005) and are capable of self-renewal (Dhawan and Rando, 2005; Collins 2006).

Across phyla the evolutionary conservation of mstn function is evident: in almost all vertebrates studied, null mutation leads to a hypermuscular phenotype.

In invertebrates only few proteins sharing a significant sequence homology with myostatin have been found: *myoglianin*, expressed in the embryonic muscle of *Drosophila* (Lo and Frasch, 1999) and *sMSTN*, cloned from the bay scallop *Argopecten irradians*, expressed at high levels in muscle as well as in many other tissues and organs (Kim et al. 2004). The authors who reported the latter finding have also identified a mstn-like gene from the *Ciona intestinalis* genome confirming that this gene has been conserved throughout evolution.

In the subsequent paragraphs some historical and recent discoveries are introduced with regard to mstn biology, in fish, birds and mammals.

4.2. Myostatin and Fish Development

The bulk of fish muscles are made up of metameric myotomal lateral muscles. The typical W shape and the complex anatomical arrangement of the myomeres adopted by all the gnathostome fishes have been considered an improvement in the mechanical efficiency of the muscles in relation to the dorsal position of the spinal cord and the limited range of body flexions in the lateral plane. Inside each myomere, the fibres are arranged according to a highly complex pattern, the most superficial fibres running parallel to the long body axis, and the innermost fibres running obliquely. As observed by Alexander (1969), this peculiar pattern of fibre orientation is strictly related to the need of all the different fibres to contract at about the same rate as the body flexes. In addition, the suitable combination of structurally and functionally different fibre types in each myomere (red slow fibres and white fast fibres, plus pink intermediate fibres in some cases) gives rise to two distinct and specialized motor systems (Bone et al. 1995).

As regards the development of skeletal muscles, fish differ from other vertebrates. In mammals muscle fibres, derived respectively from primary and secondary myotubes, are produced mainly in two steps. In contrast, in fish axial muscle, the different fibre types develop in distinct proliferative zones; the formation of different muscle fibres is spatially and temporally separated (Rowlerson et al. 1995). Moreover, in mammals, hyperplastic growth as a source of new muscle fibres usually stops at the perinatal stage, whereas in most fish species hyperplasia is important during larval as well post-larval life. In most teleost fish, muscle development of the yolk-sac larva gives rise to two different larval muscle types, the inner and the superficial muscles which are well-developed at hatching (Veggetti et al. 1993; Mascarello et al. 1995). During larval life, the formation and the growth of differentiated red and white fibre types is mainly due to the mechanism of hyperplastic growth that occurs by apposition of new fibres along proliferative zones, principally under the lateral line and in the apical myomere regions, but also immediately beneath the superficial monolayer (Rowlerson et al. 2001). In the early larva of some species the inner white muscle fibres grow by hypertrophy, whereas in other species the inner white muscle fibres grow also by hyperplasia. Generally, species that reach a large adult size show a new hyperplastic process disseminated throughout the fast white muscle layer also during post-larval life. In contrast, in those species that never reach a large size (Veggetti et al. 1993) no such disseminated hyperplastic process is found throughout the inner white muscle. In all species, hypertrophic muscle fibre growth occurs at all stages of growth/development (until the maximum somatic size is reached), but it is predominant during juvenile and adult life.

Our group investigated the distribution of myostatin mRNA and protein during the development and growth of different teleost species in view of its importance in accurately determining the proper muscle mass in mammals. However, few studies have shown mstn function in fish (Xu et al. 2003; Amali et al. 2004) and one has demonstrated that mstn suppression, achieved by RNA interference, increases the size of zebrafish due to an increase in muscle mass (Acosta et al. 2005). In fish two distinct myostatin genes have been characterised and, unlike in mammals, fish myostatin is expressed in several tissues other than in muscle (Kocabas et al. 2002; Maccatrozzo et al. 2001a, 2001b); therefore, other possible functions cannot be ruled out. Moreover, the great bulk of mstn studies have been carried out on mRNA expression while information on the cellular localization of mstn protein is scarce (Radaelli et al. 2003).

The fish mstn gene was duplicated during the evolution and two clades are present: the mstn-1 and the mstn-2 paralog genes. In salmonids, a second gene duplication occurred and four mstn paralogs are present in these fish (Garikipati et al. 2007). The majority of fish studies have been conducted on the mstn-1 ortholog which, unlike mstn-2, has been cloned in numerous species such as the sea bream, the trout, the Atlantic salmon, the channel catfish, the sea bass, the spotted grouper and the zebrafish. The expression of mstn-1 and mstn-2 during zebrafish development differ (Helterline et al. 2007). The latter research group found that levels of zfMSTN-1 rose during the late stages of somitogenesis while in the zebrafish mstn-2 peaked rapidly at the early stages of somitogenesis and then dropped in the successive stages of development. Recently we investigated mstn expression during the post hatching phases of sea bass (*Dicentrarchus labrax*) development showing a low level of mstn mRNA as early as two days post hatching (and none before) while a significant mRNA increase was observed at the later stages of development (from day 25 post hatching). Mstn is expressed at a very low level during early post hatching development and it is extremely difficult to visualize its mRNA by in situ hybridization (ISH) at these stages; it was therefore necessary to combine several approaches in order to support the experimental hypothesis. In sea bass we performed Real-Time PCR to confirm that there is a lack of a proper mstn mRNA signal up to the third week from hatching; following this period, mRNA expression was detected in the trunk lateral muscle of young larvae only in the epiaxial and hypoaxial proliferative muscle zones (Patruno et al. 2008). ISH performed on frozen sections of developing sea bass larvae (up to 80 days post hatching) revealed the presence of mstn mRNA in the superficial monolayer (future red muscle under the skin) and in the underlying proliferative zone, adjacent to the horizontal septum. The latter, an area of hyperplastic growth, is characterized by small-diameter fibres among large ones (future red and white fibres), known as the "mosaic" (Figure 4) (Patruno et al. 1998; Rowlerson and Veggetti 2001). We also measured mstn mRNA levels in red and white muscles from juvenile and adult fish since the literature regarding the expression of myostatin gene in white and red muscle in fish is inconsistent. It appears that a differential expression is seen among fibre types (red or white) within various fish species, depending on which myostatin is studied (type 1 or 2) and, more importantly, which developmental stage is examined: juveniles, adult or old fish. It has been shown that mstn mRNA expression was higher in the red muscle of adult brook trout, king mackerel, and yellow perch by Northern Blot analysis; in small tunny mstn was expressed predominately in white muscle while in mahi-mahi MSTN was expressed equally in both muscle types (Roberts and Goetz 2001). Furthermore, Østbye et al. (2001) showed that in salmon the mRNA of mstn was detected in red muscle only (by a standard PCR assay). Radaelli et al. (2003) showed a consistent difference in the pattern of myostatin immunoreactivity in the red and white muscle of postlarval fish, being positive in red muscle and negative in white muscle (in solea, sea bream and zebrafish). It is important to note that, generally, red muscle grows more slowly than fast/white muscle and this is consistent with a role for myostatin as a negative growth regulator in this muscle type, although it does not rule out other possible functions (since the zone between the red and white muscle is also a proliferative area in large size species).

In our recent report (Patruno et al. 2008) sea bass at 80 days post hatching exhibited, by means of immunohistochemistry localization, mstn negative red fibres and mstn positive white fibres whereas adults had an opposite profile. Our immunohistochemical data were confirmed in adult fish by Real-Time PCR assay, in which the level of mRNA was much higher in the red/slow portion of lateral muscle than in the white/fast muscle (this was also the case in shi drum and sea bream). However, the lack of mstn protein in fry red muscle (against the high level of expression found in the same muscle in adults) might suggest that mstn plays a role as a negative growth regulator in this muscle type.

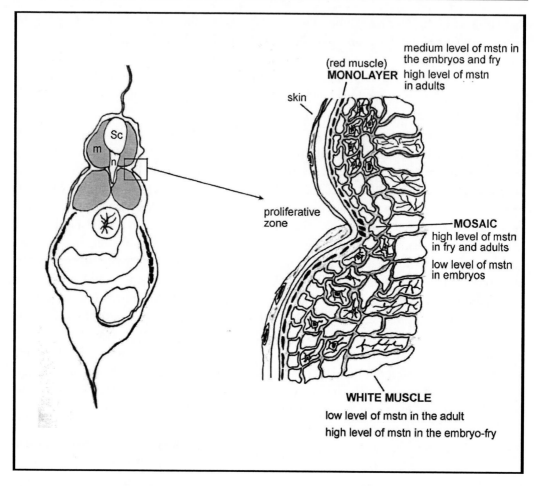

Figure 4. Expression of myostatin in the red/white muscles and mosaic of a commonly studied large size fish (the sea bass for example). The drawing shows a transversal section of a juvenile fish (left) and a enlargement of the proliferative zone (right).

The mstn mRNA levels observed in the two sea bass muscle types, together with previous results obtained in other fish species are probably due to the fact that mstn is developmentally regulated and a quick up- or down-regulation of this factor is necessary in both slow (adults) or rapid (juveniles) growing phases of fish up to their reaching a large adult size.

This pattern of mstn expression, influenced by the developmental stage considered, is analogous to that observed in mammals (see next paragraph). Other similarities were found by Funkenstein and Rebhan (2007) who, by means of an elegant approach, showed that the presence of mstn prodomain is essential for refolding fish mstn *in vitro* and also prevents its precipitation during the prolonged refolding process. The same authors observed that following the cleavage by endopeptidase furin, the mature fish mstn dimer remains in a latent form, which is activated by acidic or thermal treatment in order to elicits its biological activity; this fact suggests that fish mstn can probably signal through the same receptor and signaling pathway as mammalian mstn.

Mstn is also expressed in cardiac fish musculature as observed in solea and zebrafish (Radaelli et al. 2003), indicating that this molecule may play an important role during cardiac muscle development since, as in mammal cardiomyocytes, there may be a direct link between IGF-I and mstn.

4.3. Myostatin and Bird Development

The first comprehensive report of myostatin mRNA patterns in chicken embryos was carried out by Kocamis et al (1999a, 1999b). In chicks mstn is already expressed at the blastoderm stage and its expression fluctuates during different phases of myogenesis. Many subsequent studies have shown that mstn functions as a negative regulator of muscle mass also in birds; findings made on poultry indicated the need for further studies on the skeletal muscle growth of meat-producing animals by blocking myostatin activity. This type of study has recently been performed using monoclonal antibodies that increased the skeletal muscle mass of post hatch broilers treated with *in ovo* administration of the antibody (Kim et al. 2006).

Moreover, it was demonstrated that mstn can induce the arrest Pax7[+] cell proliferation and the down-regulation of MRFs and Pax3 if beads coated with high concentrations of myostatin are implanted in developing limb buds of chick embryos (Amthor et al. 2004, 2006). These results, also obtained in mammalian satellite cells using different approaches (McCrosckery et al. 2003), suggest that the activation of myostatin signaling in the embryo, as well as in adults, results in the quiescence of muscle progenitors and the arrest of their myogenic differentiation programme. More recent results, however, appear to indicate the phenomenon is more complex, since mstn might also promote the terminal differentiation of chick embryonic muscle progenitors. These observations indicate that mstn may be a pivotal molecule in balancing the proliferative/differentiative ratio of embryonic muscle progenitors, and not merely a negative regulator (Manceau et al. 2008).

In a recent report, our research group set up chick cocultures between embryonic spinal cord and myotubes isolated from adductor muscle in order to obtain a useful tool to study the cascade of myogenic positive and negative signals activated by paracrine neuronal factors (Martinello and Patruno 2008). The findings obtained indicated that one of the factors affected by innervations was myostatin (see point 4.5).

4.4. Myostatin and Mammalian Development

The development of skeletal muscle fibres starts at the level of embryonic somites which develop at the paraxial mesoderm (the closest mesoderm layer to the notochord). The signalling from the surrounding tissues (neural tube, nothochord, lateral plate mesoderm) patterns the somites into compartments, which give rise to different cell types. In the somites it is initially possible to identify a dorsal part, called the dermomyotome and a ventral portion, the sclerotome. The latter give rise to the the vertebrae and the ribs. The dermomyotome is subdivided into a medial, epaxial (or more correctly primaxial), and lateral (hypaxial, or rather abaxial) regions. Cells from the primaxial region inwards and form the myotome (between the dermomyotome and sclerotome), from which myoblast (muscle progenitor cells) develop in the axial muscles of the body. Another group also originates from the hypaxial somite or hypaxial dermomyotomes, but then migrates and builds up the limb, diaphragm and tongue muscles (migratory hypaxial muscles).

The mesodermal somitic cells from the dermomyotome receive inductive signals from the surrounding tissues (Wnt, Sonic hedgehog, Nogging) or inhibitory hints (Notch, Frzb, Sfrp, BMP4); afterwards, the expression of some myogenic factors such as Pax3 and primary myogenic factors, Myf5 and MyoD (belonging to the myogenic regulatory factors family, MRFs), commit the cells to their myogenic lineage (Figure 5). Wnt and Shh signalling play an important role in promoting and maintaining the expression of myogenic regulatory factors (MRFs); they therefore promote and maintain myogenesis. MRFs, a family of basic helix loop helix (bHLH) proteins, induce the transcription of many muscle-specific genes). The MRF family consists of four structurally related transcription factors (Myf5, MyoD, myogenin and MRF4 or Myf6) that participate in the regulation of both skeletal muscle development and postnatal hypertrophic growth or the regeneration process. During muscle development, each MRF has a distinct temporal pattern of expression (Buckingham, 2003, 2006). Myf5, the earliest to be expressed before MyoD; both play an important role in myoblast activation, proliferation, and subsequent differentiation. In the differentiation program are more implicated Myogenin and MRF4 that drive the myoblasts to fuse in myotubes till mature fibres. After birth, MRFs continue to be expressed even if their expression pattern differs and tends to decrease; however, MRFs appear to be involved in the regulative modulation of certain mature muscle fibre types.

Once muscle progenitor cells are committed, they develop into myoblasts that fuse to originate myotubes. Mammal muscle fibres (derived from primary and secondary myotubes) are produced in two main steps. Early or embryonic myoblasts fuse to form primary myotubes independently of innervation; these form a type of scaffold around which fetal myoblasts develop and fuse to form secondary myotubes in a nerve dependent process (Figure 6). Consequently, the different muscle fibre components that form the adult muscle derive from different generations of myotubes interspersed in the developing muscle. Innervation regulates not only mitosis in populations of secondary generation myoblasts but also influences the expression of the different myosin isoforms. Among large mammals, porcine skeletal muscles are peculiar for their development and their "unique" fibre type composition. Indeed, the porcine hindlimb and trunk muscles express all four MHC isoforms, including 2B MHC (Toniolo et al. 2004). In pigs primary myotubes form during the first month of gestation, while secondary myotubes develop during the second half of gestation (Figure 6); a third generation of myotubes, also observed in other species, only appears in the pig one month after birth (Mascarello et al. 1992; Lefaucheur et al. 1995, Picard et al. 2002; de Jonge et al. 2006).

Masticatory, eye and laryngeal muscles, which are considered specialized muscles belonging to head muscles, have a different embryonic origin from trunk and limb muscles. In the head region muscles arise from a non-somitic paraxial mesoderm that does not follow the same pathways as skeletal muscle myogenesis. The regulation of muscle development from the pre-otic mesoderm is not yet well understood (Noden and Francis-West 2006).

Caudal to the otic vesicle are the so-called occipital somites, the most cranial of somite series that, during evolution, have been incorporated into the head.

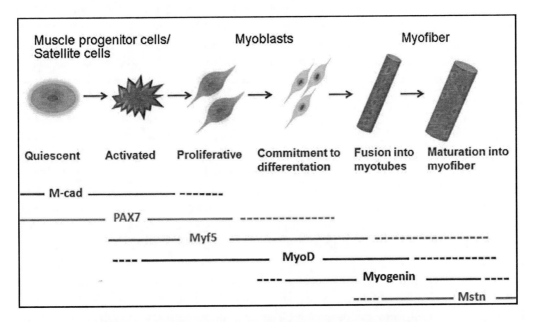

Figure 5. Growth factors and other molecules involved in the myogenic program.

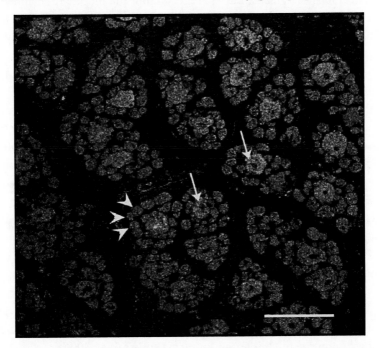

Figure 6. The picture shows primary myotubes (arrows) surrounded by secondary myotubes (arrowheads) during pig foetal muscle development. Scale bar, 75μm.

They provide epaxial and hypaxial muscles of the neck (pharyngeal and laryngeal muscles that develop in the caudal branchial arches) and the tongue musculature.

Cranial to the otic vesicle, the head muscles origin from pre-otic paraxial mesoderm, incompletely segmented in seven mesenchymal structures termed somitomeres, and further cranially from the pre-chordal mesoderm, which does not form somites. Extraocular muscles originate from the pre-chordal mesoderm while the masticatory muscles develop in the core of the first three branchial arches (Noden and Francis-West 2006).

Large size animals need a massive activation of the satellite cells, which are located beneath the basal lamina of the myofibre, and are necessary for the accretion of muscle protein within growing fibres (Mauro, 1961). It is generally accepted that myostatin and Pax7 are crucial growth factors for the physiology of satellite cells (McCroskery et al. 2003; Zammit et al. 2006a, 2006b), the majority of which originate in the central area of the dermamyotome (Gros et al. 2005); they are, moreover, capable of self-renewal (Rando 2006). Cattle (Belgian blue) with a natural mutation of myostatin showed increased muscling by muscle hyperplasia unlike from myostatin-null mice (Joulia-Ekaza and Cabello 2006). These observations confirmed that myostatin negatively regulates muscle growth. In fact, some results have demonstrated that this growth factor may inhibit myoblast proliferation (Thomas et al. 2000). In their study, Kambadur et al. (2004) showed that myostatin is expressed early in the development its expression apparently being restricted to muscle tissue (McPherron et al. 1997). Furthermore, myostatin transcript and/or protein expression have been shown to be regulated during different physiological and pathological situations which affect muscle mass, such as muscle atrophy, heart infarct, muscle unloading, HIV infection, microgravity exposure, chemical muscle damage, muscle regeneration and muscle re-loading (reviewed by Matsakas and Diel 2005; Matsakas et al. 2006). All these findings have raised the possibility that myostatin plays an important role also in muscle growth during postnatal life (Kambadur et al. 2004).

During the development of different muscle types myostatin seems to act in several ways or, at least, is expressed in different patterns. The expression studies published on myostatin have often failed to investigate the relation between the developmental expression and the muscle type composition. Myogenic cells (myoblasts) synthesise myostatin especially during the early phases of development (Thomas et al. 2000) but few data are available on the post-natal stages of growth, also with regard to mstn expression in "allotype" muscles (Joulia-Ekaza and Cabello 2006). The latter term, which refers to a concept introduced by Hoh and Hughes (1988), is based on some of the characteristics of muscle, such as the expression of peculiar MHC isoforms, a different embryological origin and the type of innervation.

Our group showed, by Real Time PCR and Western blotting experiments, that the expression of myostatin in skeletal muscles is developmentally regulated (Patruno et al. 2008). During early foetal development, the mstn precursor is the most abundant form expressed in pig (as observed in cattle) although its putative role played in the regulation of mstn activity difficult to explain. It may be speculated that high precursor levels promote a decrease in processing and therefore in secreting mature mstn, probably participating in the negative feedback loop recently proposed for mstn (Forbes et al. 2006).

The Real time PCR assay carried out by us on allotype muscles showed that mstn expression was more abundant and followed a different age-dependent evolution in extraocular compared to skeletal muscles at both mRNA and protein level. In particular, mstn mRNA was very abundant in almost all developmental stages examined while the mature mstn were present only in adult muscles. Mstn processing and synthesis never appear to be ceaseless/arrested in these muscles, and this is in line with with the intense production of MRF. Our group showed for the first time that extraocular muscles express higher levels of both mstn mRNA and protein than skeletal muscles, and, among adult extraocular muscles, the scleral and orbital layers of the rectus showed a lower level of mstn mRNA than that found in the Rb muscle. Attempts to quantify variations in mstn mRNA expression between skeletal (Longissimus dorsi) and extraocular (Rectus superior) muscles were also performed in cell culture experiments. Myogenic cells isolated from Ld and Rs muscles in piglets sampled at 30 days postpartum revealed that during myoblastic differentiation mstn mRNA levels were reduced in both groups of cells, confirming in vitro that mstn retained its role as an inhibitor of proliferation in both sets of cells.

Towards the end of gestation, myostatin immunostaining was observed in type 1 fibres only. Likewise, in the double-muscled bovine the inactivating mutations of myostatin are associated with fast fibre hyperplasia and in mstn knocked out mice a a deficiency in type 1 fibres is found (Joulia-Ekaza and Cabello 2006). Other authors have observed a different localisation of mstn, this depending on the physiological/differentiative state of the cells; in vitro a nuclear or perinuclear staining was observed in proliferating myoblasts (Artaza et al. 2002) while a cytoplasmic antibody reactivity was described after their differentiation (McFarlane et al. 2005). The same authors suggested that the perinuclear localization found during proliferative phases of myogenesis probably reflects the myostatin synthesis and processing which occur in the endoplasmic reticulum and Golgi. More puzzling is the mstn staining observed in postnatal stages of development; it might concerns also satellite cells given the co-localization with Pax7 found in a previous study (McCrosckery et al. 2003) and since both factors play similar roles in protecting satellite cells and myoblasts from apoptosis (Buckingam 2006).

The findings made by us bear out the idea that molecular determinants have a strong influence in defining the terminal differentiation of slow, fast and special allotype muscles. Indeed, each group of muscles examined by us was found to have a specific expression pattern for each molecule investigated, thus confirming that a totally reliable experimental animal model to be used for therapeutic purposes in humans and/or animals has yet to be found.

The mstn pattern found in EO muscles, which are considered an example of continual tissue remodelling, strengthens the idea that mstn is strictly linked to myogenesis regulation and that it may be involved in the delay of terminal fibre maturation or in the block of hypertrophic mechanisms. In addition, EO muscles are spared in diseases such as Duchenne dystrophy and the different expression of myogenic genes might be an essential requisite (Kaminski et al. 1992).

Regarding skeletal muscles, three developmental profiles of mstn expression emerge (Figure. 7): i) high level of mstn mRNA combined with high level of the precursor protein ("foetal profile"); ii) intermediate level of mRNA combined with low level of precursor and mature peptide ("birth profile"); iii) very low mRNA level combined with the presence of mature peptide ("post-natal/adult profile"). Allotype muscles show a different expression pattern (Figure 7).

Skeletal muscles

Mstn foetal profile
- -High level of mRNA and high level of the precursor protein
- -Absence of mature peptide?

Mstn birth profile
- -Decrease of mRNA expression
- -Low level of precursor and mature peptide

Mstn adult profile
- -Very low level of mRNA
- -Presence of mature peptide

Extraocular muscles

Mstn foetal profile

Mstn birth profile
- -Higher level of mRNA compared to skeletal muscle
- -Higher level of precursor protein compared to skeletal muscle

Mstn adult profile
- -Higher level of mRNA, precursor protein and mature petide compared to skeletal muscle

Masseter muscle

Mstn birth profile
- -Higher mRNA levels compared to skeletal muscles
- -Expression of the mature peptide from birth (not seen in skeletal and eye muscles)

Mstn adult profile
- -Low levels of mRNA

Figure 7. Myostatin expression "profile" in skeletal, extraocular and masseter mammalian muscles.

4.5. Myostatin Expression in Co-Culture, Exercise Training and NMES Experiments

Our group is actively investigating the expression of myostatin in several experimental conditions and in different animal species. The *in vitro* model used by us consisted of a chick co-culture system of neuronal and muscular tissues, and an exercise protocol performed in rats (acute and endurance swimming training) was employed in order to follow the fluctuation of myostatin after a synaptic contact or a physical activity. More recently, we have used NMES protocols in healthy humans, the aim being to investigate if the recruitment of satellite cells detected after this treatment might involve the increasing expression of myostatin.

Findings made by us in a co-culture experiment using embryonic spinal cord fragments placed on myotubes from chick embryos, confirmed that innervations accelerate the differentiation of skeletal muscles (Martinello and Patruno 2008). The main aim of this study was to investigate the muscle response to neuron contacts. In our model, the developing neurons showed exploratory microspikes which test the microenvironment in order to send signals back to the soma. Once the functional contacts were established between chick spinal cord and myotubes, most of the adventitious sprouts were retracted, probably limiting the extent of multiple innervation. Our findings also showed that axonal connections accelerate myogenesis, since the presence of spinal cord segments was found to cause earlier contractions and therefore an earlier maturation of myoblasts. Real Time PCR carried out on these co-cultures showed that myostatin increases in the presence of innervation and therefore inhibits myoblast proliferation in favor of a differentiative muscle program; this confirm that it has a developmental role, controlling the myoblast differentiation mechanism (Martinello et al. unpublished observations).

Regarding myostatin expression after training (therefore in a stressed microenvironment), a decreasing trend has been found in the muscles of rodents exposed to exercise. This is expected, since a trained muscle increases its satellite cell turnover (or other muscle progenitor cells), thus modifying its mass (usually it increases but this depends on the protocol followed) and therefore the level of mstn must be low. Several authors have previously found lower muscle RNA transcripts of myostatin in response to muscle exercise through different types of muscle loading [e.g. acute and short-term swimming training, chronic wheel running (Matsakas et al. 2005), muscle loading through treadmill running (Wehling et al. 2000) or isometric resistance training after atrophy induced by hindlimb unloading (Haddad et al. 2006)]. Although several studies have been conducted to verify the effect of muscle loading on myostatin expression, findings reported in literature are somewhat contradictory/ there is still some controversy among researchers. In humans both decreasing and increasing levels of myostatin mRNA and protein have been observed, as one bout of resistance exercise as well as heavy resistance training for several weeks have been reported to be followed by a reduced myostatin mRNA expression (Roth et al. 2003; Kim et al. 2005) and a decreased concentration of circulating myostatin (Walker et al. 2004). However, in other studies it has been reported that a heavy resistance training in healthy individuals for twelve weeks is accompanied by an increase of myostatin mRNA and protein expression and subsequent increase in serum level (Willoughby 2004). Such observations indicate that the up- and/or down-regulation of the myostatin molecule is influenced by several external factors, including training or disuse, although in only one study (Walker et al. 2004) has a link between changes in myostatin expression pattern and muscle mass increase been found. With the aim of assessing whether heavy training might impact myostatin expression and whether distinct changes might occur in relation to fibre type composition, we chose to study the gastrocnemius muscle, which is heavily involved in swimming and is composed of fast-glycolytic (superficial portion) and slow-oxidative (deep portion) fibres. On the basis of our previous experience (Matsakas et al. 2005), we expected a significant decrease in myostatin mRNA expression, especially after intense and long lasting training. Our starting premise was that the changes in myostatin mRNA expression induced by training would be different in fast and slow muscle fibres since the basal level of this molecule is different in the two muscle types and the muscle response to both strength and endurance training protocols is fibre type-dependent. Indeed, our results showed that in control conditions myostatin expression was higher in white than in red fibres and its expression was significantly reduced following intense and long lasting training, while brief exercise did not significantly modify its mRNA level (Matsakas et al. 2006). These results suggest that swimming training induces myostatin suppression in order to increase satellite cell turnover (proliferation), supplying nuclei to enlarging fibres or repairing damaged fibres due to the intense effort. This hypothesis is supported by the observation that the cells positive to Pax7, a satellite cell marker, are increased after an intensive period of training (Patruno et al. unpublished results). Moreover, our study was the first to provide evidence of a myostatin mRNA increase in the heart after long-term endurance training, thus supporting the view that myostatin also plays a role in the regulation of cardiac muscle (Matsakas et al. 2006). Cook et al. (2002) reported an up-regulation of myostatin in hypertrophic hearts of transgenic mice, and Shyu et al. (2005) proposed that myostatin represents a chalone of the insulin-like growth factor-I pathway in the cardiac muscle. However, the biological significance of these findings has yet to be elucidated. Another plausible hypothesis is that higher myostatin mRNA levels in the heart of trained rats might be linked to cardiac tissue damage, for example stress-induced infarctions (Sharma et al. 1999).

We are recently using protocols of transcutaneous neuromuscular electrical stimulation (NMES), which exerts significant effects on the skeletal muscle phenotype and function by increasing muscle mass, and its force and exercise capacity. NMES is used as a tool for muscle strength training in athletes and subjects undergoing rehabilitation. The basic principle underlying NMES is the application of electrical stimuli as intermittent trains through surface electrodes followed by activation of motor axons and their terminal branches over sensory axons. In NMES, intermittent train frequencies range between 20 and 50Hz and the stimulation regimens have usually been categorized in low (LF, <40Hz) or high frequencies (HF, >40Hz). Both programs have been found to increase muscle maximal voluntary strength (MVC) and/or endurance. In a research project in which we investigated the possible clinical application of NMES in the treatment of sarcopenia of ageing and the prevention of secondary atrophy of clinical relevance we also examined putative cellular damage caused by NMES and the growth factors effected by this treatment. Preliminary results obtained in healthy men undergoing LF stimulations indicated an increasing level of mstn, although in fast muscles only (Caliaro et al. unpublished observations). This confirms that when muscle mass enlargement occurs in "stressful" conditions (as is the case in long-term endurance training in rat hearts), the expression of this molecule must be modified (usually it increases) but only in specific group of muscles. Future research will indicate the potential of this technique in association with mstn or TGF-β blockade in the attempt to increase the regenerative capabilities of skeletal muscle.

5. Conclusive Remarks

Developmental regulatory genes follow evolutionary changes in a stochastic and/or in a deterministic manner (in concert); it is therefore very difficult to predict/identify their role(s). Moreover, gene duplication has enhanced the complexity involved in explaining the ancient role of a primordial TGF-β factor. Indeed, "primitive" animals appear to have had just one or two components of the TGF-β superfamily but, since the exceptions found in this research field are quite numerous, our predictive capacity is greatly compromised. However, one fascinating process can be explained: common developmental gene systems may function in a completely diverse environment (even in adult mechanisms), thus leading to unrelated morphogenetic pathways. The BMP pathway is functionally conserved in patterning basic body axes in Protostomes as well as in Deuterostomes; nevertheless, these master regulator genes have produced divergent ontogenies.

As explained earlier in this chapter, although the numerous factors belonging to the TGF-β that have been identified control many phases of development and regenerative mechanisms, much has yet to be discovered. The molecule myostatin, found in mammals, may represent an "evolutionary BMP committed molecule" towards a muscle check point system. Thanks to the few changes that have occurred to the ancestral TGF-β gene (and its receptor), we can still trace the evolutionary history of myostatin through the animal kingdom, observing its expression in many tissues other than muscle (including regenerating ones). This applies to invertebrates, amphiouxus and fish/bird mstn homologs, in which this molecule is expressed in several tissues and organs. At a certain point during mammalian evolution, mstn became "specialized" for muscle, but this link which connects genes with the developmental process and morphological evolution, is both intriguing and still evasive.

Overall, both TGF-β and myostatin appear valuable molecules to study, as our improved understanding of them may enable us to boost the regenerative capability of skeletal muscle; recent reports, moreover, have examined the link between these factors and regeneration, highlighting the fact that, where mstn is absent (blocked in different ways), muscle regeneration in muscle is enhanced and muscle fibrosis is reduced (Wagner et al. 2005; Li et al. 2008). The authors of another recent report showed that i) TGF-β is essential for muscle regeneration in both young and adult mice muscles and ii) indicated that TGF-β, but not mstn, is the main age-specific local inhibitor in the skeletal muscle niche, since a change in the balance between ligand/receptor and downstream genes interferes with muscle stem cells renewal and, therefore, with a "high-quality" regeneration (Carlson et al. 2008).

It is therefore plausible that the clue to morphological and functional changes occurring in different phyla lies in the modulation of the cascade of signaling factors, within a developmental pathway (performed by gradients, antagonist balance, miRNAs influence or other unknown mechanisms), and not in the mere mutations of single nucleic acid molecules. If this is so, will be far more simple to understand why invertebrates and mammals share so many similar biochemical pathways.

"Biodiversity is a wonderful phenomenon. So-called misleading genetic mechanisms can create marvellous individuals. I write this with Luca, my Down's syndrome nephew, in mind"

M.P.

Acknowledgments

I wish to thank my colleagues Prof. G. Radaelli and Dr. A.Matsakas for reading the manuscript and improving upon the discussion and Dr. Elena Negrato for the drawings in Figures 3 and 4. I am also very grateful to all the researcher teams that I collaborated with in the past (Prof. M. D. Candia Carnevali, Prof. M. Thorndyke and Prof. A. Graham, in particular). My thanks also go to the colleagues with whom I am currently working especially Dr. L. Maccatrozzo, Prof. F. Mascarello and Prof. C. Reggiani.

References

Acosta J, Carpio Y, Borroto I, González O, Estrada MP (2005) Myostatin gene silenced by RNAi show a zebrafish giant phenotype. *J. Biotechnol.* 119:324-31

Alexander RM (1969) Orientation of muscle fibres in the myomers of fishes. *J. Mar. Biol. Assoc. UK* 49:263–290

Amali AA, Lin CJ, Chen YH, Wang WL, Gong HY, Lee CY, Ko YL, Lu JK, Her GM, Chen TT, Wu JL (2004) Up-regulation of muscle-specific transcription factors during embryonic somitogenesis of zebraWsh (*Danio rerio*) by knock-down of myostatin-1, *Dev. Dyn.* 229, 847–856.

Amthor H, Nicholas G, McKinnell I, Kemp CF, Sharma M, Kambadur R, Patel K (2004) Follistatin complexes Myostatin and antagonises Myostatin-mediated inhibition of myogenesis. *Dev. Biol* .270:19-30

Amthor H, Otto A, Macharia R, McKinnell I, Patel K (2006) Myostatin imposes reversible quiescence on embryonic muscle precursors. *Dev. Dyn.* 235:672-80

Angerer LM, Oleksyn, Logan CY, McClay DR, Dale L, Angerer RC (2000) A BMP pathway regulates cell fate allocation along the sea urchin animal-vegetal embryonic axis. *Development* 127, 1105-1114

Artaza JN, Bhasin S, Mallidis C, Taylor W, Ma K, Gonzalez-Cadavid NF (2002) Endogenous expression and localization of myostatin and its relation to myosin heavy chain distribution in C2C12 skeletal muscle cells. *J. Cell. Physiol* .190:170-179

Baguña J, Saló E, Romero R, Garcia-Fernàndez J, Bueno D, Muñoz-Mármol AM, Bayascas-Ramirez JR, Casali A (1994) Regeneration and pattern formation in planarians: cells, molecules and genes. *Zool. Sci* .11, 781-795

Bannister R, McGonnell IM, Graham A, Thorndyke MC, Beesley PW (2005) Afuni, a novel transforming growth factor-beta gene is involved in arm regeneration by the brittle star Amphiura filiformis. *Dev. Genes Evol.* 215:393-401

Bannister R, McGonnell IM, Graham A, Thorndyke MC, Beesley PW (2008) Coelomic expression of a novel bone morphogenetic protein in regenerating arms of the brittle star Amphiura filiformis. *Dev. Genes Evol.* 218:33-38

Bayascas JR, Castillo E, Muñoz-Mármol AM, Saló E (1997) Planarian Hox genes: novel patterns of expression during regeneration. *Development* 124:141-148

Bogdanovich, S., Krag T.O., Barton E.R., Morris, L.D., Whittemore, L.A., Ahima, R.S., Khurana, T.S., (2002) Functional improvement of dystrophic muscle by myostatin blockade. *Nature* 420, 418-421

Bollner T, Howalt S, Thorndyke MC, Beesley PW (1995) Regeneration and post-metamorphic development of the central nervous system in the prtochordate *Ciona intestinalis*: a study with monoclonal antibodies. *Cell Tissue Res.* 279, 421-432

Bone Q, Marshall NB, Blaxter JHS (1995) Biology of fishes. Chapman and Hall London Weinheim New York, pp 1–328

Brockes JP (1997) Amphibian limb regeneration: rebuilding a complete structure. *Science* 276, 82-87

Buckingham, M., Bajard, L., Chang, T., Daubas, P., Hadchouel, J., Meilhac, S., Montarras, D., Rocancourt, D., and Relaix, F. (2003) The formation of skeletal muscle: from somite to limb. *J. Anat* .202:59-68.

Buckingham, M. (2006) Myogenic progenitor cells and skeletal myogenesis in vertebrates. *Curr Opin Genet Dev* 16:525-532

Burnett, AL (1966) A model of growth and cellular differentiation in *Hydra Amer. Nat.* 100:165-189

Candia Carnevali MD, Bonasoro F, Lucca E, Thorndyke MC (1995) Pattern of cell proliferation in the early stages of arm regeneration in the feather star *Antedon mediterranea. J. Exp. Zool.* 272:464-474

Candia Carnevali MD, Bonasoro F, Biale A (1997) Pattern of bromodeoxyuridine incorporation in the advanced stages of arm regeneration in the feather star *Antedon mediterranea. Cell Tissue Res.* 289:363-374

Candia Carnevali MD, Bonasoro F, Patruno M, Thorndyke MC (1998) Cellular and molecular mechanisms of arm regeneration in crinoid echinoderms: the potential of arm explant. *Dev. Genes Evol* .208:421-430

Candia Carnevali MD, Bonasoro F, Welsch U, Thorndyke MC (1998b). Arm regeneration and growth factors in Crinoids. In: Mooi R, Telford M (eds) Echinoderms: San Francisco. Balkema, Rotterdam, pp. 145-150

Candia Carnevali MD, Bonasoro F, Patruno M, Thorndyke MC, Galassi S (2000) PCB-induced abnormal regeneration in crinoid echinoderms. *Mar. Ecol. Proc. Ser.* 215:155-167

Cannata SM, Bernardini S, Filoni S (1992) Regenerative responses in cultured hindlimb stumps of larval *Xenopus laevis. J. Exp. Zool.* 262:446-453

Carlson ME, Hsu M, Conboy IM (2008) Imbalance between pSmad3 and Notch induces CDK inhibitors in old muscle stem cells. *Nature* 454:528-532

Collins CA (2006) Satellite cell self-renewal. *Curr Opin Pharmacol* 6:1-6

Connes R (1974) L'origine des cellules blastogénétiques chez *Suberitus domoncula*: L'euilibre choanocytes-archéocytes chez les les spongiaires *Ann .Sci. Nat. Zool* .16:111-118

Cook SA, Matsui T, Li L and Bosenzweig A (2002) Transcriptional effects of chronic Akt activation in the heart. *J. Biol. Chem* .277:22528-22533

Couso JP, Bate M, Martinez-Arias Al (1993) A wingless-dependent polar coordinate system in *Drosophila mealanogaster. Science* 259:484-489

Davidson EH, Cameron RA, Ransick A (1998) Specification of cell fate in the sea urchin embryo: summary and some proposed mechanisms. *Development* 125:3269-3290

de Jonge HW, van der Wiel CW, Eizema K, Weijs WA, Everts ME (2006) Presence of SERCA and calcineurin during fetal development of porcine skeletal muscle. *J. Histochem. Cytochem.* 54:641-648

De Robertis EM and Sasay Y (1996) A common plan for dorsoventral patterning in Bilateria. *Nature* 380:37-40

Draeger A, Weeds AG, Fitzsimons RB (1987) Primary, secondary and tertiary myotubes in developing skeletal muscle: a new approach to the analysis of human myogenesis. *J. Neurol. Sci.* 81:19-43

Dhawan J and Rando TA (2005) Stem cells in postnatal myogenesis: molecular mechanisms of satellite cell quiescence, activation and replenishment. *Trends Cell Biol* .15:666-673

Feng and Derynck, (2005) Specificity and versatility in TGF signaling through Smads. *Annu Rev. Cell Dev. Biol* .21:659-693

Ferretti P and Gèraudie J (1998) Cellular and molecular basis of regeneration: from invertebrates to humans. Chichester: John Wiley and Sons Ltd.

French V, Bryant PJ, Bryant SV (1976) Pattern regulation in epimorphic fields. *Science* 193:969-980

Forbes D, Jackman M, Bishop A, Thomas M, Kambadur R, Sharma M (2006) Myostatin auto-regulates its expression by feedback loop through Smad7 dependent mechanism. *J. Cell Physiol.* 206: 264-272

Funkenstein B and Rebhan Y (2007) Expression, purification, renaturation and activation of Wsh myostatin expressed in *Escherichia coli*: Facilitation of refolding and activity inhibition by myostatin prodomain. *Protein Expr. Purif* 54:54-65

Garikipati DK, Gahr SA, Roalson EH, Rodgers BD (2007) Characterization of rainbow trout myostatin-2 genes (rtMSTN-2a and -2b): genomic organization, differential expression, and pseudogenization. *Endocrinology* 148:2106-2115

Gentry and Nash (1990) The pro domain of pre-pro-transforming growth factor beta 1 when independently expressed is a functional binding protein for the mature growth factor. *Biochemistry* 29:6851-6857

Goss RJ (1969) Principle of regeneration. New York: Academic Press

Goss RJ (1987) Why mammals don't regenerate-or do they? *News Physiol Rev* 2:112-115

Gros J, Manceau M, Thome V, Marcelle C (2005) A common somitic origin for embryonic muscle progenitors and satellite cells. *Nature* 435:954-958

Haddad F, Adams GR, Bodell PW and Baldwin KM (2006) Isometric resistance exercise fails to counteract skeletal muscle atrophy processes during the initial stages of unloading. *J Appl. Physiol* .100:433-441.

Herpin A, Lelong C, Becker T, Rosa F, Favrel P, Cunningham C (2005) Structural and functional evidence for a singular repertoire of BMP receptor signal transducing proteins in the lophotrochozoan Crassostrea gigas suggests a shared ancestral BMP/activin pathway. *FEBS J.* 272:3424-3440

Herpin A, Lelong C, Becker T, Favrel P, Cunningham C (2007) A tolloid homologue from the Pacific oyster Crassostrea gigas. *Gene Expr. Patterns* 7:700-708

Helterline DL, Garikipati D, Stenkamp DL, Rodgers BD (2007) Embryonic and tissue-specific regulation of myostatin-1 and -2 gene expression in zebrafish. *Gen. Comp. Endocrinol.* 151:90-97

Hoh JFY and Hughes S (1988) Myogenic and neurogenic regulation of myosin gene expression in cat jaw-closing muscles regenerating in fast and slow limb muscle beds. *J. Muscle Res. Cell Motil.* 9:59-72

Holland LZ and Holland ND (1999) Chordate origins of the vertebrate central nervous system. *Curr. Opin. Neur* .9:596-602

Hwang SPL, Chen CA, Chen CP (1999) Sea urchin TgBMP2/4 gene encoding a bone morphogenetic protein closely related to vertebrate BMP2 and BMP4 with maximal expression at the later stages of embryonic development. *Biochem. Biophys. Res. Comm.* 258:457-463

Joulia-Ekaza D, and Cabello G (2006) Myostatin regulation of muscle development: molecular basis, natural mutations, physiopathological aspects. *Exp. Cell Res.* 312:2401-2414

Kambadur R, Bishop A, Salerno MS, McCroskery S, Sharma M (2004) Role of myostatin in muscle growth. In: Te Pas, M.F.W., Everts, M.E. and Haagsmann, H.P. (ed) Muscle Development of Livestock Animals: Physiology, Genetics and Meat Quality. CABI Publishing, Wallingoford, Oxfordshire, pp 297-316

Kaminski HJ, al-Hakim M, Leigh RJ, Katirji MB, Ruff RL. (1992) Extraocular muscles are spared in advanced Duchenne dystrophy. *Ann. Neurol.* 32:586-588

Kim HW, Mykles DL, Goetz FW, Roberts SB (2004) Characterization of a myostatin-like gene from the bay scallop, *Argopecten irradians. Biochim. Biophys. Acta* 1679:174–179

Kim JS, Cross JM and Bamman MM (2005) Impact of resistance loading on myostatin expression and cell cycle regulation in young and older men and women. *Am. J. Physiol.* 288:E1110-E1119

Kim YS, Bobbili NK, Paek KS, Jin HJ (2006) Production of a monoclonal anti-myostatin antibody and the effects of in ovo administration of the antibody on posthatch broiler growth and muscle mass. *Poult. Sci.* 85:1062-1071

Kingsley DM (1994) The TGF-β superfamily: new members, new receptors, and new genetic tests of function in different organisms. *Genes and Dev.* 8:133-146

Kocabas AM, Kucuktas H, Dunham RA, Liu Z (2002) Molecular characterization and differential expression of the myostatin gene in channel catfish (*Ictalurus punctatus*). *Biochim. Biophys. Acta* 1575:99-107

Kocamis H, Kirkpatrick-Keller DC, Richter J, Killefer J. (1999a) The ontogeny of myostatin, follistatin and activin-B mRNA expression during chicken embryonic development. *Growth. Dev .Aging* .63:143-150

Kocamis H, Yeni YN, Kirkpatrick-Keller DC, Killefer J. (1999b) Postnatal growth of broilers in response to in ovo administration of chicken growth hormone. *Poult. Sci* .78:1219-1226

Lefaucheur L, Edom F, Ecolan P, Butler-Browne GS. (1995) Pattern of muscle fiber type formation in the pig. *Dev. Dyn.* 203:27-41

Lo PC and Frasch M (1999) Sequence and expression of myoglianin, a novel Drosophila gene of the TGF-beta superfamily. *Mech. Dev.* 86:171–175

Li ZB, Kollias HD, Wagner KR (2008) Myostatin directly regulates skeletal muscle fibrosis. *J. Biol .Chem* .283:19371-19378

Lindholm D, Castren E, Kiefer R, Zafra F, Thoenen H (1992) Transforming growth factor-ß in the rat brain: increase after injury and inhibition of astrocyte proliferation. *J. Cell Biol.* 117:395-400

Loosli F, Kmita-Cunisse M, Gehring WJ (1996) Isolation of a Pax-6 homolog from the ribbonworm *Lineus sanguineus. PNAS* 93:2658-2663

Maccatrozzo L, Bargelloni L, Cardazzo B, Rizzo G, Patarnello T (2001a) A novel second myostatin gene is present in teleost fish. *FEBS Lett* 509:36-40

Maccatrozzo L, Bargelloni L, Radaelli G, Mascarello F, Patarnello T (2001b) Characterization of the myostatin gene in the gilthead sea bream (*Sparus aurata*): sequence, genomic structure, and expression pattern. *Mar. Biotechnol.* 3:224-230

Manceau M, Gros J, Savage K, Thomé V, McPherron A, Paterson B, Marcelle C (2008) Myostatin promotes the terminal differentiation of embryonic muscle progenitors. *Genes Dev.* 22:668-81

Martello G, Zacchigna L, Inui M, Montagner M, Adorno M, Mamidi A, Morsut L, Soligo S, Tran U, Dupont S, Cordenonsi M, Wessely O, Piccolo S (2007) MicroRNA control of Nodal signalling. *Nature* 449:183-188

Martinello and Patruno (2008) Embryonic chick cocultures of neuronal and muscle cells. *Neurol .Res.* 30:179-82.

Martin, P. D'Souza D, Martin J, Grose R, Cooper L, Maki R, McKercher SR (2003) Wound healing in the PU.1 null mouse tissue repair is not dependent on inflammatory cells. *Curr. Biol.* 13:1122–1128

Massagué J (1998) TGF-β signal transduction. *Annu. Rev. Biochem.* 67:753-791

Massagué J (2000) How cells read TGF-ß signals. *Nature Rev. Mol. Cell Biol.* 1:169-178

Massague J and Wotton D (2000) Transcriptional control by the TGF-beta / Smad signaling system. *EMBO J .*19:1745–1754

Mascarello F, Stecchini ML, Rowlerson A, Ballocchi E (1992) Tertiary myotubes in postnatal growing pig muscle detected by their myosin isoform composition. *J. Anim. Sci.* 70:1806-1813

Mascarello F, Rowlerson A, Radaelli G, Scapolo PA, Veggetti A (1995) Differentiation and growth of muscle in *Sparus aurata* (L.): I. Myosin expression and organisation of fibre types in lateral muscle from hatching to adult. *J. Muscle Res. Cell Motil .*16:213–222

Matsakas A and Diel P (2005). The growth factor myostatin, a key regulator in skeletal muscle growth and homeostasis. *Int. J. Sports Med .*26:83-89

Matsakas A, Friedel A, Hertrampf T and Diel P (2005). Short-term endurance training results in a muscle-specific decrease of myostatin mRNA content in the rat muscle. *Acta Physiol. Scand* 183:299-307

Matsakas A, Bozzo C, Cacciani N, Caliaro F, Reggiani C, Mascarello F, Patruno M (2006) Effect of swimming on myostatin expression in white and red gastrocnemius muscle and in cardiac muscle of rat. *Exp. Physiol.* 91:983-994

Mauro A (1961) Satellite cell of skeletal muscle fibers. *J Biophys Biochem Cytol* 9:493-495

McCroskery S, Thomas M, Maxwell L, Sharma M, Kambadur R (2003) Myostatin negatively regulates satellites cell activation and self-renewal. *J. Cell Biol.* 162:1135-1147

McFarlane C, Langley B, Thomas M, Hennebry A, Plummer E, Nicholas G, McMahon C, Sharma M, Kambadur R (2005) Proteolytic processing of myostatin is auto-regulated during myogenesis. *Dev. Biol.* 283:58-69

McPherron AC, Lawler AM, Lee S-J (1997) Regulation of skeletal muscle mass in mice by a new TGF-β superfamily member. *Nature* 387:83-90

McPherron AC and Lee S-J (1997) Double muscling in cattle due to mutations in the myostatin gene. *Proc. Natl. Acad Sci .*USA 94:12457-12461

McPherron AC, Lawler AM, Lee SJ (1999) Regulation of anterior/posterior patterning of the axial skeleton by growth/differentiation factor 11. *Nat. Genet*.22:260-264

Mita K Fujiwara S (2007) Nodal regulates neural tube formation in the Ciona intestinalis embryo. *Dev. Genes Evol.* 217:593-601

Molina MD, Saló E, Cebrià F (2007) The BMP pathway is essential for re-specification and maintenance of the dorsoventral axis in regenerating and intact planarians. *Dev. Biol.* 311:79-94

Moss C, Hunter AJ, Thorndyke MC (1998) Patterns of bromodeoxyuridine incorporation and neuropeptide immunoreactivity in the regenerating arm of the starfish *Asterias rubens. Phil. Trans. Royal. Soc. U.K.* 353:421-436

Muneoka K and Sasson D (1992) Molecular aspects of regeneration in developing vertebrates limbs. *Dev. Biol*.152:37-49

Newfeld SJ, Wisotzkey RG, Kumar S (1999) Molecular evolution of developmental pathway: phylogenetic analyses of transforming growth factor- β family ligands, receptors and smad signal transducers. *Genetics* 152:783-795

Noble NA, Harper JR, Border WA (1992) In vivo mechanism of TGF-β and extracellular matrix. *Prog. Growth. Factor Res*.4:369-382

Nodder S and Martin P (1997) Wound healing in embryos: a review. *Anat. Embryol.* 195:215-228

Noden DM and Francis-West P (2006) The differentiation and morphogenesis of craniofacial muscles. *Dev. Dynam*.235:1194-1218

O'Kane S and Ferguson MWJ (1997) Transforming growth factor βs and wound healing. *Int J Biochem. Cell Biol.* 29:63-78

Orii and Watanabe (2007) Bone morphogenetic protein is required for dorso-ventral patterning in the planarian Dugesia japonica. *Dev. Growth Differ.* 49:345-349

Østbye TK, Falck Galloway T, Nielsen C, Gabestad I, Bardal T, Andersen Ø (2001) The two myostatin genes of Atlantic salmon (*Salmo salar*) are expressed in a variety of tissues. *Eur. J. Biochem.* 268:5249-5257

Padgett RW and Reiss M (2007) TGFβ superfamily signalling: notes from the desert. *Development* 134:3565-3569

Parker MH, Seale P, Rudnicki MA (2003) Looking back to the embryo: defining transcriptional networks in adult myogenesis. *Nat. Rev. Genetic.* 4:497-507

Patruno M, Radaelli G, Mascarello F, Candia Carnevali MD (1998) Muscle growth in response to changing demands in the teleost *Sparus aurata* (L) during development from hatching to juvenile. *Anat. Embryol.* 198:487-504

Patruno M, Thorndyke MC, Candia Carnevali MD, Bonasoro F, Beesley PW. (2001) Growth factors, heat-shock proteins and regeneration in echinoderms. *J. Exp. Biol.* 204:843-848

Patruno M, Smertenko A, Candia Carnevali MD, Bonasoro F, Beesley PW, Thorndyke MC (2002) Expression of transforming growth factor beta-like molecules in normal and regenerating arms of the crinoid Antedon mediterranea: immunocytochemical and biochemical evidence. *Proc. Biol. Sci*.269:1741-1747

Patruno M, McGonnell I, Graham A, Beesley P, Candia Carnevali MD, Thorndyke M (2003) Anbmp2/4 is a new member of the transforming growth factor-beta superfamily isolated from a crinoid and involved in regeneration. *Proc. Biol. Sci*.270:1341-1347

Patruno, M., Maccatrozzo, L., Funkenstein, B., and Radaelli, G. (2006) Cloning and expression of insulin-like growth factor I and II in the Shi drum (*Umbrina cirrosa*). *Comp. Biochem. Physiol. B* 144:137-151

Patruno M, Caliaro F, Maccatrozzo L, Sacchetto R, Martinello T, Toniolo L, Reggiani C, Mascarello F (2008) Myostatin shows a specific expression pattern in pig skeletal and extraocular muscles during pre-natal and post-natal growth. *Differentiation* 76:168-181

Patruno M, Sivieri S, Poltronieri C, Sacchetto R, Maccatrozzo L, Martinello T, Funkenstein B, Radaelli G. (2008) Real-time polymerase chain reaction, in situ hybridization and immunohistochemical localization of insulin-like growth factor-I and myostatin during development of Dicentrarchus labrax (Pisces: Osteichthyes). *Cell Tissue Res.* 331:643-58

Picard B, Lefaucheur L, Berri C, Duclos MJ (2002) Muscle fibre ontogenesis in farm animal species. *Reprod Nutr. Dev.* 42:415-431

Ponce MR, Micol JL, Peterson KJ, Davidson EH (1999) Molecular characterization and phylogenetic analysis of SPBMP5-7, a new member of the TGF-β superfamily expressed in sea urchin embryos. *Mol. Biol. Evol.* 16:634-645

Rando TA (2006) Stem cells, ageing and the quest for immortality. *Nature* 441:1080-1086

Radaelli G, Rowlerson A, Mascarello F, Patruno M, Funkenstein B (2003) Myostatin precursor is present in several tissues in teleost fish: a comparative immunolocalization study. *Cell Tiss. Res.* 311:239-250

Raftery LA and Sutherland D J (1999) TFGβ family signal transduction in *Drosophila* development: from MA to Smads. *Dev. Biol.* 210:251-268

Reber-Müller S, Streitwolf-Engel R, Yanze N, Schmid V, Stierwald M, Erb M, Seipel K (2006) BMP2/4 and BMP5-8 in jellyfish development and transdifferentiation. *Int. J. Dev. Biol.* 50:377-384

Reinhardt B, Broun M, Blitz IL, Bode HR. (2004) HyBMP5-8b, a BMP5-8 orthologue, acts during axial patterning and tentacle formation in hydra *Dev. Biol.* 267:43-59

Rentzsch F, Guder C, Vocke D, Hobmayer B, Holstein TW. (2007) An ancient chordin-like gene in organizer formation of Hydra. *Proc. Natl. Acad. Sci. USA.* 104:3249-3254

Rios, R., Carneiro, I., Arce, V.M., Devesa, J. (2002) Myostatin is an inhibitor of myogenic differentiation. *Am. J. Physiol. Cell Physiol.* 282:C993-999

Roberts SB, Goetz FW (2001) Differential skeletal muscle expression of myostatin across teleost species, and the isolation of multiple myostatin isoforms. *FEBS Lett.* 491:212-216

Roth SM, Martel GF, Ferrell RE, Metter EJ, Hurley BF and Rogers MA (2003). Myostatin gene expression is reduced in humans with heavy-resistance strength training: a brief communication. *Exp. Biol. Med.* 228:706-709

Rowlerson A, Mascarello F, Radaelli G, Veggetti A. (1995) Differentiation and growth of muscle in the fish Sparus aurata (L): II. Hyperplastic and hypertrophic growth of lateral muscle from hatching to adult. *J. Muscle Res. Cell Motil.* 16:223-236

Rowlerson A, Veggetti A (2001) Cellular mechanisms of post-embryonic muscle growth in aquaculture species. In: Johnston IA (ed) Fish Physiology, vol 18. Academic Press, London

Saharinen J, Hyytianen M, Taipale J, Keski-Oja J (1999) Latent transforming factor-b binding proteins (LTBPs) structural extracellular matrix proteins for targeting TGF-β action. *Cytokine and Growth Factor Reviews* 10:99-117

Sariola H and Saarma M 2003 Novel functions and signalling pathways for GDNF. *J. Cell Sci* .116:3855-3862

Schmierer and Hill, (2007) TGFβ-SMAD signal transduction: molecular specificity and functional flexibility. *Nat. Rev. Mol .Cell Biol.* 8:970-982

Schmidt V and Adler H (1984) Isolated, mononucleated, striated muscle can undergo pluripotent transdifferentiation and form a complex regenerate. *Cell* 38:801-809

Schuelke M, Wagner KR, Stolz LE, Hübner C, Riebel T, Kömen W, Braun T, Tobin JF, Lee SJ. (2004) Myostatin mutation associated with gross muscle hypertrophy in a child. *N. Engl. J. Med* .350:2682-8

Shah M, Foreman DM and Ferguson MWJ, (1992) Control of scarring in adult wounds by neutralising antibody to transforming growth factor beta. *Lancet*. 339:213-214

Sharma M, Kambadur R, Matthews KG, Somers WG, Devlin GP, Conaglen JV, Fowke PJ and Bass JJ (1999). Myostatin, a transforming growth factor-beta superfamily member, is expressed in heart muscle and is upregulated in cardiomyocytes after infarct. *J. Cell Physiol* .180:1-9

Shyu KG, Ko WH, Yang WS, Wang BW and Kuan P (2005). Insulin-like growth factor-I mediates stretch-induced upregulation of myostatin expression in neonatal rat cardiomyocytes. *Cardiovasc Res.* 68:405-414

Singer M and Craven L (1948) The growth and morphogenesis of the regenerating forelimb of adult *Triturus* following denervation at various stages of development. *J. Exp. Zool.* 108:279-308

Stenzel P, Angerer LM, Smith BJ, Angerer RC, Vale, WW (1994) The univin gene encodes a member of the transforming growth factor-ß superfamily with restricted expression in the sea urchin embryo. *Dev .Biol* .166:149-158

Sporn MB and Roberts AB (1992) Transforming growth factor-ß: recent progress and new challenges. *J. Cell Biol.* 119:1017-1021

Tamaki T, Akatsuka A, Ando K, Nakamura Y, Matsuzawa H, Hotta T, Roy RR, Edgerton VR (2002) Identification of myogenic-endothelial progenitor cells in the interstitial spaces of skeletal muscle. *J. Cell Biol* .4:571-577

Tamaki T, Akatsuka A, Okada Y, Matsuzaki Y, Okano H, Kimura M (2003) Growth and differentiation potential of main and side-population cells derived from murine skeletal muscle. *Exp. Cell Res.* 291:83-90

Theisen H, Syed A, Nguyen BT, Lukacsovich T, Purcell J, Srivastava GP, Iron D, Gaudenz K, Nie Q, Wan FY, Waterman ML, Marsh JL. (2007) Wingless directly represses DPP morphogen expression via an armadillo/TCF/Brinker complex. *PLoS ONE.* 2(1), e142.

Thomas M, Langley B, Berry C, Sharma M, Kirk S, Bass J, Kambadur R (2000) Myostatin, a negative regulator of muscle growth, functions by inhibiting myoblast proliferation. *J. Biol. Chem.* 275:40235-40243

Thorndyke MC, Patruno M, Chen WC, Beesley PW. (2001a) Stem cells and regeneration in invertebrate Deuterostomes. In "Brain Stem Cells", J.A. Miyan, M. Thorndyke, P.W. Beesley, C.M. Bannister eds. pp.107-119; ISBN: 185996222X. Oxford: Bios Scientific Publishers Ltd (United Kingdom).

Thorndyke MC, Chen WC, Beesley PW, Patruno M. (2001b) Molecular approach to echinoderm regeneration. *Microsc Res. Tech* .55:474-485

Tomoda T, Shirasawa T, Yahagi Y, Ishii K, Takagi H, Furiya Y, Arai K, Mori H, Muramatsu M (1996) Transforming growth factor-beta is a survival factor for neonate cortical neurones: Coincident expression of type I receptors in developing cerebral cortices. *Dev. Biol.* 179:79-90

Toniolo L, Patruno M, Maccatrozzo L, Pellegrino MA, Canepari M, Rossi R, D'Antona G, Bottinelli R, Reggiani C, Mascarello F (2004) Fast fibres in a large animal: fibre types, contractile properties and myosin expression in pig skeletal muscles. *J. Exp. Biol.* 207:1875-1886

Truby P (1985) Separation of wound healing from regeneration in the cockroach leg. *J. Embryol. Exp. Morphol.* 85:177-190

Veggetti A, Mascarello F, Scapolo PA, Rowlerson A, Candia Carnevali MD (1993) Muscle growth and myosin isoform transitions during development of a small teleost fish, *Poecilia reticulata* (Peters) (Atheriniformes, Poeciliidae): a histochemical, immunohistochemical, ultrastructural and morphometric study. *Anat Embryol.* 187:353–361

Wagner KR, Liu X, Chang X, Allen RE (2005) Muscle regeneration in the prolonged absence of myostatin. *Proc. Natl. Acad Sci. USA* 102:2519-2524

Walker KS, Kambadur R, Sharma M and Smith HK (2004). Resistance training alters plasma myostatin but not IGF-1 in healthy men. *Med Sci Sports Exerc* 36:787-93

Wallace H Vertebrate limb regeneration. Chichester and New York: John Wiley, 1981

Wehling M, Cai B and Tidball JG (2000). Modulation of myostatin expression during modified muscle use. *FASEB J.* 14, 103-110.

Willoughby DS (2004). Effects of heavy resistance training on myostatin mRNA and protein expression. *Med. Sci. Sports Exerc.* 36:574-582

Wolpert L (2002) Principle of development. New York: Oxford University Press Inc.

Wozney JM, Rosen V, Celeste AJ, Mitsock LM, Whitters MJ, Kriz RW, Hewick RM, Wang EA (1988) Novel regulators of bone formation: molecular clones and activities. *Science* 242:1528-1534

Xu C, Wu G, Zohar Y, Du SJ, (2003) Analysis of *myostatin* gene structure, expression and function in zebraWsh, *J. Exp. Biol.* 206:4067–4079

Zammit PS, Partridge TA, Yablonka-Reuveni Z (2006a) The Skeletal Muscle Satellite Cell: The stem cell that came in from the cold. *J. Histochem. Cytochem.* 54:1177-1191

Zammit PS, Relaix F, Nagata Y, Ruiz AP, Collins CA, Partridge TA, Beauchamp JR (2006b) Pax7 and myogenic progression in skeletal muscle satellite cells. *J. Cell Sci.* 119:1824-1832

In: Development Gene Expresión Regulation
Editor: Nathan C. Kurzfield

ISBN: 978-60692-794-6
©2009 Nova Science Publishers, Inc.

Chapter VI

Alpha-Foetoprotein: It's all About Timing

De Mees Christelle[a] and Streel Emmanuel[b]

[a]Université Libre de Bruxelles, Institut de Biologie et Médecine Moléculaires,
12 rue des prof. Jeener et Brachet, 6041 Gosselies, Belgium.
[b]Université Libre de Bruxelles, ISM, Campus Erasme , Brussels, Belgium.

Abstract

Alpha-foetoprotein (AFP) is a well known diagnostic biomarker used in medicine to detect foetal developmental anomalies such as neural tube defects or Down's syndrome, or to follow the development of tumors such as hepatocellular carcinomas. However, the role of AFP goes way further than that. AFP is involved at least in rodents in the correct differentiation of the female brain, through its estrogen binding capacity. This chapter present an overview of what is known about the regulation of the *Afp* gene, describes the phenotype of the AFP knock-out (AFP KO) mouse and offers an overview of other mouse models available to study estrogen function.

Being in the right place, at the right time, is a key factor for alpha-foetoprotein (AFP). Firstly, because it is involved in major events occurring during narrow time-windows, such as sexual differentiation of the female brain. Secondly, because AFP is expressed in an onco-foetal way.

AFP: A Brief Introduction

Alpha-foetoprotein is a highly glycosylated protein of 65-70 kDa, massively produced during foetal life by the liver hepatocytes and the visceral endoderm of the yolk sac and to a lesser extent, by the developing intestines, pancreas and kidney [1,2,3]. It is the major foetal serum protein, with a concentration reaching the order of the mg/ml.

The AFP produced by the embryo is spotted in the amniotic fluid but is also able to cross the placental barrier to reach the maternal blood flow, where its titer is routinely used as biomarker to detect developmental anomalies of the foetus. AFP is part of the triple test (now quadruple test) for antenatal Down's syndrome screening, undertaken at 14-22 weeks of each human pregnancy. Abnormally low levels of AFP, along with an altered combination of values for unconjugated estradiol, human chorionic gonadotropin and inhibin A, are indicative of an increased risk of developing this pathology [4]. On the other hand, abnormally high levels of AFP in the maternal serum are indicative of an increased risk for neural tube defects of the foetus such as Spina Bifida or anencephaly [5]. Other foetal pathologies are also associated with aberrant AFP values (for review, see [6]). However, the link between AFP and those pathologies is still, at this date, not understood and additional studies will be needed to find out if it is a causal one or merely a consequence of other deregulations.

Afp transcription is dramatically repressed perinatally, resulting in a 10^{-4} fold reduction in liver AFP mRNA levels within a few weeks after birth [7]. In normal conditions, AFP is only present in trace amounts during adulthood and is then produced only by the liver. Its synthesis can however resume in case of liver pathologies or liver regeneration (cirrhosis, hepatitis, partial hepatectomy) or in case of tumours such as hepatocellular carcinoma, hepatoblastomas, germ cell tumours (embryonic carcinoma and teratocarcinoma) and some pancreatic and renal tumours [8-13]. The exact mechanisms that control AFP gene silencing or expression are still largely not understood. A brief overview of what is known is given below.

AFP Gene Regulation

The *Afp* gene is a member of the albumin gene family along with albumin, alpha-albumin (afamin) and vitamin D binding protein [14]. These four genes have evolved from a common ancestor by a series of duplication events and still share some regulatory elements such as enhancers [15]. In addition to amino acid sequence similarities, members of the albumin family also share a characteristic pattern of disulfide bridges in their polypeptide chains, resulting in a protein structure composed of three domains. They have a nearly identical intron/exon splitting pattern, are positioned near each other on the chromosome and have a common direction of transcription. However, the four genes show different developmental regulations. The expression of albumin and vitamin D binding protein starts in the foetal liver and is maintained during adulthood. The expression of alfa-albumin starts after birth and continues in the adult stage. Finally, the expression of alpha-foetoprotein, as previously described, starts during foetal life and is turned off after birth.

The regulatory regions of the rodent AFP gene have been extensively studied in cultured cells and transgenic mouse models, and contain five distinct elements: a promotor region, three distal enhancers, and a silencer element.

The promotor consists of a genomic fragment of 250 bp upstream of exon 1.

This region contains binding sites for numerous positive factors, including HNF1, FoxA, FTF, HNF6, AP1 (complex formed by the oncoproteins Jun and Fos), NF1 and Nhx2.8 [16-22]. Nhx2.8 is expressed only in embryonic liver and AFP-expressing hepatoma lines [23] and seems to be necessary for the stimulation of the AFP promotor by the remote enhancers through interactions with regulatory protein complexes. HNF1, NF1 and C/EBP all bind a sequence situated at -120 bp from the transcription start site of the *Afp* gene and seem crucial for AFP promoter activity [24-26]. NF1 however seems to exert a dual effect as it is weakly stimulating at low concentrations but suppresses the AFP promotor activity at higher doses [26]. Several factors on the other hand exert a negative action on the AFP promoter, mainly by interfering with HNF1 binding. ATBF1 for example suppresses the activity of both AFP promoter and enhancer, apparently by competing with HNF1 for common binding sites [27]. The heterodimer Ku on the other hand binds a segment able to adopt a peculiar secondary structure, probably a cruciform one, that eliminates the binding sites of HNF1 [28]. Ku is thus probably involved in the maintenance of this transcritpionnaly inactive structure inside the AFP promotor. The mode of action of other transcriptional repressors of the AFP gene is less understood. There is a binding site located at -135 bp of the AFP promoter for the transcriptional repressor COUP-TF. However, its physiological significance in the regulation of AFP transcription is still unknown [29,30]. Also, overexpression of c-Jun in hepatoma cells inhibits AFP promoter activity in a DNA binding-independent manner [31].

Recently a much more serious candidate for AFP gene transcription repression has been discovered [32]. ZBTB20, also named DPZF, HOF or ZNF288 is a transcription factor belonging to the BTB/POZ zinc finger family. It has two isoforms due to alternative translation initiation, both containing an intact N-terminal BTB domain and a C-terminal zinc finger domain [33]. The expression of ZBTB20 is developmentally regulated and displays an inverse correlation with AFP in the postpartum liver, increasing while AFP drastically decreases. Several lines of evidence suggest that ZBTB20 acts as a dominant transcriptional repressor on the *Afp* gene. Firstly, hepatocyte-specific disruption of the ZBTB20 gene in the mouse leads to a dramatic derepression of AFP in the adult liver, with expression levels close to that observed in the foetal liver. Secondly, in vitro studies showed that the forced expression of ZBTB20 in AFP producing cells represses AFP-driven transcriptional activity of reporter genes in a dose-dependent manner. The ZBTB20 responsive region seems to be located within the -162 bp to +27 bp region of the AFP promoter. Thirdly, ZBTB20 binds to the AFP promotor both *in vitro* and *in vivo*, in the adult liver. The most likely scenario for ZBTB20 action is interference with transcriptional activators or recruitment of transcriptional repressors. Further studies will be needed to discover exactly witch ones are involved.

Finally, nuclear hormone receptors modulate AFP gene expression in various ways. Gluticorticoids and retinoic acid have various effects, that are detailed in review [34]. The diversity of their action can be explained by competition for overlapping binding sites of other transcription factors, protein-protein interactions with those factors, and the existence of several nuclear receptors with different transactivation properties.

The AFP promotor has a tissue-specific activity and although displaying substantial activity in hepatoma cells, it is not active in vivo in the absence of a linked enhancer element [35]. Three distant enhancers (E1-E2-E3), each 200-300 bp in length, have been described. E1 and E2 have similar sequences, suggesting that they arose by duplication from a common ancestor [36]. E3 is the most studied element and binds several factors, such as C/EBP, Foxa, ROR and COUP-TF [37-40]. Like the AFP promotor, the enhancers contribute to the tissue-specific expression of the gene [41]. Their additive properties and the exact interactions with the promoter however differs according to species and *in vivo/in vitro* status. In addition, E3 is also active in brain cells [41].

Each enhancer is able to stimulate the heterologous albumin promoter and remains active in the adult liver if it is coupled with this promotor in transgenic constructions [42]. Interestingly, each enhancer is zonally active in the adult liver. E1 and E2 are active in all hepatocytes but their activity is highest in cells surrounding the central vein of a liver lobule, and decreases in cells of the periportal region [41]. E3 on the other hand is exclusively active in a single layer of hepatocytes surrounding the central vein [41]. This restricted activity seems to be based on an active repression occurring during the perinatal period, as the activity of E3 seems dominant over the activity of E2 [43]. As AFP repression in the perinatal liver follows a portal-central gradient, i-e is first seen in cells surrounding the portal triad and last experienced by the cells surrounding the central vein, it rises the interesting possibility that the negative action on E3 could contribute to the postnatal AFP repression [43].

Several mechanisms, in addition to those mentioned above, are likely to be involved in the perinatal gene silencing of AFP. The AFP gene contains a silencer element spanning in the mouse from 250 to 850 bp upstream of the start of transcription. This repressor region is required for AFP post-natal repression in pericentral hepatocytes but is not essential for complete AFP repression in the intermediate zone and periportal hepatocytes. p53 and FoxA both regulate AFP expression through mutually exclusive binding of a sequence centered around -850 bp [44,45]. Fox A has a stimulatory effect that can be abolished by p53: these results are confirmed by the delayed repression of the AFP gene in p53 deficient mice [46]. p53 repression can be amplified by p300. p53 can also act together with p73 to modify the chromatine structure at the AFP locus [47]. Another site, situated at -165 bp, is also involved in AFP gene repression. FoxA is unable to bind this site but seems to inhibit AFP via interactions with other binding proteins [17]. Finally, both ING1b and ING2 can repress AFP through mechanisms that might involve interactions with p53, HNF1 and chromatin remodelling [48].

The differences in the AFP levels among several inbred strains of mice provide further insight into the AFP gene regulation. BALB/cJ mice have AFP levels that are 20 fold higher than the other strains tested [49]. They also have higher steady-state AFP mRNA levels in the adult liver. This continued expression is a recessive trait. The gene modulating AFP repression was called *Afr1*, for alpha-fetoprotein regulator 1. It is a liver-specific repressor that acts in a manner that couples both transcriptional and post-transcriptionnal mechanisms [42, 50, 51]. In addition to *Afp*, *Afr1* regulates *H19*, an imprinted gene also expressed in the foetal liver and silenced after birth. BALB/cJ mice also show abnormal, elevated levels of *H19* in the adult liver. A recent study using positional cloning identified the zinc-fingers and homeoboxes 2 (*Zhx2*) gene as *Afr1* [52]. *Zhx2* is a member of a small gene family that includes *Zhx1* and *Zhx3* [53, 54]. The proteins coded by these three genes are able to form homodimers and heterodimers with each other and with NF-YA [54, 55]. Their predicted structure contains two zinc fingers domains of the C2-H2 type and four homeodomains (five in the case of *Zhx1*). ZHX2 acts as a transcriptional repressor [53] and regulates both *Afp* and *H19*.

The *Zhx2* allele is mutated in the BALB/cJ mouse line. The insertion of an endogenous retroviral element in the first intron disturbs the splicing pattern of the gene, leading to a dramatic reduction of the functional (wild type) mRNA transcripts [52]. However, other actors must be involved in AFP gene silencing, as *Zhx2* is ubiquitely expressed and the *Afr1* phenotype is liver-specific. Interactions with other liver-specific factors may occur. Alternatively, the higher levels of ZHX2 in other organs might be sufficient to induce a correct regulation of the target genes.

Interesting information can also be deduced from human cases of hereditary persistence of AFP expression. These were revealed by the results of the screening tests performed during pregnancy to detect foetal development defects. The persons affected still express AFP in the adult liver, although at a much lesser dose than in the foetal liver. Several mutations have been found in the AFP promotor, each one affecting the binding of HNF1 in a positive way [56, 57, 58]. Only one case remains unexplained and probably involves mutations in other regulatory regions or in trans-acting factors [59].

Finally, transcription of the *Afp* gene generates multiple mRNAs. The main product of AFP gene transcription in foetal liver is a 2,1 kb mRNA. Besides, 1,7, 1,4, and 1 kb mRNAs were detected in foetal and regenerating liver, and in carcinogenesis [60, 61]. The shorter forms (1,4 and 1 kb) are dominant in adult liver [62]. All these forms can be translated, but the exact function of the resulting proteins (if indeed different from the classical AFP, translated from the 2,1 kb form) is still unknown [62]. In addition, a second promotor has been discovered inside the first intron of the AFP gene. Transcription from this promotor generates an mRNA lacking the first exon of the gene [63]. This mRNA is detected in the yolk sac and fetal liver. The protein it encodes (named AFP2), if translated, might have a quite different function than AFP, as AFP2 lacks the secretory peptide signal coded by the first exon of the gene and should therefore be intracellular.

The AFP Knock-Out Mouse

In order to better understand the exact function of AFP, AFP knock-out mice were generated by replacing a genomic fragment in the *Afp* gene extending from exon 1 to intron 3 (KO1, *Afp*^*tm1lbmm* allele) or extending from exon 2 to intron 3 (KO2, *Afp*^*tm2lbmm* allele) by an IRES-LacZ-neo selection cassette [64]. Both constructions generate a null allele, meaning that AFP production is abolished at both the mRNA and protein level.

The difference between the two knock-outs is that KO2 places the lacZ gene under the control of both *Afp* promoters (the main one and the one situated inside the first intron) whereas KO1 destroys the intronic promotor. However, the two knock-outs are of similar use in the study of AFP function as they both effectively destroy the gene. The AFP KO1 mouse line has been recently archived in the EMMA (European Mouse Mutant Archive) database and is thus available to laboratories wanting to use it in their research.

Both invalidations gave rise to viable homozygous animals with no apparent phenotypic anomaly. All studied functions, including embryonic development, liver function and liver regeneration were undistinguishable between AFP KO animals and wild type littermates. The only dysfunction observed lies in the fertility field.

AFP KO female mice are sterile and suffer from anovulation. AFP KO males are fertile. Reciprocal ovary transplantation experiments demonstrated that AFP KO ovaries are functional: AFP KO ovaries transplanted in normal mice were able to ovulate and the transplanted females generated pups from the mutated parental oocytes. AFP KO ovaries also contain follicles at different stages of maturation, including the last Grafiaan follicle stage. However no corpora lutea, indicative of ovulation, could be detected which is in accordance with the smooth exterior aspect of the ovaries and the abnormally low levels of progesterone in the serum. Ovulation in AFP KO mice can be induced by the injection of gonadotropins. However, although ova can be released and fertilized, blastocysts are unable to implant in the uterine horns. These are badly damaged by chronic overstimulation by estrogens, as a result of the absence of corpus luteum and thus the lack of normal cycles of proliferation and regression of the uterus linings.

Anovulation in AFP KO mice is caused by a defect in the HPG axis (hypothalamic-pituitary-gonadal axis) which provides the adequate hormonal environment necessary for ovulation. In the proestrus phase of the sexual cycle, the hypothalamus located in the brain responds to a stimulatory signal of estrogens by the secretion of the GnRH decapeptide which reaches the pituitary, binds to its receptor and triggers the release of the LH and FSH hormones responsible in fine for ovulation. In AFP KO female mice, GnRH neurons are deficient. This conclusion was reached in two complementary studies [65, 66].

At the mRNA level, a microarray analysis of the pituitaries revealed a down-regulation of the GnRH pathway in AFP KO female mice. The expression levels of the GnRH gene, the GnRH receptor gene and several other genes activated by the GnRH receptor (*cFos*, *Egr2*, *Tgfbli4*, *Ptp4a1*) were affected [65]. Furthermore, several other genes previously implicated in female fertility (*Egr1*, *Cish2*, *Ptprf*, *Psa*, *Tkt*) were down-regulated.

At a functional level, no activation of the GnRH neurons could be detected after a stimulatory treatment with both estradiol benzoate and progesterone, or with estradiol benzoate alone. As a consequence, no steroid-induced preovulatory LH surge could be observed in the plasma of the mutant females [66]. This proves that even with the right hormonal priming, GnRH neurons don't function correctly. Furthermore, the neurons controlling the GnRH neurons also fail to behave properly. The Kiss/GPR54 system is a key upstream regulator of the GnRH system: in fact, it is the most potent activator of GnRH neuron firing yet to be discovered [67-69]. It is composed of Kisspeptin coded by the Kiss1 gene and its receptor, the G-protein coupled receptor 54 (GPR54) present among others at the surface at the GnRH neurons. In AFP KO female mice, the amount of kisspeptin-immunoreactive neurons as well as the level of their activity was severely diminished. Furthermore, their distribution was abnormal. Kisspeptin-immunoreactive neurons have a sexually dimorphic distribution pattern in some hypothalamic nuclei, being more numerous in females than in males. In AFP KO females however, the amount of kisspeptin immunoreactive neurons was similar to that observed in males, both in neutral and in stimulatory conditions. This 'defeminization' of the AFP KO female brains could also be observed through other aspects. Other neuronal populations have a sexually dimorphic pattern of distribution. Among these, the tyrosine hydroxylase expressing neurons are more numerous in females than in males in some hypothalamic nuclei such as the AVPV. Again, AFP KO females display a male pattern of distribution of these neurons [70]. Behaviour is also affected. AFP KO female mice fail to exhibit the female typical behaviour of lordosis (posture with raised head and rump, and deflected tail, to facilitate copulation) in the presence of a sexually active male and show an increase of male-typical behaviours such as mounting and intromission like behaviours [70]. However, odour preferences remained intact [71].

The phenotype of the AFP KO mice can be rescued by ante-natal treatment with anti-estrogen. If a pregnant mouse is given anti-estrogen during the last half of her gestation, the born AFP KO pups will show restored fertility when they reach adulthood. The fertility of the pups does not need to be sustained by any treatment in adulthood and several litters can be obtained from them, thereby proving that the cause of anovulation has indeed been solved. Furthermore, normal typical female behaviour, normal distribution of tyrosine hydroxylase expressing neurons and normal gene expression in the hypothalamic-pituitarian-gonadal axis is restored [65,70].

Link between AFP and Fertility: Estrogens

AFP is able, at least in rodents, to inhibit estrogen responsiveness. AFP can bind estrogens (but not androgens) at its C-terminal extremity with a KD of 10^{-8} M, indicating that it can act as an estrogen carrier in the blood [72-74]. AFP also seems to exert anti-estrogenic effects independent of its binding or sequestering properties [75].

Estrogens are known to exert masculinizing effects on the female developing brain in the peri-natal period. Perinatal exposure to estrogens in rodent females results in anovulatory sterility in adulthood, associated with altered gonadotropin production and absence of female typical behaviour [76,77]. It is thus classically assumed that the function of AFP is to shield the developing female brain from estrogens, by sequestrating them in the foetal serum [78]. In the males case, testosterone produced by the foetal testes is free to reach the brain, as AFP is unable to bind androgens. Testosterone is then converted to estrogens by an enzyme called aromatase and both hormones can exert their masculinizing effects. The AFP KO mice support this model and weakens the alternative idea that AFP specifically delivers estrogens to targeted brain cells in order to ensure correct female differentiation [79,80] as AFP KO female mice that developed in an embryonic environment almost depleted in estrogens showed restored fertility in adulthood.

However, this model does not explain why AFP is found inside neurons without being produced there locally, as no AFP mRNA could be detected [81]. An intracellular pool of AFP of unknown function has been found in the cytosol of neurons of both sexes, at all stages of development of the central nervous system, from the postmitotic neuroblast to more differentiated neurons [80]. AFP is not present in the adult mouse brain, though. This observation led to the theory that AFP could deliver estrogens to specific brain cells through receptor-mediated endocytosis. Since there is a difference of several orders of magnitude between the affinity constants for estradiol binding by AFP (KD 10^{-8} M [62]) and by the estrogen receptor (KD10^{-11}M), the subsequent intracytoplasmic dissociation of the AFP/estradiol complex in estrogen receptor containing neurons could liberate the steroid and lead to its immediate binding to estrogen receptors. AFP might thus act as a reservoir of estrogen. Although possible, this mechanism is unlikely to be crucial for the female brain sexual differentiation, at least for the reproduction linked aspects. Firstly, the results obtained with the anti-estrogenic prenatal treatment of the AFP KO mice question this hypothesis, but furthermore there is a mismatch between the areas in which AFP-containing neurons are found, and areas with estrogen-receptor expressing neurons. No AFP immunoreactivity has been found in the diagonal band of Broca, the medial preoptic area, the arcuate nuclei, the ventral premammillary nuclei of the hypothalamus, and the medial and cortical amygdaloid nuclei. However AFP is able, like albumin, to bind other steroids than estrogens as well as endogenous and exogenous substances such as fatty acids, bilirubin and various pharmaceutical agents [82]. AFP could play an active role in neural development through its binding capacity of teratogens and polyunsaturated free fatty acids such as arachidonic, docosahexaenoic and docosatetraenoic acids, all of them important in brain development.

It is also possible that the intracellular uptake of estrogens though endocytosis of an AFP-estrogen complex might be important in other cognitive functions such as learning and memory [83].

There are several useful models beside AFP KO mice for the study of the impact of estrogen on the brain development and differentiation. Among these, the ArKO mouse model is of particular interest [84,85]. ArKO mice are deficient in aromatase activity due to a targeted mutation of the *Cyp19* gene. The aromatase cytochrome P450 catalyses the final step in the biosynthesis of estrogens from C19 steroids [86]. The mutated animals are thus unable to produce estrogens but still have functional estrogen receptors: their genetic deficiency can thus be corrected at the phenotype level by an exogenous administration of the hormone. It is thus possible to study the consequences of the absence of estrogen biosynthesis and cellular uptake early in life and then correct the default to discriminate organizational and activational effects of estradiol on the brain. ArKO mice are born phenotypically normal and reach adulthood [87]. Impairments begin to appear as they grow up: at three weeks of age, their coat seems to be duller than their wild type littermates. At sexual maturity, females present underdeveloped external genitalia, mammary glands, and uteri. Ovaries contain numerous follicles with evidence of antral formation but no corpora lutea could be found, suggesting anovulation. Furthermore, stroma cells are hypertrophied and ovaries contain structures suggestive of atretic follicles. The lordosis behaviour of the female ArKO mice is also severely impaired. Male ArKO mice are fertile early in age, and present an enlargement of the male accessory sex glands because of the increased content of secreted material. However, male ArKO mice develop progressive infertility after four months of age, due to spermatogenesis disruptions [88]. Both sexes show a weight increase of the gonadal fat pads. Both sexes also have increased levels of LH, FSH and testosterone.

Bakker and colleagues proposed a 'time-window' model for estrogen action on female brain differentiation and behaviour after the combined analysis of the two knock-out mouse models, AFP KO and ArKO [83]. In that model, estrogen are not needed in the prenatal period and are harmful for the developing female brain. Estrogen are sequestered by AFP in the blood and their defeminisation action on the brain is blocked. In males, testosterone is able to reach the brain, is transformed into estrogen by the aromatase enzyme and acts along them to defeminize the brain. After birth however, as AFP levels dramatically decrease, estrogen might contribute to the development of the female brain. AFP levels decrease by about 50% in the first 24h after birth, and only trace amounts (about 0,01% of fetal levels) are detected after three weeks after birth in rats. AFP thus no longer plays a significant role when the ovaries begin to secrete estrogens, at about one week after birth. Estrogens are then needed to exert a feminization action between birth and puberty (postnatal days 40-50) on the female brain. Estrogen thus exert dual effects during the prenatal and the postanatal period. The blocking of the prenatal, defeminizing effects, and the correct happening of the postnatal, feminizing effects, are needed to ensure correct female functioning.

Finally, the estrogen receptors (ER) knock-out mouse models could also provide further information about estrogen action, although phenotype rescue is much harder to achieve. Three different models exist: the estrogen receptor α knock-out mouse [89,90], the estrogen receptor β knock-out mouse [91,92], and the knock-out for both receptors [93]. However, things might be complicated by the suspicion of a membrane estrogen receptor and by the combined genomic and non-genomic actions of estrogens (for review see [94]).

Genomic actions group the transcriptional regulation of target genes through direct binding of the estrogen-receptor, or an estrogen-receptor containing protein complex, on regulatory sites inside the promotor. They can be subdivided in classical and non classical estrogen receptor pathways. In the classical pathway, estrogen bind to their intracellular receptors. This binding induces a conformational change that allows the receptors to dimerize and interact with diverse coactivators and corepressor molecules. This complex then binds with high affinity to specific estrogen-response elements (ERE) in the regulatory regions of target genes to either activate or repress them. Non classical estrogen pathways include protein-protein interactions of liganded estrogen receptors with other transcription factors bound to their own response elements such as AP-1, Sp1 and NB-KB [95-98]. They also include ligand-independent estrogen receptor signalling, in witch estrogen receptors are activated through phosphorylation [99]. This phosphorylation results from an altered intracellular kinase and phosphatase activity caused by the upstream activation of second messenger pathways.

Non-genomic actions of estrogens (or MISS, membrane-initiated steroid signalling) are faster. They include the activation of several signal transduction pathways after the binding of estrogens on membrane-bound or intracellular receptors and can also be seen as non-classical estrogen receptor pathways.

The various estrogen knock-out and knock-in mouse models provide insight in the cellular and molecular mechanisms of estradiol action. Genomic actions can be deduced from the phenotype of the estrogen-receptor knock-out mice.

Estrogen receptor alpha knock-out mice appear normal but experience fertility problems in both sexes [89,90]. Females have hypoplastic uteri with diminished size of the stromal, epithelial and myometrial tissue compartments. Ovaries have cystic and hemorrhagic follicles containing few granulosa cells and lack corpus luteum. This could be a consequence of overstimulation of the follicles by the LH hormone that is abnormally elevated in ERKOα mice. Androgen and estrogen secretion is also increased. The antenatal development of the mammary gland is normal but the differentiation of the epithelial ducts is incomplete in adulthood. Finally, ERKOα females do not display lordosis, are aggressive and lack maternal behaviour. The phenotype of the estrogen receptor beta knock-out mice is milder [91,92]. Mice lacking this receptor develop normally and also appear normal. Females are fertile and exhibit normal sexual behavior, but have fewer and smaller litters than their wild type littermates. This is due to a reduced ovarian efficiency. On the other hand, ERKOβ females have normal breast development and lactate normally.

The generation of double estrogen knock-out mice lead to surprising results. Both sexes are infertile. ERKO$\alpha\beta$ females have hypoplastic uteri but furthermore, seem to endure post-natal sex regression in the ovaries [93]. Ovaries contain structures resembling seminiferous tubules of the testis and many differentiation markers of Sertoli cells such as mullerian-inhibiting substance, sulphated glycoprotein 2 and sox9 are expressed. Both receptors seem thus to be needed to insure correct ovarian function.

The non-classical estrogen receptor pathways can be studied in another transgenic mouse model. A single mutation within the first zinc finger of the DNA-binding domain of the estrogen receptor alpha (E207A/G208A, 'AA' mutation) completely abolishes ERE binding and activation of ERE containing reporter genes [96,100]. However, the receptor still has the ability to interact with Jun and can drive the expression of genes containing AP1 response elements. The mutation does not alter the structure of the receptor nor its expression and activity, and does not exert dominant-negative effects in vitro [96,100]. The E207A/G208A mutation was introduced in the mouse genome by targeted insertion (knock-in). Heterozygous females were obtained. They are acyclic and suffer from anovulation, uterine defects, reduced serum progesterone levels and inhibited mammary gland development, which is a way more pronounced phenotype than that observed in heterozygous ERαKO females, that are fertile. This observation stresses the importance of the non-genomic effects of estrogen but also raises the question of an antagonism of the wild type allele by the AA mutation. Heterozygous ER$\alpha^{WT/AA}$ males are fertile and were used to introduce the AA mutation in the ERαKO mice background, thereby generating ER$\alpha^{KO/AA}$ mice lacking ERE-dependent signalling. Physiological roles of the non-classical estrogen pathway could then be deduced from the comparison of the ERαKO and ER$\alpha^{KO/AA}$ phenotypes.

Estrogen exert negative and positive effect on the hypothalamic-pituitiarian axis, and therefore on GnRH and LH secretion. Throughout most of the ovulatory cycle, estrogen exert suppressive effects on GnRH and LH secretion. However, slowly increasing levels of estrogen produced by the developing ovarian follicles, over an extended period of time, imbalance the negative effects and transform estrogen into stimulatory hormones when a certain threshold is reached. Estrogen then exert a positive action and evoke a preovulatory GnRH surge followed by a subsequent LH surge, which triggers ovulation.

Similarly to ERαKO females, ER$\alpha^{KO/AA}$ females do not exhibit an LH surge, spontaneous ovulation, or estrous cyclicity, indicating that they do not respond to the positive feedback actions of estrogens. They respond however to the negative actions of estrogen, as ER$\alpha^{KO/AA}$ mice have reduced LH serum levels compared to ERαKO females and thereby experience less ovarian hyperstimulation and have less hemorrhagic cysts. This reduction of LH also rescues steroidogenesis, as evidenced by a complete restoration of estradiol serum levels in ER$\alpha^{KO/AA}$ females. ER$\alpha^{KO/AA}$ females also respond to the ablation of the ovaries by a normal elevation of LH levels, elevation that can be reduced by an exogenous administration of estrogens.

AFP KO females also respond to the negative effects of estrogen but are unable to perceive the positive ones. AFP KO females show an elevation of LH serum levels in response to ovariectomy: however, no LH nor GnRH surge could be obtained even in stimulatory hormonal conditions. Finally, ER$\alpha^{KO/AA}$ females are, like ERαKO females, unreceptive to males and do not show lordosis behaviour. However, they do display significantly less rejective behaviour (eg kicking, fleeing and rearing) and more proceptive behaviour (approach to males) than their ERαKO counterparts.

In conclusion, several mouse models exist and can be used as tools to better understand the role of estrogens. AFP KO mice are a model for prenatal and perinatal estrogen overexposure. ArKO mice are devoid of estrogens. ERKO mice produce estrogens, but lack the estrogen receptor or the ERE binding ability of the estrogen receptor. All those mice could be used separately or together, in the context of double knock-outs, to deepen our understanding of estrogen action. Consequently, an exciting field lies wild open to new discoveries.

References

[1]   Sell S, Becker FF. Alpha-Fetoprotein. J Natl Cancer Inst. 1978;60(1):19-26.

[2]   Andrews GK, Dziadek M, Tamaoki T. Expression and methylation of the mouse alpha-fetoprotein gene in embryonic, adult, and neoplastic tissues. J Biol Chem. 1982; 257(9):5148-53.

[3]   Belayew A, Tilghman SM. Genetic analysis of alpha-fetoprotein synthesis in mice. Mol Cell Biol. 1982;2(11):1427-35.

[4]   Wald NJ, Huttly WJ, Hackshaw AK. Antenatal screening for Down's syndrome with the quadruple test. Lancet. 2003;361(9360):835-6.

[5]   Leighton PC, Kitau MJ, Chard T, Gordon YB, Leek AE. Levels of alpha-fetoprotein in maternal blood as a screening test for fetal neural-tube defect. Lancet. 1978;2(7943):1012-5.

[6]   Mizejewski GJ. Biological roles of alpha-fetoprotein during pregnancy and perinatal development. Exp Biol Med (Maywood). 2004;229(6):439-63.

[7]   Tilghman SM, Belayew A. Transcriptional control of the murine albumin/alpha-fetoprotein locus during development. Proc Natl Acad Sci U S A. 1982;79(17):5254-7.

[8]   Masopust J, Kithier K, Rádl J, Koutecký J, Kotál L. Occurrence of fetoprotein in patients with neoplasms and non-neoplastic diseases. Int J Cancer. 1968;3(3):364-73.

[9]   Chiu JF, Huang DP, Burkhardt AL, Cote G, Schwartz CE.The alteration of gene expression in rat liver during chemical carcinogenesis. Arch Biochem Biophys. 1983;222(1):310-20.

[10]   Abelev GI, Eraiser TL. Cellular aspects of alpha-fetoprotein reexpression in tumors. Semin Cancer Biol. 1999;9(2):95-107.

[11]   Labdenne P, Heikinheimo M. Clinical use of tumor markers in childhood malignancies.

[12]   Ann Med. 2002;34(5):316-23.

[13]   Yuen MF, Lai CL. Serological markers of liver cancer. Best Pract Res Clin Gastroenterol. 2005;19(1):91-9.

[14]   Ishigami S, Natsugoe S, Nakashima H, Tokuda K, Nakajo A, Okumura H, Matsumoto M, Nakashima S, Hokita S, Aikou T. Biological aggressiveness of alpha-fetoprotein (AFP)-positive gastric cancer. Hepatogastroenterology. 2006;53(69):338-41.

[15]   Gibbs PE, Witke WF, Dugaiczyk A. The molecular clock runs at different rates among closely related members of a gene family. J Mol Evol. 1998;46(5):552-61.

[16] Jin JR, Wen P, Locker J. Enhancer sharing in a plasmid model containing the alpha-fetoprotein and albumin promoters. DNA Cell Biol. 1995;14(3):267-72.

[17] Feuerman MH, Godbout R, Ingram RS, Tilghman SM.Tissue-specific transcription of the mouse alpha-fetoprotein gene promoter is dependent on HNF-1. Mol Cell Biol. 1989;9(10):4204-12.

[18] Huang MC, Li KK, Spear BT. The mouse alpha-fetoprotein promoter is repressed in HepG2 hepatoma cells by hepatocyte nuclear factor-3 (FOXA). DNA Cell Biol. 2002;21(8):561-9.

[19] Galarneau L, Paré JF, Allard D, Hamel D, Levesque L, Tugwood JD, Green S, Bélanger L. The alpha1-fetoprotein locus is activated by a nuclear receptor of the Drosophila FTZ-F1 family. Mol Cell Biol. 1996;16(7):3853-65.

[20] Zhang XK, Dong JM, Chiu JF. Regulation of alpha-fetoprotein gene expression by antagonism between AP-1 and the glucocorticoid receptor at their overlapping binding site. J Biol Chem. 1991;266(13):8248-54.

[21] Bernier D, Thomassin H, Allard D, Guertin M, Hamel D, Blaquière M, Beauchemin M, LaRue H, Estable-Puig M, Bélanger L. Functional analysis of developmentally regulated chromatin-hypersensitive domains carrying the alpha 1-fetoprotein gene promoter and the albumin/alpha 1-fetoprotein intergenic enhancer. Mol Cell Biol. 1993;13(3):1619-33.

[22] Apergis GA, Crawford N, Ghosh D, Steppan CM, Vorachek WR, Wen P, Locker J. A novel nk-2-related transcription factor associated with human fetal liver and hepatocellular carcinoma. J Biol Chem. 1998;273(5):2917-25.

[23] Nacer-Cherif H, Bois-Joyeux B, Rousseau GG, Lemaigre FP, Danan JL. Hepatocyte nuclear factor-6 stimulates transcription of the alpha-fetoprotein gene and synergizes with the retinoic-acid-receptor-related orphan receptor alpha-4. Biochem J. 2003;369(Pt 3):583-91.

[24] Wen P, Locker J. A novel hepatocytic transcription factor that binds the alpha-fetoprotein promoter-linked coupling element. Mol Cell Biol. 1994;14(10):6616-26.

[25] Feuerman MH, Godbout R, Ingram RS, Tilghman SM. Tissue-specific transcription of the mouse alpha-fetoprotein gene promoter is dependent on HNF-1. Mol Cell Biol. 1989;9(10):4204-12.

[26] Zhang DE, Ge X, Rabek JP, Papaconstantinou J. Functional analysis of the trans-acting factor binding sites of the mouse alpha-fetoprotein proximal promoter by site-directed mutagenesis. J Biol Chem. 1991;266(31):21179-85.

[27] Bois-Joyeux B, Danan JL. Members of the CAAT/enhancer-binding protein, hepatocyte nuclear factor-1 and nuclear factor-1 families can differentially modulate the activities of the rat alpha-fetoprotein promoter and enhancer. Biochem J. 1994;301 (Pt 1):49-55.

[28] Yasuda H, Mizuno A, Tamaoki T, Morinaga T. ATBF1, a multiple-homeodomain zinc finger protein, selectively down-regulates AT-rich elements of the human alpha-fetoprotein gene. Mol Cell Biol. 1994;14(2):1395-401.

[29] Liénard P, De Mees C, Drèze PL, Dieu M, Dierick JF, Raes M, Szpirer J, Szpirer C. Regulation of the alpha-fetoprotein promoter: Ku binding and DNA spatial conformation. Biochimie. 2006;88(10):1409-17.

[30] Liu Y, Chiu JF.Transactivation and repression of the alpha-fetoprotein gene promoter by retinoid X receptor and chicken ovalbumin upstream promoter transcription factor. Nucleic Acids Res. 1994;22(6):1079-86.

[31] Leng X, Cooney AJ, Tsai SY, Tsai MJ. Molecular mechanisms of COUP-TF-mediated transcriptional repression: evidence for transrepression and active repression. Mol Cell Biol. 1996;16(5):2332-40.

[32] Bois-Joyeux B, Denissenko M, Thomassin H, Guesdon S, Ikonomova R, Bernuau D, Feldmann G, Danan JL. The c-jun proto-oncogene down-regulates the rat alpha-fetoprotein promoter in HepG2 hepatoma cells without binding to DNA. J Biol Chem. 1995;270(17):10204-11.

[33] Xie Z, Zhang H, Tsai W, Zhang Y, Du Y, Zhong J, Szpirer C, Zhu M, Cao X, Barton MC, Grusby MJ, Zhang WJ. Zinc finger protein ZBTB20 is a key repressor of alpha-fetoprotein gene transcription in liver. Proc Natl Acad Sci U S A. 2008;105(31):10859-64.

[34] Mitchelmore C, Kjaerulff KM, Pedersen HC, Nielsen JV, Rasmussen TE, Fisker MF, Finsen B, Pedersen KM, Jensen NA.Characterization of two novel nuclear BTB/POZ domain zinc finger isoforms. Association with differentiation of hippocampal neurons, cerebellar granule cells, and macroglia. Biol Chem. 2002;277(9):7598-609.

[35] Chen H, Egan JO, Chiu JF. Regulation and activities of alpha-fetoprotein. Crit Rev Eukaryot Gene Expr. 1997;7(1-2):11-41.

[36] Hammer RE, Krumlauf R, Camper SA, Brinster RL, Tilghman SM. Diversity of alpha-fetoprotein gene expression in mice is generated by a combination of separate enhancer elements. Science. 1987;235(4784):53-8.

[37] Godbout R, Ingram RS, Tilghman SM. Fine-structure mapping of the three mouse alpha-fetoprotein gene enhancers. Mol Cell Biol. 1988;8(3):1169-78.

[38] Long L, Davidson JN, Spear BT. Striking differences between the mouse and the human alpha-fetoprotein enhancers. Genomics. 2004;83(4):694-705.

[39] Groupp ER, Crawford N, Locker J. Characterization of the distal alpha-fetoprotein enhancer, a strong, long distance, liver-specific activator. J Biol Chem. 1994;269(35):22178-87.

[40] Millonig JH, Emerson JA, Levorse JM, Tilghman SM. Molecular analysis of the distal enhancer of the mouse alpha-fetoprotein gene. Mol Cell Biol. 1995;15(7):3848-56.

[41] Bois-Joyeux B, Chauvet C, Nacer-Chérif H, Bergeret W, Mazure N, Giguère V, Laudet V, Danan JL. Modulation of the far-upstream enhancer of the rat alpha-fetoprotein gene by members of the ROR alpha, Rev-erb alpha, and Rev-erb beta groups of monomeric orphan nuclear receptors. DNA Cell Biol. 2000;19(10):589-99.

[42] Ramesh TM, Ellis AW, Spear BT. Individual mouse alpha-fetoprotein enhancer elements exhibit different patterns of tissue-specific and hepatic position-dependent activities. Mol Cell Biol. 1995;15(9):4947-55.

[43] Camper SA, Tilghman SM. Postnatal repression of the alpha-fetoprotein gene is enhancer independent. Genes Dev. 1989;3(4):537-46.

[44] Peyton DK, Ramesh T, Spear BT. Position-dependent activity of alpha -fetoprotein enhancer element III in the adult liver is due to negative regulation. Proc Natl Acad Sci U S A. 2000;97(20):10890-4.

[45] Lee KC, Crowe AJ, Barton MC. p53-mediated repression of alpha-fetoprotein gene expression by specific DNA binding. Mol Cell Biol. 1999;19(2):1279-88.

[46] Crowe AJ, Sang L, Li KK, Lee KC, Spear BT, Barton MC. Hepatocyte nuclear factor 3 relieves chromatin-mediated repression of the alpha-fetoprotein gene. J Biol Chem. 1999;274(35):25113-20.

[47] Nguyen TT, Cho K, Stratton SA, Barton MC. Transcription factor interactions and chromatin modifications associated with p53-mediated, developmental repression of the alpha-fetoprotein gene. Mol Cell Biol. 2005;25(6):2147-57.

[48] Cui R, Nguyen TT, Taube JH, Stratton SA, Feuerman MH, Barton MC. Family members p53 and p73 act together in chromatin modification and direct repression of alpha-fetoprotein transcription. J Biol Chem. 2005;280(47):39152-60.

[49] Kataoka H, Bonnefin P, Vieyra D, Feng X, Hara Y, Miura Y, Joh T, Nakabayashi H, Vaziri H, Harris CC, Riabowol K. ING1 represses transcription by direct DNA binding and through effects on p53. Cancer Res. 2003;63(18):5785-92.

[50] Olsson M, Lindahl G, Ruoslahti E. Genetic control of alpha-fetoprotein synthesis in the mouse. J Exp Med. 1977;145(4):819-27.

[51] Peyton DK, Huang MC, Giglia MA, Hughes NK, Spear BT. The alpha-fetoprotein promoter is the target of Afr1-mediated postnatal repression. Genomics. 2000;63(2):173-80.

[52] Vacher J, Camper SA, Krumlauf R, Compton RS, Tilghman SM. raf regulates the postnatal repression of the mouse alpha-fetoprotein gene at the posttranscriptional level. Mol Cell Biol. 1992;12(2):856-64.

[53] Perincheri S, Dingle RW, Peterson ML, Spear BT. Hereditary persistence of alpha-fetoprotein and H19 expression in liver of BALB/cJ mice is due to a retrovirus insertion in the Zhx2 gene. Proc Natl Acad Sci U S A. 2005;102(2):396-401.

[54] Kawata H, Yamada K, Shou Z, Mizutani T, Yazawa T, Yoshino M, Sekiguchi T, Kajitani T, Miyamoto K. Zinc-fingers and homeoboxes (ZHX) 2, a novel member of the ZHX family, functions as a transcriptional repressor. Biochem J. 2003;373(Pt 3):747-57.

[55] Yamada K, Kawata H, Shou Z, Hirano S, Mizutani T, Yazawa T, Sekiguchi T, Yoshino M, Kajitani T, Miyamoto K. Analysis of zinc-fingers and homeoboxes (ZHX)-1-interacting proteins: molecular cloning and characterization of a member of the ZHX family, ZHX3. Biochem J. 2003;373(Pt 1):167-78.

[56] Kawata H, Yamada K, Shou Z, Mizutani T, Miyamoto K. The mouse zinc-fingers and homeoboxes (ZHX) family; ZHX2 forms a heterodimer with ZHX3. Gene. 2003;323:133-40.

[57] McVey JH, Michaelides K, Hansen LP, Ferguson-Smith M, Tilghman S, Krumlauf R, Tuddenham EG. A G-->A substitution in an HNF I binding site in the human alpha-fetoprotein gene is associated with hereditary persistence of alpha-fetoprotein (HPAFP). Hum Mol Genet. 1993;2(4):379-84.

[58] Alj Y, Georgiakaki M, Savouret JF, Mal F, Attali P, Pelletier G, Fourré C, Milgrom E, Buffet C, Guiochon-Mantel A, Perlemuter G. Hereditary persistence of alpha-fetoprotein is due to both proximal and distal hepatocyte nuclear factor-1 site mutations. Gastroenterology. 2004;126(1):308-17.

[59] Nagata-Tsubouchi Y, Ido A, Uto H, Numata M, Moriuchi A, Kim I, Hasuike S, Nagata K, Sekiya T, Hayashi K, Tsubouchi H. Molecular mechanisms of hereditary persistence of alpha-fetoprotein (AFP) in two Japanese families A hepatocyte nuclear factor-1 site mutation leads to induction of the AFP gene expression in adult livers. Hepatol Res. 2005;31(2):79-87.

[60] Yeh SH, Kao JH, Chen PJ. Heterogeneity of hereditary persistence of alpha-fetoprotein. Gastroenterology. 2004;127(2):687.

[61] Petropoulos CJ, Yaswen P, Panzica M, Fausto N. Cell lineages in liver carcinogenesis: possible clues from studies of the distribution of alpha-fetoprotein RNA sequences in cell populations isolated from normal, regenerating, and preneoplastic rat livers. Cancer Res. 1985;45(11 Pt 2):5762-8.

[62] Wan YJ, Chou JY. Expression of the alpha-fetoprotein gene in adult rat liver. Arch Biochem Biophys. 1989;270(1):267-76.

[63] Lemire JM, Fausto N. Multiple alpha-fetoprotein RNAs in adult rat liver: cell type-specific expression and differential regulation. Cancer Res. 1991;51(17):4656-64.

[64] Scohy S, Gabant P, Szpirer C, Szpirer J. Identification of an enhancer and an alternative promoter in the first intron of the alpha-fetoprotein gene. Nucleic Acids Res. 2000;28(19):3743-51.

[65] Gabant P, Forrester L, Nichols J, Van Reeth T, De Mees C, Pajack B, Watt A, Smitz J, Alexandre H, Szpirer C, Szpirer J. Alpha-fetoprotein, the major fetal serum protein, is not essential for embryonic development but is required for female fertility. Proc Natl Acad Sci U S A. 2002;99(20):12865-70.

[66] De Mees C, Laes JF, Bakker J, Smitz J, Hennuy B, Van Vooren P, Gabant P, Szpirer J, Szpirer C. Alpha-fetoprotein controls female fertility and prenatal development of the gonadotropin-releasing hormone pathway through an antiestrogenic action. Mol Cell Biol. 2006;26(5):2012-8.

[67] González-Martínez D, De Mees C, Douhard Q, Szpirer C, Bakker J. Absence of gonadotropin-releasing hormone 1 and Kiss1 activation in alpha-fetoprotein knockout mice: prenatal estrogens defeminize the potential to show preovulatory luteinizing hormone surges. Endocrinology. 2008;149(5):2333-40.

[68] Gottsch ML, Clifton DK, Steiner RA. Kisspepeptin-GPR54 signaling in the neuroendocrine reproductive axis. Mol Cell Endocrinol. 2006;254-255:91-6.

[69] Dungan HM, Gottsch ML, Zeng H, Gragerov A, Bergmann JE, Vassilatis DK, Clifton DK, Steiner RA. The role of kisspeptin-GPR54 signaling in the tonic regulation and surge release of gonadotropin-releasing hormone/luteinizing hormone. J Neurosci. 2007;27(44):12088-95.

[70] Li C, Chen P, Smith MS. Morphological evidence for direct interaction between arcuate nucleus neuropeptide Y (NPY) neurons and gonadotropin-releasing hormone neurons and the possible involvement of NPY Y1 receptors. Endocrinology. 1999;140(11):5382-90.

[71] Bakker J, De Mees C, Douhard Q, Balthazart J, Gabant P, Szpirer J, Szpirer C. Alpha-fetoprotein protects the developing female mouse brain from masculinization and defeminization by estrogens. Nat Neurosci. 2006;9(2):220-6.

[72] Bakker J, De Mees C, Szpirer J, Szpirer C, Balthazart J. Exposure to oestrogen prenatally does not interfere with the normal female-typical development of odour preferences. J Neuroendocrinol. 2007;19(5):329-34.

[73] Uriel J, Bouillon D, Aussel C, Dupiers M. Alpha-fetoprotein: the major high-affinity estrogen binder in rat uterine cytosols. Proc Natl Acad Sci U S A. 1976;73(5):1452-6.

[74] Savu L, Benassayag C, Vallette G, Christeff N, Nunez E. Mouse alpha 1-fetoprotein and albumin. A comparison of their binding properties with estrogen and fatty acid ligands. J Biol Chem. 1981;256(18):9414-8.

[75] Nishi S, Matsue H, Yoshida H, Yamaoto R, Sakai M. Localization of the estrogen-binding site of alpha-fetoprotein in the chimeric human-rat proteins. Proc Natl Acad Sci U S A. 1991;88(8):3102-5.

[76] Mizejewski GJ, Vonnegut M, Jacobson HI. Estradiol-activated alpha-fetoprotein suppresses the uterotropic response to estrogens. Proc Natl Acad Sci U S A. 1983 May;80(9):2733-7.

[77] Gorski RA. Modification of ovulatory mechanisms by postnatal administration of estrogen to the rat. Am J Physiol. 1963;205(5):842-4.

[78] Whalen RE, Nadler RD. Suppression of the development of female mating behavior by estrogen administered in infancy. Science. 1963;141:273-4.

[79] McEwen BS, Plapinger L, Chaptal C, Gerlach J, Wallach G. Role of fetoneonatal estrogen binding proteins in the associations of estrogen with neonatal brain cell nuclear receptors. Brain Res. 1975;96(2):400-6.

[80] Döhler KD, Hancke JL, Srivastava SS, Hofmann C, Shryne JE, Gorski RA. Participation of estrogens in female sexual differentiation of the brain; neuroanatomical, neuroendocrine and behavioral evidence. Prog Brain Res. 1984;61:99-117.

[81] Toran-Allerand CD.On the genesis of sexual differentiation of the general nervous system: morphogenetic consequences of steroidal exposure and possible role of alpha-fetoprotein. Prog Brain Res. 1984;61:63-98.

[82] Schachter BS, Toran-Allerand CD. Intraneuronal alpha-fetoprotein and albumin are not synthesized locally in developing brain. Brain Res. 1982;281(1):93-8.

[83] Gillespie JR, Uversky VN. Structure and function of alpha-fetoprotein: a biophysical overview. Biochim Biophys Acta. 2000;1480(1-2):41-56.

[84] Bakker J, Baum MJ. Role for estradiol in female-typical brain and behavioral sexual differentiation. Front Neuroendocrinol. 2008;29(1):1-16.

[85] Takagi Y, Maeda S. Disruption of sexual behavior in male aromatase-deficient mice lacking exons 1 and 2 of the cyp19 gene. Biochem Biophys Res Commun. 1998;252(2):445-9.

[86] Toda K, Takeda K, Okada T, Akira S, Saibara T, Kaname T, Yamamura K, Onishi S, Shizuta Y. Targeted disruption of the aromatase P450 gene (Cyp19) in mice and their ovarian and uterine responses to 17beta-oestradiol. J Endocrinol. 2001;170(1):99-111.

[87] Nelson DR, Kamataki T, Waxman DJ, Guengerich FP, Estabrook RW, Feyereisen R, Gonzalez FJ, Coon MJ, Gunsalus IC, Gotoh O, et al. The P450 superfamily: update on new sequences, gene mapping, accession numbers, early trivial names of enzymes, and nomenclature. DNA Cell Biol. 1993;12(1):1-51.

[88] Fisher CR, Graves KH, Parlow AF, Simpson ER. Characterization of mice deficient in aromatase (ArKO) because of targeted disruption of the cyp19 gene. Proc Natl Acad Sci U S A. 1998;95(12):6965-70 .

[89] Robertson KM, O'Donnell L, Jones ME, Meachem SJ, Boon WC, Fisher CR, Graves KH, McLachlan RI, Simpson ER.Impairment of spermatogenesis in mice lacking a functional aromatase (cyp 19) gene. Proc Natl Acad Sci U S A. 1999;96(14):7986-91.

[90] Lubahn DB, Moyer JS, Golding TS, Couse JF, Korach KS, Smithies O. Alteration of reproductive function but not prenatal sexual development after insertional disruption of the mouse estrogen receptor gene. Proc Natl Acad Sci U S A. 1993;90(23):11162-6.

[91] Schomberg DW, Couse JF, Mukherjee A, Lubahn DB, Sar M, Mayo KE, Korach KS. Targeted disruption of the estrogen receptor-alpha gene in female mice: characterization of ovarian responses and phenotype in the adult. Endocrinology. 1999;140(6):2733-44.

[92] Krege JH, Hodgin JB, Couse JF, Enmark E, Warner M, Mahler JF, Sar M, Korach KS, Gustafsson JA, Smithies O. Generation and reproductive phenotypes of mice lacking estrogen receptor beta. Proc Natl Acad Sci U S A. 1998;95(26):15677-82.

[93] Ogawa S, Chan J, Chester AE, Gustafsson JA, Korach KS, Pfaff DW. Survival of reproductive behaviors in estrogen receptor beta gene-deficient (betaERKO) male and female mice. Proc Natl Acad Sci U S A. 1999;96(22):12887-92.

[94] Couse JF, Hewitt SC, Bunch DO, Sar M, Walker VR, Davis BJ, Korach KS. Postnatal sex reversal of the ovaries in mice lacking estrogen receptors alpha and beta. Science. 1999;286(5448):2328-31.

[95] Osborne CK, Schiff R. Estrogen-receptor biology: continuing progress and therapeutic implications. J Clin Oncol. 2005;23(8):1616-22.

[96] Gaub MP, Bellard M, Scheuer I, Chambon P, Sassone-Corsi P.Activation of the ovalbumin gene by the estrogen receptor involves the fos-jun complex. Cell. 1990;63(6):1267-76.

[97] Jakacka M, Ito M, Weiss J, Chien PY, Gehm BD, Jameson JL. Estrogen receptor binding to DNA is not required for its activity through the nonclassical AP1 pathway. J Biol Chem. 2001;276(17):13615-21.

[98] Ray A, Prefontaine KE, Ray P. Down-modulation of interleukin-6 gene expression by 17 beta-estradiol in the absence of high affinity DNA binding by the estrogen receptor. J Biol Chem. 1994;269(17):12940-6.

[99] Safe S. Transcriptional activation of genes by 17 beta-estradiol through estrogen receptor-Sp1 interactions.Vitam Horm. 2001;62:231-52.

[100] Weigel NL, Zhang Y. Ligand-independent activation of steroid hormone receptors. J Mol Med. 1998;76(7):469-79.

[101] Jakacka M, Ito M, Martinson F, Ishikawa T, Lee EJ, Jameson JL. An estrogen receptor (ER)alpha deoxyribonucleic acid-binding domain knock-in mutation provides evidence for nonclassical ER pathway signaling in vivo. Mol Endocrinol. 2002;16(10):2188-201.

In: Development Gene Expresión Regulation
Editor: Nathan C. Kurzfield

ISBN: 978-60692-794-6
©2009 Nova Science Publishers, Inc.

Chapter VII

Molecular Mechanisms of Cell-Specific Expression of Neuropeptide Hormone Genes, *DH-PBAN* and *PTTH* in the Silkworm, *Bombyx Mori*

Kunihiro Shiomi

Division of Applied Biology, Faculty of Textile Science and Technology, Shinshu
University, Ueda, Nagano 386-8567, Japan

Abstract

Neurosecretory cells play critical roles in different and specific physiological and behavioral processes via spatiotemporal regulation of neurohormone secretion. In insects, diapause and metamorphosis are induced by neuropeptide hormones secreted from a few neurosecretory cells that project intrinsic axons to intrinsic neurohemal sites. Diapause hormone (DH) is responsible for induction of embryonic diapause in *Bombyx mori*. The diapause hormone-pheromone biosynthesis activating neuropeptide gene, *DH-PBAN,* is expressed exclusively in seven pairs of DH-PBAN-producing neurosecretory cells (DHPCs) on the terminally differentiated processes of the subesophageal ganglion (SG). On the other hand, prothoracicotropic hormone (PTTH) plays a central role in controlling molting and metamorphosis in *Bombyx mori* by stimulating the prothoracic glands to synthesize and release the molting hormone, ecdysone. The *PTTH* gene is constantly expressed during larval-pupal development, and the peptide is produced exclusively in two pairs of lateral PTTH-producing neurosecretory cells (PTPCs) in the brain. To help reveal the regulatory mechanisms of cell-specific expression of *DH-PBAN* and *PTTH*, we identified *cis*-regulatory elements that regulate expression in DHPCs and PTPCs, respectively, using a recombinant baculovirus (AcNPV)-mediated gene transfer system and a gel-mobility shift assay. Interestingly, *Bombyx mori* Pitx (BmPitx), a bicoid-like homeobox transcription factor, binds the 5'-upstream sequence of both *DH-PBAN* and *PTTH* and activates gene expression. This article describes the regulatory mechanisms of cell-specific expression of the neuropeptide hormone genes involved in diapause and metamorphosis in *Bombyx*.

1. Neuropeptide Hormones and Neurosecretory Cells Involved in Diapause and Metamorphosis of the Silkworm, *Bombyx Mori*

Neurosecretory cells play critical roles in different and specific physiological and behavioral processes via spatiotemporal regulation of neurohormone secretion. In insects, diapause and metamorphosis are induced by neuropeptide hormones secreted from a few neurosecretory cells that project intrinsic axons to intrinsic neurohemal sites [1] [2]. The silkworm, *Bombyx mori*, is a typical insect whose neurohormones have been systematically studied, especially the neuropeptides regulating diapause and metamorphosis (Fig. 1) [3] [4]. Diapause hormone (DH) is a unique neurohormone responsible for the induction of embryonic diapause in the silkworm [3]. DH is a 24-amino-acid peptide which is post-translationally processed from a precursor polypeptide, DH-PBAN. DH belongs to the FXPRLamide peptide family, which also includes the pheromone biosynthesis activating neuropeptide (PBAN), myotropin, and pyrokinin [5] [6] [7] [8] [9]. *DH-PBAN* encodes a polyprotein precursor containing DH; PBAN; α-, β-, and γ-SGNP; and is exclusively expressed in seven pairs of neurosecretory cells, known as the DH-PBAN-producing neurosecretory cells (DHPCs), in the subesophageal ganglion (SG) (Fig. 1) [10] [11] [12] [13]. *DH-PBAN* expression is first detected at the histogenesis stage and gradually increases as the embryo develops to the hatching stage. Furthermore, somata recognized by anti-FXPRLamide antibody detection are found in the SG at the blastokinesis stage and these cells are already morphologically identical to those of larval and pupal SG [14]. In addition, immunoreactive axons that have varicosities and neurohemal organ corpus cardiacum (CC) are observed at the complete embryonic reversal stage, which occurs just after blastokinesis (Fig. 1) [14]. *DH-PBAN* and DH are spatiotemporal-expressed in terminal differentiation processes of DHPCs, such that the neuroendocrine system is responsible for DH secretion and subsequent, induction of diapause by activating the DH receptor [15] in developing ovaries during pupal-adult development. However, it is not known if the molecular mechanism(s) of diapause induction including DHPCs differentiation and specification relies on *DH-PBAN* expression and DH release, despite the fact that many attempts have been made to identify a role for *DH-PBAN* expression in this process [16] [17].

In addition, prothoracicotropic hormone (PTTH) stimulates the prothoracic glands to synthesize and release ecdysone, the steroid necessary for molting, metamorphosis, and the termination of pupal diapause (Fig. 1) [18] [19] [20]. PTTH was first purified and sequenced from the silkworm, *Bombyx mori* [21] [22]. The *PTTH* gene is constantly expressed during larval-pupal development [23], and the peptide is produced exclusively in two pairs of lateral PTTH-producing neurosecretory cells (PTPCs) in the brain (Fig. 1) [24]. From there it is transported *via* axons to the corpora allata and then released into the hemolymph. The PTTH titer in the hemolymph has been shown to correlate closely with the ecdysteroid titer [19]

[25]. Fluctuations of PTTH titer in hemolymph consequentially act as a pacemaker in the neuroendocrine regulation of development by varying the secretion of ecdysone. The timing of the increase in hemolymph PTTH titer on the day of wandering is photoperiodically controlled in *Bombyx mori* [19]. In addition, in larvae of *Heliothis virescens*, the expression of the *PTTH* gene declines sharply at the onset of larval wandering behavior and remains low during pupal diapause [26]. Thus, analysis of the molecular mechanisms controlling PTTH secretion in PTPCs is important for understanding the termination of pupal diapause as well as the induction of molting and metamorphosis.

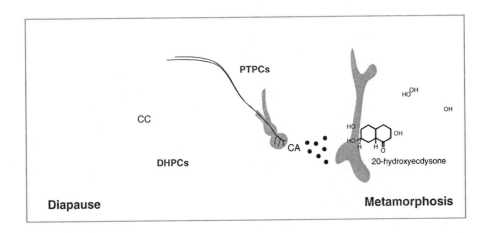

Figure 1. Neuropeptide hormone and neurosecretory cells involved in diapause and metamorphosis of the silkworm, *Bombyx mori*. *DH-PBAN* encodes a polyprotein precursor containing diapause hormone (DH) and is exclusively expressed in seven pairs of neurosecretory cells, the DH-PBAN-producing neurosecretory cells (DHPCs), in the subesophageal ganglion (SG). DHPCs consists of three neuromeres; four mandibular cells [SMd]; six maxillary cells [SMx]; two labial cells [SLb]) along the ventral midline; and the lateral cells (two cells of lateral [SL]). Furthermore, the DHPCs project into the corpora cardiacum (CC), a putative neurohemal organ of DH so that DH is transported via axons to CC, and released into hemolymph. Axons of DHPCs are represented in the hemisphere in green lines. Furthermore, released DH acts upon the receptor in developing ovaries, and is responsible for diapause induction. On the other hand, *PTTH* is exclusively expressed in two pairs of lateral PTTH-producing neurosecretory cells (PTPCs) in the brain. PTTH is transported via axons to the corpora allata (CA) and then released into hemolymph. Axons of PTPCs are represented in the hemisphere in red lines. Fluctuations of PTTH titer in hemolymph consequentially act as a pacemaker in the neuroendocrine regulation of development by varying the secretion of ecdysone (20-hydroxyecdysone) in the prothoracic gland, and are responsible for metamorphosis.

The limited expression of *DH-PBAN* and *PTTH* genes in specific sets of neurosecretory cells in the central nervous system (CNS) at a specific stage of development should be under the control of the integrated machinery of regulation involving the *cis*-regulatory elements and cellular factors. The stringent regulation of neuropeptide secretion seems to be essential for the neuropeptide to exert its own physiological function. To pursue the molecular mechanisms of neuropeptide gene expression and release involved in diapause and metamorphosis, we first developed a baculovirus, *Autographa californica* nucleopolyhedrovirus (AcNPV)-mediated gene transfer system.

2. Acnpv-Mediated Gene Transfer System in the Central Nervous System of the Silkworm, *Bombyx Mori*

There are several gene transfer techniques that can be used to elucidate the regulatory mechanism of gene expression at the cellular level in *Bombyx mori*. Transient gene transfer has been achieved by electroporation of a reporter gene into the intact brain of *Bombyx* [27], but the efficiency is too low for most laboratory applications. A more effective gene transfer method has been achieved by injection of a vector derived from the *piggyBac* transposon [28] [29]. This system requires a highly technical micro-injection through the hard egg shell at a limited stage of embryogenesis, so it cannot deal with a large number of transgenic reporters at once. Furthermore, a virus-mediated gene transfer system has been developed in *Bombyx mori*. Although sindbis virus is an alphavirus that enables systemic infection in many insects, and has been reported to be an expression vector for *B. mori* [30], this virus-based vector results in instability of recombinant clones after multiple passages, due to infection barriers in reproductive organs [31]. Furthermore, *Bombyx mori* nucleopolyhedrovirus (BmNPV) has been used as a vector, which achieved efficient delivery of the enhanced green fluorescence protein (EGFP) gene in neural tissues of the larvae. However, recipient animals died rapidly due to their permissiveness for BmNPV [32].

We have tried to develop another system for transient gene transfer into *Bombyx* using the recombinant baculovirus, *Autographa californica* nucleopolyhedrovirus (AcNPV). AcNPV baculovirus vectors are extensively used in numerous biotechnology applications [33] [34] [35] [36], because the budded form of the virus is harmless to humans and to the environment. In a wide variety of organisms, the gene transfer system using recombinant AcNPV has been developed to study the effects of the ectopic expression of genes [37]. Furthermore, AcNPV replicates the viral DNA and produces infectious budded viruses with no polyhedrin synthesis in *Bombyx* cells [38]. When the budded virion was injected into silkworm larvae, no symptoms of nuclear polyhedrosis appeared and no abnormal metamorphoses of pupae occurred [39] [40]. These facts suggest that the budded virion can transfer viral DNA into the host nucleus, and the damage to recipients should be less than with BmNPV.

We first constructed recombinant AcNPV containing the *EGFP* gene driven by the *Bombyx actin A3* promoter using the Bac-to-bac baculovirus expression system (Fig. 2A). This recombinant AcNPV (v[A3/EGFP]) was prepared by transposition of the expression cassette from a recombinant plasmid (pA3/EGFP) (Fig. 2A). Cultured media containing the virus was injected into newly ecdysed fifth-instar larvae of the N4 strain. Under the experimental conditions, no developmental disturbance occurred until 5 days after injection, when the non-injected larvae ceased feeding to begin cocoon spinning. While the virus-infected larvae continued feeding for 5 days further, the body became swollen and the cuticle appeared glossy. Finally, all larvae died in a few days. When the recombinant AcNPV was injected into pupae just after larval-pupal ecdysis, the pupal-adult development appeared to be arrested at the early stage, because neither the coloration of adult eyes nor morphological change of the brain-SG complex occurred. The infected pupae eventually died within 10 days post-injection, 3 days after eclosion in the non-injected animal (at 7 days after pupation).

Recombinant AcNPV, v[A3/EGFP], was injected into newly ecdysed fifth-instar larvae of the N4 strain, and observed every day under a microscope equipped for epifluorescence detection. Figure 2B shows the whole-body signal 5 days after injection with intense signals at intersegments in the anterior and posterior regions of the body [13]. Likewise, a fluorescence signal was detected in the whole body of pupae 5 days after injection (Fig. 2C), and a signal was also visible in the excised CNS (Fig. 2D). The accumulation of the EGFP protein in the brain-SG complex was detected by western blot analysis using an anti-GFP antibody [13]. The EGFP protein was detected in the epidermis, fat body, ovary, testis, and hemocytes, in addition to the CNS, 5 days after injection by western blot analysis (data not shown). Thus, recombinant AcNPV entered into the cells in a variety of tissues and expressed the reporter gene, *EGFP*, under the control of the *BmA3* promoter. Notably, a very strong fluorescence signal was detectable in CNS by day 3 after AcNPV injection, and the signal was confirmed by immunoblotting.

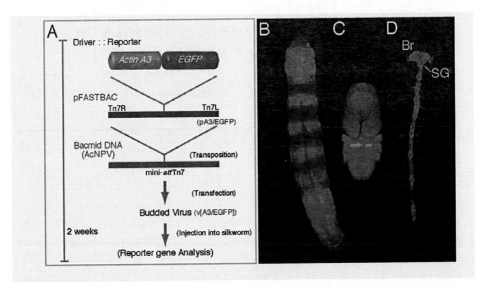

Figure. 2. AcNPV-mediated gene transfer system. (A); Recombinant AcNPV containing *EGFP* cDNA driven by the *Bombyx actin A3* promoter was prepared by transposition of expression cassettes from recombinant plasmid (pA3/EGFP) according to the instruction manual of the Bac-to-bac baculovirus expression system (Invitrogen). Recombinant bacmid, which is a baculovirus shuttle vector propagated in *E. coli*, was constructed by transposing a mini-Tn7 element from a recombinant pFASTBAC plasmid to the mini-*att*Tn7 attachment site on the bacmid, with the Tn7 transposition functions provided in trans by a helper plasmid. *Spodoptera frugiperda* (SF9) cells were transfected with recombinant bacmid DNA. Cells were incubated at 27°C for 4 days, after which the medium was collected and briefly centrifuged and the virus-containing supernatant was injected into silkworm. Five days after injection, reporter gene analysis was performed using fluorescence microscopy. This system was complete in the short time of 2 weeks from construction of the recombinant plasmid. (B - D); Expression of EGFP driven by an *actin A3* promoter throughout the body (B and C) and in CNS (D) of the N4 strain. Newly molted larvae (B) and day 0 pupae (C) were injected with recombinant AcNPV (v[A3/GFP]) and, 5 days later, subjected to epifluorescence microscopy to detect EGFP. EGFP expression in isolated CNS of a pupa 5 days after injection (D). Br, brain; SG, subesophageal ganglion.

So far, it has been believed that *B. mori* is permissive to BmNPV, but nonpermissive to AcNPV, since the original host of AcNPV is *Trichoplusia ni.* The difference in permissiveness between two experiments may be due to the silkworm strain used. To address this possibility, we examined twenty bivoltine strains in the reporter gene experiment under the same conditions as above. We identified six AcNPV-permissive strains. Furthermore, Guo *et al.* demonstrated that the existence of only one dominant host anti-AcNPV gene or a set of genetically linked genes prevent AcNPV infection in nonpermissive silkworm strain Qingsong and are absent in permissive silkworm strain Haoyue [41]. Therefore, we concluded that the N4 strain is highly susceptible to AcNPV infection, so that we selectively used the N4 strain for subsequent experiments. In particular, this system allows highly reproducible detection of the reporter gene, *EGFP*, expressed under the control of various promoters in the short time of 2 weeks from construction of the recombinant plasmid (Fig. 2A).

3. EGFP Expression in a Specific Set of Neuropeptide Hormone-Producing Neurosecretory Cells of the CNS

To study molecular mechanisms of neuropeptides gene expression of *DH-PBAN* and *PTTH*, we performed reporter gene analyses using two recombinant AcNPV strains, v[DH7/EGFP] and v[PT/EGFP], containing the *EGFP* gene under the control of *DH-PBAN* and *PTTH* upstream regions, respectively (Fig. 3) [13]. In the reporter gene assay of the *DH-PBAN* promoter, we first asked if an *EGFP* reporter gene is expressed in the somata and neurites of DHPCs when under the control of nucleotides -7113 to +94 of the *DH-PBAN* promoter (v[DH7/EGFP]; Fig. 3A - C). Observation of fluorescence clearly showed that EGFP expression was detected in 12 cells of the SG (Fig. 3A, B). The DHPCs consist of three neuromeres; four mandibular cells [SMd]; six maxillary cells [SMx]; two labial cells [SLb]) along the ventral midline; and the lateral cells (two cells of lateral [SL]) so that DHPCs are known to consist of 14 cells in the SG (Fig. 1) [11] [12] [13]. These cells were localized near the ventral midline of the SG and were aggregated into three clusters in the pattern expected for DHPCs. We confirmed that EGFP is expressed in DHPCs by colocalization of EGFP fluorescence with endogenous DH detected by immunohistochemistry. Merging the Cy3 (anti-DH[N]) and EGFP (v[DH7/EGFP]) signals showed that these signal-positive somata correspond to somata of DHPCs, but no EGFP signals were observed in SL (Fig. 3C). Furthermore, the Cy3 signal in DHPCs projected into the CC, where most of the signal overlapped with the EGFP signal [42]. Thus, the *DH-PBAN* promoter directed the expression of the reporter gene in DHPCs, ensuring that this reporter system is under the control of the inserted promoter. Therefore, the results suggest that the *DH-PBAN* promoter region that extends from nucleotides -7113 to +94 contains one or more *cis*-regulatory elements that direct *DH-PBAN* expression in six pairs of DHPCs [13] [42].

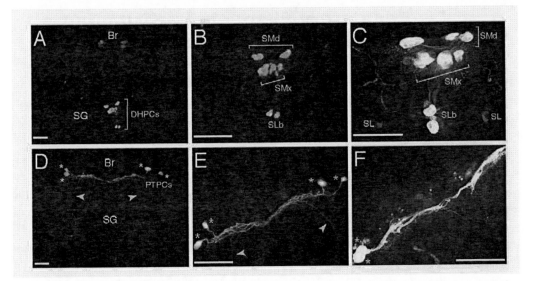

Figure 3. Reporter gene assay of *Bombyx* neurohormone gene in the brain-SG complex. EGFP expression in the brain-SG complexes of pupae injected with v[DH7/EGFP] (A - C) and v[PTG/EGFP] (D - F) were visualized by fluorescence microscopy. B and E show magnified views of EGFP-expressing cells, taken from different animals from those shown in A and D, respectively. The arrowheads indicate the axon emanating from the somata running contralateral after crossing the midline, and then projecting into the corpus allatum (D and E). Co-expression of neuropeptides and EGFP driven by a *DH-PBAN* and *PTTH* promoter in the SG and brain, respectively (C and F). Pupae were injected with v[D7DH/EGFP] (C) and v[PT/EGFP] (F), then the SG and brain were subjected to whole-mount immunostaining with anti-DH[N] antibody and anti-PTTH antibody, respectively. EGFP fluorescence was visualized by green colors. Immunofluorescence signals were visualized by Cy3 labeling (magenta colors). The merged image shows that DHPCs (C) and PTPCs (F) co-express EGFP and DH, or EGFP and PTTH, respectively. Br, brain; SG, subesophageal ganglion: DHPCs, DH-PBAN-producing neurosecretory cells; SMd, mandibular neuromere cell of the SG; SMx, maxillary neuromere cell of the SG; SLb, labial neuromere cell of the SG; PTPCs, PTTH-producing neurosecretory cells. Scale bar = 100 μm.

In the reporter gene assay of the *PTTH* promoter, we tried to examine the reporter gene construct containing *EGFP* under control of nucleotides -879 to +52 of the *PTTH* promoter (v[PT/EGFP]). The fluorescence signals in the brain-SG complex injected with v[PT/EGFP] appeared in two pairs of lateral cells in the brain (Fig. 3D and E). The axon arising from the somata runs towards the midline of the brain with some arborization, and then run contralateral after crossing the midline. Immunohistochemical staining with an anti-PTTH antibody to visualize endogenous PTTH produced by PTPCs identified two pairs of lateral cells in the protocerebrum (Fig. 3F). Furthermore, the Cy3 signals in the PTPCs projected into the corpus allatum-corpus cardiacum complex where most of the signal overlapped with the EGFP signal [43]. Thus, the neurosecretory cells in the brain of *Bombyx* expressing v[PT/EGFP]-derived EGFP corresponded to PTPCs. The results also suggest that the sequence of the *PTTH* promoter from nucleotides -879 to +52 contains *cis*-regulatory elements that drive *PTTH* gene expression in PTPCs. In summary, we have successfully developed an AcNPV-mediated gene transfer for gene expression of the neuropeptide hormone *in situ* in the silkworm.

4. *Cis*-Regulatory Elements Participating with *DH-PBAN* Expression

To help reveal the regulatory mechanisms of cell-specific DH-PBAN expression, we identified a cis-regulatory element that regulates expression in DHPCs using the recombinant AcNPV-mediated gene transfer system and a gel-mobility shift assay. To better understand the cis-regulatory elements that direct DH-PBAN expression, we constructed several recombinant AcNPVs varied in terms of the upstream regions of DH-PBAN, which was fused with an EGFP reporter gene, and tested their expression patterns (Fig. 4A). In order to evaluate the efficiency of infection, the constructs also included the DsRed reporter gene under the control of the ActinA3 promoter. Progressive deletion of the 5'-upstream region of DH-PBAN from nucleotides -4233 to +94 had no significant effect on EGFP expression, for which fluorescence signals were strongly observed in DHPCs. However, recombinant AcNPVs carrying nucleotides -1123 to +94 of the DH-PBAN promoter showed a slight decrease in expression of EGFP in DHPCs, particularly in SLb (Figs. 3B, compare to 4B). Using a recombinant AcNPV with nucleotides -1103 to +94 of the DH-PBAN promoter, no EGFP signal was observed in DHPCs (Fig. 4D), although DsRed signals were observed throughout the SG (Fig. 4E). This was also the case for a different recombinant AcNPV that included the 5'-upstream region from nucleotides -1123 to +94 of the DH-PBAN promoter (Fig. 4C). From the results of this stepwise deletion of the DH-PBAN promoter, we speculate that a 20-base sequence from -1123 to -1103 of DH-PBAN contains the cis-regulatory element(s) that participates in the decision to switch on or off expression of DH-PBAN. In addition, we speculate that a region from -5,421 to -4,233 of DH-PBAN contains the cis-regulatory element(s) that participates in enhancement of DH-PBAN expression (Fig. 4A).

To identify the trans-activating factors that control DH-PBAN expression, we searched the 35-base sequence from nucleotide -1123 to -1088 of DH-PBAN for transcription factor binding sites using MatInspector (*http://www.genomatix.de/*). We found that the DNA sequence reported to be recognized by bicoid-like homeobox transcription factor, Pituitary homeobox 1 (Pitx), 5'-G(T/G)(A/G)*GATT(A/C)G(G*/T/A)GTCCA-3' (the Pitx binding core sequence has been underlined), is conserved at the 5'-upstream region of the fragment we examined; that is, from nucleotides -1117 to -1088 of DH-PBAN.

Next, we performed further reporter gene analysis and a gel-mobility shift assay using a 30 base-pair, double-stranded oligonucleotide encoding nucleotides -1117 to -1088 of the DH-PBAN promoter [42]. We revealed that a Bombyx Pitx (BmPitx) analog exists in the pupal Br-SG, and further, that BmPitx binds the Pitx consensus binding sequence we identified in the DH-PBAN promoter.

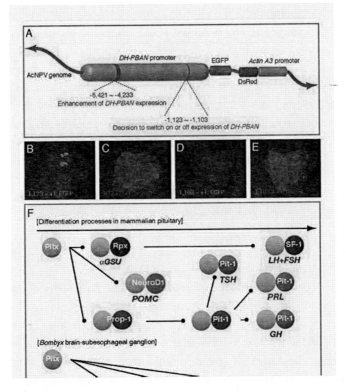

Figure 4. Reporter gene analysis and *cis*-regulatory elements of *DH-PBAN* expression. (A); *Cis*-regulatory elements of *DH-PBAN* expression. Several recombinant AcNPVs varied in terms of the upstream regions of *DH-PBAN*, which was fused with an *EGFP* reporter gene, were constructed with the *DsRed* reporter gene under the control of the *Actin A3* promoter to evaluate the efficiency of infection. Two *cis*-regulatory elements were identified that (i) in a 20-base sequence from -1123 to -1103 of *DH-PBAN* contain the *cis*-regulatory element that participates in switching on or off expression of *DH-PBAN*, and that (ii) in a region from -5,421 to -4,233 of *DH-PBAN* contain the *cis*-regulatory element(s) that participates in enhancement of *DH-PBAN* expression. (B - E); Fluorescence microscopy was used to visualize EGFP expression by green colors (B, D; under the *DH-PBAN* promoter) and DsRed by red colors (C, E; under the *ActA3* promoter) in the SG injected with recombinant AcNPVs carrying nucleotides -1,123 to +94 (B, C) and 1,103 to +94 (D, E) of *DH-PBAN*. (F); Pitx functions in both pituitary development in mammals and the *Bombyx* brain-SG complex. In the mammalian pituitary, a pan-pituitary regulator, Pitx1 achieves cell- and neuropeptide promoter-specific transcriptional activation via cooperation with lineage-restricted transcription factors. Pitx1 cooperates physically with Pit1 on the PRL, TSH, and GH promoters, with NeuroD1 on the POMC promoter, and with SF-1 on the LHβ and FSH promoters, and with Rpx on the αGSU promoter [49]. On the other hand, Pitx also acts as a transcription factor in *Bombyx*, which regulates neuropeptide gene expression of *DH-PBAN* and *PTTH*.

5. Function of the *Pitx* Homeobox Gene in the Central Nervous System of *Bombyx*

We cloned *BmPitx* cDNA from brain-SG complex mRNA using a PCR-based strategy. A 1599 bp sequence containing the 5'- and 3'-untranslated regions of *BmPitx* (AB162107) was obtained by RT-PCR and rapid amplification of cDNA ends [42]. The open reading frame corresponding to positions +296 to +1456 of the cDNA has the ability to encode a 387 aa protein. Both a homeodomain and an OAR domain [44] [45] [46] are encoded by a 60 aa region (homeodomain) and a 16 aa region (OAR domain) found in the predicted protein sequence. The two regions are highly conserved in Pitx proteins from a variety of organisms.

To determine if expression of BmPitx affects *DH-PBAN* expression, we injected recombinant AcNPVs, v[DH7/Pitx-s] and v[DH7/Pitx-i], for overexpression and silencing of the *BmPitx*, respectively, into a diapause-egg producer of the 'Kosetsu' strain, which was selected from among twenty bivoltine strains that are semi-permissive for AcNPV [42]. When v[DH7/Pitx-s] was injected, the level of *BmPitx* mRNA detected in pupae via northern blot analysis (3.7 kb band) was higher than that in animals that were not injected or that were injected with v[DH7/EGFP] [42]. The amount of *DH-PBAN* mRNA increased after injection with v[DH7/Pitx-s], and v[DH7/Pitx-i] appeared to suppress expression of *DH-PBAN* mRNA [42]. Thus, the two recombinant AcNPVs, v[DH7/Pitx-s] and v[DH7/Pitx-i], were able to induce overexpression and to suppress expression of the *DH-PBAN*, respectively. We have thus provided evidence that *BmPitx* encodes a transcription factor involved in *DH-PBAN* neuropeptide hormone gene expression in DHPCs.

Since BmPitx is highly conserved both in the 60 aa homeodomain region and in the 16 aa OAR domain [44] [45] [46], BmPitx appears to be a structural analog of Pitx family proteins which are members of the *bicoid*-like *Pitx* homeobox transcription factor family and include three vertebrate paralogues, *Pitx1, Pitx2*, and *Pitx3* (also called *Ptx1-3*) [47] [48]. The *Pitx1/2* genes are upstream regulators of transcription factors required for progression of pituitary development and cell differentiation. Pitx1 is present in all pituitary lineages and is a strong activator of several pituitary-specific promoters; thus, Pitx1 is considered to be a pan-pituitary transcriptional activator (Fig. 4F) [49]. In addition, mammalian Pitx1 acts to control expression at promoters of hormone-encoding genes, including the promoters of growth hormone (GH), prolactin (PRL), pro-opiomelanocortin (POMC), luteinizing hormone β-subunit (LHβ), glycoprotein hormone α-subunit (αGSU), follicle stimulating hormone, b (FSH), and thyroid-stimulating hormone (TSH) (Fig. 4F) [49]. In the insect *Drosophila melanogaster*, only one member of this *Pitx* gene family, *D-Ptx1*, has been identified [50]. Interestingly, the lack of apparent morphological phenotypes in *D-Ptx1* gain- and loss-of-function animals has implied a role for this gene in controlling physiological cell functions, as opposed to roles in pattern formation. It is tempting to speculate, then, that D-Ptx1 controls physiological functions such as transmitter or hormone production [50], similar to control of UNC-30 in *C. elegans* [51].

Currently, we also identified the *cis*-regulatory element participating in the decision to switch on or off expression of *PTTH* expression using an AcNPV-mediated gene transfer system and gel-mobility shift assay. We also demonstrated that BmPitx or Pitx-like homeodomain protein binds this *cis*-regulatory element (unpublished data) (Fig. 4F). Thus, it appears that Pitx function contributes to neuropeptide hormone gene expression in neurosecretory cells, not in only the vertebrate pituitary, but also in the insect central nervous system (Fig. 4F).

As described above, Pitx1 achieves cell- and promoter-specific transcriptional activation via cooperation with lineage-restricted transcription factors (Fig. 4F). Indeed, Pitx1 cooperates physically with Pit1, a POU domain transcriptional factor, on the PRL and GH promoters [49] [52], with the basic helix-loop-helix heterodimer NeuroD1/Pan1 on the

POMC promoter [53], with the orphan nuclear receptor SF-1 [49] [54], with Tpit [55], and with the immediate early factor Egr-1 [56], in order to achieve regulation at the LHβ promoter. These interactions are considered to be part of a framework of a combinational code that sets up lineage-specific gene expression in the pituitary.

Interestingly, binding of the POU transcription factor POU-M2 to the proximal promoter of *DH-PBAN* directly activates the *DH-PBAN* promoter [17]. Thus, BmPitx and POU-M2 may also cooperate to achieve transactivation of *DH-PBAN,* an effect that may occur through a "combinational code mechanism" [57] [58] such as that described for the vertebrate neuroendocrine system. However, as *BmPitx* is expressed in various tissues (including in the neurosecretory cells of SG) [42] and as *POU-M2* is also widely distributed in a variety of tissues, it is possible that these regulators exert effects not just in the SG, but also in other cell types, such that the cell-specific expression observed for *DH-PBAN* includes the participation of as yet unidentified factor(s) expressed in DHPCs. In addition, the results are consistent with the idea that the insect neurosecretory cells and SG-CC are functionally comparable to the vertebrate neuroendocrine organ (that is, the pituitary gland). Indeed, as both the function and the developmental regulation of the CC and the pituitary are similar [59], it is perhaps not surprising to find that the two systems share a common regulatory mechanism; that is, a gene expression cascade.

Thus, the present study is the first to identify a neuropeptide-encoding gene as a target of the Pitx transcriptional regulator in invertebrates. It is tempting to speculate that functional conservation of Pitx family members in neuropeptide gene expression occurs through a "combinational code mechanism" in both vertebrate and invertebrate in neuroendocrine systems.

6. Regulatory Mechanisms of *PTTH* Expression

We also used the AcNPV-mediated gene transfer system to investigate the molecular mechanisms controlling *PTTH* expression by PTPCs [43]. We constructed various recombinant AcNPVs carrying different upstream regions of the *PTTH* fused with the *EGFP* reporter gene. EGFP fluorescence was observed in PTPCs. We also measured the fluorescence intensity in somata and compared it to the intensity of the recombinant AcNPV (v[PT/EGFP]) carrying nucleotides -879 to +52 of the *PTTH* promoter (Fig. 3D, E). Progressive deletion of the 5'-upstream region, from nucleotides -180 to +52, had no significant effect on EGFP expression. However, recombinant AcNPVs carrying nucleotides -167 to +52 and/or -119 to +52 of the *PTTH* promoter caused an abrupt decrease in the expression of EGFP in the PTPCs. Using a recombinant AcNPV carrying nucleotides -105 to +52 of the *PTTH* promoter, EGFP expression was faint, and no fluorescence signal was observed in some pupa. No expression was observed when nucleotides -60 to +52 of the *PTTH* promoter were used. Thus, using EGFP reporter gene analysis and serial deletion of the *PTTH* promoter, we identified two potential *cis*-regulatory elements: (i) a 61-base sequence from nucleotides -180 to -119 that participates in the enhancement of *PTTH* expression; and (ii) a proximal region of the transcriptional start site that helps direct the expression of *PTTH* in PTPCs.

As described above, we determined that BmPitx or Pitx-like homeodomain protein binds to the proximal region of transcriptional start site of *PTTH*, and acts as a decision switch for turning on or off expression of *PTTH* (unpublished data). Furthermore, we tried to identify the *trans*-activating factors that enhance *PTTH* gene expression. We searched the 61-base sequence from nucleotides -180 to -119 of *PTTH* for transcription factor-binding sites using MatInspector (*http://www.genomatix.de/*). As shown in Fig. 5A, we found that the DNA sequence bound by myocyte enhancer factor 2 (MEF2), C/TTA(A/T)$_4$TAG/A, is conserved at the 5'-upstream region from nucleotides -180 to -151 of the *PTTH* gene (cacaatggtt CTATTTTAAGgatttatcac; MEF2 binding consensus in capital letters; Fig. 5A). MEF2 belongs to the family of MADS box transcription factors, which bind to DNA as homo- and heterodimers through the consensus MEF2 binding sequence. This sequence is found in the upstream regions of numerous genes including muscle-specific genes, and it plays a critical role in the differentiation of cells during the development of multicellular organisms [60].

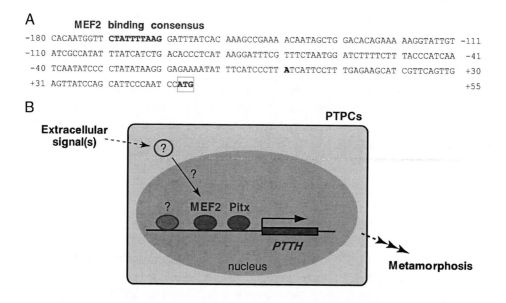

Figure 5. Regulatory mechanisms of *PTTH* expression. (A); 5'-upstream region of the *PTTH* gene from nucleotides -180 to 55. MEF2 binding consensus is shown in blue letters. The transcriptional initiation site is shown in bold letters. The translational initiation site is shown boxed. (B); Putative regulatory mechanisms of *PTTH* expression. In the *PTTH* promoter sequence, we identified two *cis*-regulatory elements that (i) participate in the enhancement of PTTH expression, and (ii) help direct *PTTH* expression in PTPCs, although there may be other *cis*-regulatory elements participating in *PTTH* expression. Each element binds to the transcription factors BmMEF2 and BmPitx, or Pitx-like homeodomain protein, respectively, and transactivates *PTTH* expression. Subsequently, PTTH stimulates the prothoracic gland to secrete 20-hydroxyecdysone, the steroid necessary for metamorphosis. Furthermore, we speculate that extracellular signal(s) activates both MEF2 and *PTTH* expression through the growth factor signaling pathway.

A gel mobility shift assay using a 30-base pair double-stranded oligonucleotide encoding nucleotides -180 to -151 of the *PTTH* promoter as a probe revealed that a protein in *Bombyx* brain bound to the MEF2 consensus binding sequence, and its binding was prevented by an antiserum that recognizes the MADS box of BmMEF2. In addition, the reporter gene assay showed that the expression of EGFP was altered by mutation of the MEF2 consensus binding sequence in the *PTTH* promoter, a region important for enhancing reporter gene expression. These findings suggest that the *Bombyx* MEF2 homologue binds to the MEF2 consensus binding sequence in the *PTTH* promoter, enhancing *PTTH* gene expression [43].

There are four isoforms (A to D) of mammalian MEF2, and they have high homology within the 56 amino acid MADS box at their N-termini and within an adjacent 29 amino acid region referred to as the MEF2 domain. The MADS box is essential for DNA binding and dimerization, and the MEF2 domain plays an important role in DNA binding affinity as well as an indirect role in dimerization. The C-terminal portion of MEF2C is required for its transcriptional activation [61]. We next cloned *MEF2* cDNA from the brain-SG complex in *Bombyx* using a PCR-based strategy with degenerate primers corresponding to the MADS-box and the MEF2 domain, regions that are highly conserved across a variety of organisms. A 2716 base-pair sequence containing the 5'- and 3'-untranslated regions of *Bombyx mori MEF2* (*BmMEF2*; accession number AB121093) was obtained by reverse transcription polymerase chain reaction (RT-PCR) and rapid amplification of cDNA ends. The open reading frame was from nucleotides +748 to +1965 and encoded a predicted 404 amino acid protein [43]. A MADS box and an adjacent MEF2 domain are highly conserved in MEF2s from various organisms. The *Bombyx* sequence is most similar to that of *Drosophila melanogaster* (D-MEF2), with 96% amino acid sequence identity in the MADS box and MEF2 domain.

We examined the developmental expression of *BmMEF2* in various tissues during embryonic and post-embryonic development by RT-PCR. *BmMEF2* mRNA was expressed in various tissues, including muscle and neural tissues, as well as in *Drosophila melanogaster* and various vertebrates [43]. Furthermore, we determined the localization of *BmMEF2* mRNA in brain by whole-mount *in situ* hybridization. In brain, the *BmMEF2* gene was expressed at elevated levels in two types of lateral neurosecretory cells, namely PTPCs and corazonin-like immunoreactive lateral neurosecretory cells (CLI-LNCs) [62]. Consequently, the correlation between structure and gene expression profiles suggests that the BmMEF2 is a structural and functional analogue of MEF2 proteins in various organisms. Furthermore, it has been speculated that BmMEF2 is responsible for the regulation of fundamental cellular processes in various tissues.

To investigate whether the expression of BmMEF2 affects *PTTH* gene expression, we constructed two recombinant AcNPVs, v[PT/MEFs] and v[PT/MEFi], which were designed to overexpress and silence *BmMEF2* mRNA under control of the *PTTH* promoter, respectively. Using RT-PCR, we first investigated the effect of infection with AcNPV on the amounts of mRNAs transcribed from the *BmMEF2*, *PTTH*, and *actin A3* genes. When v[PT/MEFs] was injected, there was a higher level of BmMEF2 mRNA than in pupae that were not injected or that were injected with v[PT/EGFP] [43]. Injection of v[PT/MEFi] caused a great reduction of the BmMEF2 mRNA compared to non-injected and v[PT/EGFP]-injected pupae. Thus, the two recombinant AcNPVs, v[PT/MEFs] and v[PT/MEFi], were able to induce overexpression and suppression of the *BmMEF2* gene, respectively. In addition, the amount of PTTH mRNA was also increased by injection with v[PT/MEFs]. However, v[PT/MEFi] did not cause elimination of PTTH mRNA. Thus, BmMEF2 activated but was not essential for *PTTH* gene expression. BmMEF2 therefore appears to be responsible for the enhancement of *PTTH* expression.

We demonstrated that the MEF2 binding sequences in the *PTTH* promoter enhance expression of the *EGFP* reporter gene in PTPCs and that they are important for binding of the BmMEF2 protein. Furthermore, overexpression of the *BmMEF2* gene can induce *PTTH* gene expression. Thus, it appears that BmMEF2 plays a role in the enhancement of *PTTH* gene expression in PTPCs. A single *MEF2* gene, *D-mef2*, has been identified in *Drosophila melanogaster*, and isoforms of the D-MEF2 protein act as functional analogues of the vertebrate forms that participate in muscle differentiation [63] [64]. Furthermore, D-MEF2 protein is expressed in Kenyon cells in the mushroom bodies of larval and adult brains, suggesting that these proteins are responsible for the differentiation of the Kenyon cells and for morphogenesis of the mushroom body learning center [65]. However, the target genes for D-MEF2 have not been identified, and MEF2 functions have not been determined in the insect nervous system. Thus, these findings are the first identification of a gene that is a target of MEF2 in the invertebrate nervous system.

Aizono and Shirai suggested that muscarinic acetylcholine receptor-induced signal transduction is involved in the control of PTTH release in *Bombyx mori* [66]. Activation of phospholipase C and subsequent activation of both protein kinase C and calmodulin-dependent kinase are essential in this signaling pathway. Furthermore, MEF2 is known to act as an endpoint for growth factor signaling pathways [67]. Although we identified BmMEF2 as a factor that enhances *PTTH* gene expression, BmMEF2 may participate in several other cellular processes that regulate PTTH secretion through signaling pathways. Thus, it will be important to further investigate the signal transduction pathway by which extracellular signals regulate insect functions including molting, metamorphosis, and diapause. For now, we have investigated the extracellular signals participating with *PTTH* expression (Fig. 5B).

6. Conclusion

We have successfully developed an efficient gene transfer system using recombinant AcNPV in the silkworm, *Bombyx mori*. Using this system, we identified *cis*-regulatory elements and trans-activating factors participating in *DH-PBAN* and *PTTH* expression, which give us a preliminary view of the molecular mechanisms of insect neuropeptide hormone gene expression involved in diapause and metamorphosis. In particular, our work suggests that the *Pitx* homeobox gene is conserved not only in the vertebrate pituitary, but also in the insect neuroendocrine system, in both structure and function as a pan-activator. Hereafter, we will use this gene transfer system to search for additional gene(s) and neural networks related to diapause and metamorphosis in insects.

7. References

[1]   Ichikawa, T. (1991) Architecture of cerebral neurosecretory cell systems in the silkworm *Bombyx mori*. *J. Exp. Biol.* 161, 217-37.

[2]   Nassel, D. R. (2000) Functional roles of neuropeptides in the insect central nervous system. *Naturwissenschaften* 87, 439-49.

[3]   Yamashita, O. (1996) Diapause hormone of the silkworm, *Bombyx mori*: Structure, gene expression and function. *J. Insect Physiol.* 42, 669-79.

[4]   Suzuki, A. (1995) Neuropeptides of *Bombyx mori*. In *Molecular mechanisms of insect metamorphosis and diapause* (A. Suzuki, H. Kataoka, S. Matsumoto eds.). Industrial Publishing & Consulting, Inc., Tokyo, pp. 3-10.

[5]   Holman, G. M., Cook, B. J., and Nachman, R. J. (1986) Primary structure and synthesis of a blocked myotropic neuropeptide isolated from the cockroach, Leucophaea maderae. *Comp. Biochem. Physiol. C* 85, 219-24.

[6]   Imai, K., Konno, T., Nakazawa, Y., Komiya, T., Isobe, M., Koga, K., Goto, T., Yaginuma, T., Sakakibara, K., Hasegawa, K., and Yamashita, O. (1991) Isolation and structure of diapause hormone of the silkworm, *Bombyx mori*. *Proc. Jpn. Acad. Ser. B* 67, 98-101.

[7]   Kitamura, A., Nagasawa, H., Kataoka, H., Inoue, T., Matsumoto, S., Ando, T., and Suzuki, A. (1989) Amino acid sequence of pheromone-biosynthesis-activating neuropeptide (PBAN) of the silkworm, *Bombyx mori*. *Biochem. Biophys. Res. Commun.* 163, 520-6.

[8]   Raina, A. K., Jaffe, H., Kempe, T. G., Keim, P., Blacher, R. W., Fales, H. M., Riley, C. T., Ridgway, R. I., and Hayes, D. K. (1989) Identification of a neuropeptide hormone that regulates sex pheromone production in female moths. *Science* 244, 796-98.

[9]   Schoofs, L., Holman, G. M., Hayes, T. K., Tips, A., Nachman, R. J., Vandesande, F., and De Loof, A. (1990) Isolation and identification of locusta-myotropin (Lom-MT), a novel biologically active insect neuropeptide. *Peptides* 11, 427-33.

[10]   Sato, Y., Oguchi, M., Menjo, N., Imai, K., Saito, H., Ikeda, M., Isobe, M., and Yamashita, O. (1993) Precursor polyprotein for multiple neuropeptides secreted from the suboesophageal ganglion of the silkworm *Bombyx mori*: characterization of the cDNA encoding the diapause hormone precursor and identification of additional peptides. *Proc. Natl. Acad. Sci. U S A* 90, 3251-5.

[11]   Sato, Y., Ikeda, M., and Yamashita, O. (1994) Neurosecretory cells expressing the gene for common precursor for diapause hormone and pheromone biosynthesis-activating neuropeptide in the suboesophageal ganglion of the silkworm, *Bombyx mori*. *Gen. Comp. Endocrinol.* 96, 27-36.

[12]   Sato, Y., Shiomi, K., Saito, H., Imai, K., and Yamashita, O. (1998) Phe-X-Pro-Arg-Leu-NH$_2$ peptide producing cells in the central nervous system of the silkworm, *Bombyx mori*. *J. Insect Physiol.* 44, 333-42.

[13]   [Shiomi, K., Kajiura, Z., Nakagaki, M., and Yamashita, O. (2003) Baculovirus-mediated efficient gene transfer into the central nervous system of the silkworm, *Bombyx mori*. *J. Insect Biotechnol. Sericol.* 72, 149-55.

[14]   Morita, A., Niimi, T., and Yamashita, O. (2003). Physiological differentiation of DH-PBAN-producing neurosecretory cells in the silkworm embryo. *J. Insect Physiol.* 49, 1093-102.

[15]   Homma, T., Watanabe, K., Tsurumaru, S., Kataoka, H., Imai, K., Kamba, M., Niimi, T., Yamashita, O., and Yaginuma, T. (2006) G protein-coupled receptor for diapause

hormone, an inducer of *Bombyx* embryonic diapause. *Biochem. Biophys. Res. Commun.* 344, 386-93.

[16] Ishida, Y., Niimi, T., and Yamashita, O. (2000) The *cis*-regulatory region responsible for the *BomDH-PBAN* gene expression in FXPRLamide peptide producing neurosecretory cells of the transformed *Drosophila*. *J. Seric. Sci. Jpn.* 69, 111-9.

[17] Zhang, T. Y., Kang, L., Zhang, Z. F., and Xu, W. H. (2004) Identification of a POU factor involved in regulating the neuron-specific expression of the gene encoding diapause hormone and pheromone biosynthesis-activating neuropeptide in *Bombyx mori*. *Biochem. J.* 380, 255-63.

[18] Truman, J.W. & Riddiford, L.M. (1974) Physiology of insect rhythms. 3. The temporal organization of the endocrine events underlying pupation of the tobacco hornworm. *J. Exp. Biol.* 60, 371-82.

[19] Mizoguchi, A., Dedos, S.G., Fugo, H., and Kataoka, H. (2002) Basic pattern of fluctuation in hemolymph PTTH titers during larval-pupal and pupal-adult development of the silkworm, *Bombyx mori*. *Gen. Comp. Endocrinol.* 127, 181-9.

[20] Bollenbacher, W.E. & Granger, N.A. (1985) Endocrinology of the prothoracicotropic hormone. In *Comprehensive insect physiology, biochemistry and pharmacology* (Kerkut GA & Gilgert LI, eds), vol. 7, pp. 109-151. Pergamon, Oxford.

[21] Kataoka, H., Nagasawa, H., Isogai, A., Ishizaki, H., and Suzuki, A. (1991) Prothoracicotropic hormone of the silkworm, *Bombyx mori*: amino acid sequence and dimeric structure. *Agric. Biol. Chem.* 55, 73-86.

[22] Kawakami, A., Kataoka, H., Oka, T., Mizoguchi, A., Kimura-Kawakami, M., Adachi, T., Iwami, M., Nagasawa, H., Suzuki, A., and Ishizaki, H. (1990) Molecular cloning of the *Bombyx mori* prothoracicotropic hormone. *Science* 247, 1333-5.

[23] Adachi-Yamada, T., Iwami, M., Kataoka, H., Suzuki, A., and Ishizaki, H. (1994) Structure and expression of the gene for the prothoracicotropic hormone of the silkmoth *Bombyx mori*. *Eur. J. Biochem.* 220, 633-43.

[24] Mizoguchi, A., Oka, T., Kataoka, H., Nagasawa, H., Suzuki, A., and Ishizaki, H. (1990) Immunohistochemical localization of prothoracicotropic hormone-producing neurosecretory cells in the brain of *Bombyx mori*. *Dev. Growth Differ.* 32, 591-8.

[25] Mizoguchi, A., Ohashi, Y., Hosoda, K., Ishibashi, J., and Kataoka, H. (2001) Developmental profile of the changes in the prothoracicotropic hormone titer in hemolymph of the silkworm *Bombyx mori*: correlation with ecdysteroid secretion. *Insect Biochem. Mol. Biol.* 31, 349-58.

[26] Xu, W.H. & Denlinger, D.L. (2003) Molecular characterization of prothoracicotropic hormone and diapause hormone in *Heliothis virescens* during diapause, and a new role for diapause hormone. *Insect Mol. Biol.* 12, 509-16.

[27] Moto, K., Abdel Salam, S.E., Sakurai, S., and Iwai, M. (1999) Gene transfer into insect brain and cell-specific expression of bombyxin gene. *Dev. Genes Evol.* 209, 447-50.

[28] Tamura, T., Thibert, C., Royer, C., Kanda, T., Abraham, E., Kamba, M., Komoto, N., Thomas, J.L., Mauchamp, B., Chavancy, G., Shirk, P., Fraser, M., Prudhomme, J.C., and Couble, P. (2000) Germline transformation of the silkworm *Bombyx mori* L. using a *piggyBac* transposon-derived vector. *Nat. Biotechnol.* 18, 81-4.

[29] Thomas, J.L., Da Rocha, M., Besse, A., Mauchamp, B., and Chavancy, G. (2002) 3xP3-EGFP marker facilitates screening for transgenic silkworm *Bombyx mori* L. from the embryonic stage onwards. *Insect Biochem. Molec. Biol.* 32, 247-53.

[30] Uhlirova, M., Foy, B.D., Beaty, B.J., Olson, K.E., Riddiford, L.M., and Jindra, M. (2003) Use of Sindbis virus-mediated RNA interference to demonstrate a conserved role of Broad-Complex in insect metamorphosis. *Proc. Natl. Acad. Sci. U S A.* 100, 15607-12.

[31] Foy, B.D., Myles, K.M., Pierro, D.J., Sanchez-Vargas, I., Uhlírová, M., Jindra, M., Beaty, B.J., and Olson, K.E. (2004) Development of a new Sindbis virus transducing system and its characterization in three Culicine mosquitoes and two Lepidopteran species. *Insect Mol. Biol.* 13, 89-100.

[32] Moto, K., Kojima, H., Kurihara, M., Iwami, M., and Matsumoto, S. (2003) Cell-specific expression of enhanced green fluorescence protein under the control of neuropeptide gene promoters in the brain of the silkworm, *Bombyx mori*, using Bombyx *mori* nucleopolyhedrovirus-derived vectors. *Insect Biochem. Mol. Biol.* 33, 7-12.

[33] Black, B.C., Brennan, L.A., Dierks, P.M., and Gard, I.E. (1997) Commercialization of baculoviral insecticides. In *The Baculoviruses* (L.K. Miller ed.). Plenum Press, New York, pp. 341-87.

[34] Jarvis, D.L. (1997) Baculovirus expression vectors. In *The Baculoviruses* (L.K. Miller ed.). Plenum Press, New York, pp. 389-431

[35] Hajos, J.P., Vermunt, A.M., Zuidema, D., Kulcsar, P., Varjas, L., de Kort, C.A., Zavodszky, P., and Vlak J.M. (1999) Dissecting insect development: baculovirus-mediated gene silencing in insects. *Insect Mol. Biol.* 8, 539-44.

[36] Sarkis, C., Serguera, C., Petres, S., Buchet, D., Ridet, J.L., Edelman, L., and Mallet, J. (2000) Efficient transduction of neural cells *in vitro* and *in vivo* by a baculovirus-derived vector. *Proc. Natl. Acad. Sci. USA* 97, 14638-43.

[37] Oppenheimer, D.I., MacNicol A.M., and Patel, N.H. (1999) Functional conservation of the *wingless-engrailed* interaction as shown by a widely applicable baculovirus misexpression system. *Curr. Biol.* 9, 1288-96.

[38] Ikeda, M., Katou, Y., Yamada, Y., Chaeychomsri, S., and Kobayashi, M. (2001) Characterization of *Autographa californica* nucleopolyhedrovirus infection in cell lines from *Bombyx mori*. *J. Insect Biotechnol. Sericol.* 70, 49-58.

[39] Mori, H., Yamao, M., Nakazawa, H., Sugahara, Y., Shirai, N., Matsubara, F., Sumida, M., and Imamura, T. (1995) Transovarian transmission of a foreign gene in the silkworm, *Bombyx mori*, by *Autographa californica* nuclear polyhedrosis virus. *Biotechnology* 13, 1005-7.

[40] Shikata, M., Shibata, H., Sakurai, M., Sano, Y., Hashimoto, Y., and Matsumoto, T. (1998) The ecdysteroid UDP-glucosyltransferase gene of *Autographa californica* nucleopolyhedrovirus alters the moulting and metamorphosis of a non-target insect, the silkworm, *Bombyx mori* (Lepidoptera, Bombycidae). *J. Gen. Virol.* 79, 1547-51.

[41] Guo, T., Wang, S., Guo, X., and Lu, C. (2005) Productive infection of *Autographa californica* nucleopolyhedrovirus in silkworm *Bombyx mori* strain Haoyue due to the absence of a host antiviral factor. *Virology* 341, 231-7.

[42] Shiomi, K., Fujiwara, Y., Yasukochi, Y., Kajiura, Z., Nakagaki, M., and Yaginuma, T. (2008) The *Pitx* homeobox gene in *Bombyx mori*: regulation of *DH-PBAN* neuropeptide hormone gene expression. *Mol. Cell. Neurosci.* 34, 209-18.

[43] Shiomi, K., Fujiwara, Y., Atsumi, T., Kajiura, Z., Nakagaki, M., Tanaka, Y., Mizoguchi, A., Yaginuma, T., and Yamashita, O. (2005) Myocyte enhancer factor 2 (MEF2) is a key modulator of the expression of the prothoracicotropic hormone gene in the silkworm, *Bombyx mori*. *Febs J.* 272, 3853-62.

[44] Amendt, B. A., Sutherland, L. B., and Russo, A. F. (1999) Multifunctional role of the Pitx2 homeodomain protein C-terminal tail. *Mol. Cell. Biol.* 19, 7001-10.

[45] Brouwer, A., ten Berge, D., Wiegerinck, R., and Meijlink, F. (2003) The OAR/aristaless domain of the homeodomain protein Cart1 has an attenuating role in *vivo*. *Mech. Dev.* 120, 241-52.

[46] Furukawa, T., Kozak, C. A., and Cepko, C. L. (1997) *rax*, a novel paired-type homeobox gene, shows expression in the anterior neural fold and developing retina. *Proc. Natl. Acad. Sci. U S A* 94, 3088-93.

[47] Lamonerie, T., Tremblay, J. J., Lanctot, C., Therrien, M., Gauthier, Y., and Drouin, J. (1996) Ptx1, a bicoid-related homeo box transcription factor involved in transcription of the pro-opiomelanocortin gene. *Genes Dev.* 10, 1284-95.

[48] Semina, E. V., Reiter, R., Leysens, N. J., Alward, W. L., Small, K. W., Datson, N. A., Siegel-Bartelt, J., Bierke-Nelson, D., Bitoun, P., Zabel, B. U., Carey, J. C., and Murray, J. C. (1996) Cloning and characterization of a novel bicoid-related homeobox transcription factor gene, *RIEG*, involved in Rieger syndrome. *Nat. Genet.* 14, 392-9.

[49] Tremblay, J. J., Lanctot, C., and Drouin, J. (1998) The pan-pituitary activator of transcription, Ptx1 (pituitary homeobox 1), acts in synergy with SF-1 and Pit1 and is an upstream regulator of the Lim-homeodomain gene *Lim3/Lhx3*. *Mol. Endocrinol.* 12, 428-41.

[50] Vorbruggen, G., Constien, R., Zilian, O., Wimmer, E. A., Dowe, G., Taubert, H., Noll, M., and Jackle, H. (1997) Embryonic expression and characterization of a Ptx1 homolog in *Drosophila*. *Mech. Dev.* 68, 139-47.

[51] Jin, Y., Hoskins, R., and Horvitz, H. R. (1994) Control of type-D GABAergic neuron differentiation by *C. elegans* UNC-30 homeodomain protein. *Nature* 372, 780-3.

[52] Szeto, D. P., Ryan, A. K., O'Connell, S. M., and Rosenfeld, M. G. (1996) P-OTX: a PIT-1-interacting homeodomain factor expressed during anterior pituitary gland development. *Proc. Natl. Acad. Sci. U S A* 93, 7706-10.

[53] Poulin, G., Turgeon, B., and Drouin, J. (1997) NeuroD1/beta2 contributes to cell-specific transcription of the proopiomelanocortin gene. *Mol. Cell. Biol.* 17, 6673-82.

[54] Tremblay, J. J., Marcil, A., Gauthier, Y., and Drouin, J. (1999) Ptx1 regulates SF-1 activity by an interaction that mimics the role of the ligand-binding domain. *Embo J.* 18, 3431-41.

[55] Lamolet, B., Pulichino, A. M., Lamonerie, T., Gauthier, Y., Brue, T., Enjalbert, A., and Drouin, J. (2001) A pituitary cell-restricted T box factor, Tpit, activates POMC transcription in cooperation with Pitx homeoproteins. *Cell* 104, 849-59.

[56] Tremblay, J. J., and Drouin, J. (1999) Egr-1 is a downstream effector of GnRH and synergizes by direct interaction with Ptx1 and SF-1 to enhance luteinizing hormone beta gene transcription. *Mol. Cell. Biol.* 19, 2567-76.

[57] Struhl, K. (1991) Mechanisms for diversity in gene expression patterns. *Neuron* 7, 177-81.

[58] Wang, J. C., and Harris, W. A. (2005) The role of combinational coding by homeodomain and bHLH transcription factors in retinal cell fate specification. *Dev. Biol.* 285, 101-15.

[59] De Velasco, B., Shen, J., Go, S., and Hartenstein, V. (2004) Embryonic development of the *Drosophila* corpus cardiacum, a neuroendocrine gland with similarity to the vertebrate pituitary, is controlled by sine oculis and glass. *Dev. Biol.* 274, 280-94.

[60] McKinsey, T.A., Zhang, C.L., and Olson, E.N. (2002) MEF2: a calcium-dependent regulator of cell division, differentiation and death. *Trends Biochem. Sci.* 27, 40-7.

[61] Molkentin, J.D., Black, B.L., Martin, J.F., and Olson, E.N. (1996) Mutational analysis of the DNA binding, dimerization, and transcriptional activation domains of MEF2C. *Mol. Cell. Biol.* 16, 2627-36.

[62]] Roller, L., Tanaka, Y., and Tanaka, S. (2003) Corazonin and corazonin-like substances in the central nervous system of the Pterygote and Apterygote insects. *Cell Tissue Res.* 312, 393-406.

[63] Lilly, B., Galewsky, S., Firulli, A.B., Schulz, R.A., and Olson, E.N. (1994) D-MEF2: a MADS box transcription factor expressed in differentiating mesoderm and muscle cell lineages during *Drosophila* embryogenesis. *Proc. Natl. Acad. Sci. U S A* 91, 5662-6.

[64] Nguyen, H.T,, Bodmer, R., Abmayr, S.M., McDermott, J. C., and Spoerel, N.A. (1994) D-mef2: a *Drosophila* mesoderm-specific MADS box-containing gene with a biphasic expression profile during embryogenesis. *Proc. Natl. Acad. Sci. U S A* 91, 7520-4.

[65] Schulz, R.A., Chromey, C., Lu, M.F., Zhao, B., and Olson, E.N. (1996) Expression of the D-MEF2 transcription in the *Drosophila* brain suggests a role in neuronal cell differentiation. *Oncogene* 12, 1827-31.

[66] Aizono, Y. & Shirai, Y. (1995) Control of prothoracicotropic hormone release with neurotransmitters in silkworm. In *Molecular mechanisms of insect metamorphosis and diapause.* (Suzuki A, Kataoka H & Matsumoto S, eds), pp. 119-128. Industrial Publishing & Consulting, Inc., Tokyo.

[67] Cavanaugh, J.E. (2004) Role of extracellular signal regulated kinase 5 in neuronal survival. *Eur. J. Biochem.* 271, 2056-9.

In: Development Gene Expresión Regulation
Editor: Nathan C. Kurzfield

ISBN: 978-60692-794-6
©2009 Nova Science Publishers, Inc.

Chapter VIII

Gene Expression Analysis during Development by High-Throughput Methods

Francesca Amati[1], Giovanni Chillemi[2] and Giuseppe Novelli[1]

[1]Department of Biopathology and Diagnostic Imaging, Tor Vergata University, Rome, Italy; [2]CASPUR, Rome, Italy.

Abstract

Regulation of gene expression during embryogenesis and development is a crucial clue for a normal anatomy and physiology. In fact, very little is known regarding factors that influence and regulate developmental gene expression. Similarly, there is little information available concerning the effects of a coordinate expression of a group of functionally related genes.

The analysis of temporal patterns of gene expression in embryos is essential for the understanding of the molecular mechanisms that control development. This scientific field has been innovated by the combined use of experimental high-throughput methods, such as DNA microarrays, and bioinformatic methods that take advantage of the completion of the human genome sequence, along with the genomes of related species. Microarray analysis, in fact, provides a large amount of data -at molecular level- that once acquired, must be functionally integrated in order to find common patterns within a defined group of biological samples. Following the enormous number of data obtained from these experiments, a new type of comparative embryology is now emerging, and it is based on the comparison of gene expression patterns. The sequencing of several new genomes, the increasing computational power and new bioinformatic algorithms cooperate to overcome some of the intrinsic difficulties in the study of gene regulation, thus permitting, for example, to identify regulation sites located far away from the genes. Recent bioinformatic methods applied to gene regulation are reviewed that either follow the "single species, many genes" approach or the "single gene, many species" one.

In this chapter we would review the new application of DNA microarray and bioinformatics to define a new combinatorial approach for analysis of gene expression during development.

Introduction

"Between the characters that furnish the data for the theory, and the postulated genes, to which the characters are referred, lies the whole field of embryonic development."

Thomas Hunt Morgan (1926)

Morgan's sentence illustrates that the only way to get from genotype to phenotype is through developmental processes. In fact, if the genome is the same in all somatic cells within an organism (with the exception of the lymphocytes), how do the cells become different from one another? Based on the embryological evidence for genomic equivalence (and on bacterial models of gene regulation), a consensus emerged in the 1960s that cells differentiate through differential gene expression [for a review see Weiss, 2005].

This assumption has been strengthened by the completion of the sequencing of the human genome that resulted in a downward estimation of the total number of protein-coding genes to approximately 20,000 [Goodstadt, 2006]. This is much lower than previous estimates which ranged to as many as 150,000 genes and is similar to estimates for other vertebrate species including mouse, chicken and pufferfish [Aparicio, 2002; Gregory, 2002; Okazaky, 2002; Abril, 2002; Wallis, 2004], or bovine and dog [Fadiel, 2005; Lindblad-Toh, 2005]. Now it is clear that complex organisms (like mammals) exploit a variety of regulatory mechanisms to extend the functionality of their genomes [Hayashizaki, 2004], including differential promoter activation, alternative RNA splicing, RNA modification, RNA editing, localization, translation and stability of RNA, expression of non-coding RNA, antisense RNA and microRNAs [Biemont, 2006; Rassoulzadegan, 2006; Segal , 2006; Venkatesh , 2006; Werner, 2005; Heintzman 2007; Jirtle 2007]. Most of these mechanisms work together at the RNA level to produce the functional transcriptome of an organism.

The characterization of the transcriptome, by means a gene-expression profile, in oocytes and embryos is very important for the understanding of the influence of genetic and environmental factors on preimplantation and fetal development. The comprehension of the regulation of gene expression during embryogenesis and development, is particular relevant, since it is a crucial clue for a normal anatomy and physiology [Siddiqui, 2005]. In fact, very little is known regarding factors that influence and regulate developmental gene expression. Similarly, there is little information concerning the effects of a coordinate expression of a group of functionally related genes. Zhang et al. [2004] demonstrated that distinct and coordinate expression of a group of functionally related genes implies an underlying pathway-specific transcriptional regulatory mechanism.

At the most basic level, any particular cell state can be defined by a gene-expression profile and therefore by the set of transcription factors that determine its phenotype. This is mainly true during development, where the impact of specific gene expression within a cell will dictate the fate of that cell. So, the analysis of temporal patterns of gene expression in embryos is an essential component of any research program seeking to understand molecular mechanisms that control development. This analysis may be included in the area of functional genomics.

Functional genomics, in fact, focuses on understanding the function and regulation of genes and gene products on a global or genome-wide scale. Functional genomics derives from the sequencing of human genome that introduced the new postgenomic era in biology and medicine. The studies whose interest is gene expression during development may fall back in a specific area of functional genomics. In fact, while it is important to know the sequence of a gene and its temporal-spatial pattern of expression, what's really crucial is to know the functions of that gene during development.

The greatest challenge of functional genomics has been to extract useful information on genetic interactions from large data sets [Carulli, 1998; van Someren, 2002]. However, the detection of the genetic interactions that determine a phenotype, which themselves are related to protein and metabolite interactions, requires a mixture of computational and experimental approaches [Carter, 2005]. Several papers introduced the concepts and mathematical methods for reconstructing gene networks from gene expression profiles [Wagner, 2001; Brazhnik, 2002; de la Fuente, 2002; Olivieri, 2008] and in this chapter we will try to review most of these methods.

Thanks to the improvement of new technologies, global expression profiling is now possible, and has now been applied widely to the analysis of early embryos and stem cells. Moreover, global approaches have gradually been extended to the functional analysis of genes [Davidson, 2002; Smith, 2007]. A systematic genomic approach to analyze global gene expression patterns and functions during embryogenesis have recently been named developmental genomics or embryogenomics [Ko, 2001].

Embryogenomics, consists in the systematic analysis of cohorts of genes expressed during development by means of large-scale genomic methods [Ko, 2001]. These methods include various techniques such as: cDNA subtraction [Zhu, 2000], serial analysis of gene expression [SAGE; Powell, 2000], differential display [McClelland, 1995], cDNA and oligonucleotide microarray [Blohm, 2001], large scale *in-situ* hybridization [Visel, 2004], substractive hybridization [Byers, 2000].

Most researches on mammalian development have focused on the mouse embryo, since mice are relatively easy to breed throughout the year, have large litters, and can be housed easily. Thus, most of the studies discussed here will concern mouse development.

This review focuses on functional genomics in the mouse embryogenesis and provides a road map for large-scale exploration of the mouse transcriptome.

Gene Expression Profiling with Microarrays

The fate of any cell is determined largely by the subset of genes it expresses. The gene expression pattern governs the cell's behavioural characteristics, the type of tissue it constructs and whether it is normal or diseased. Activation or expression of a specific combination of genes, in precise amounts and in a timely manner, is crucial for a cell's normal activity [Ooi, 2008].

The Human Genome Project has resulted in an exponential growth in the amount of information available about the DNA sequence of the human genome [Venter, 2001; Collins, 2003]. This information is one of the most fundamental requisite for detailed expression array analysis.

Through the use of cDNA microarrays, investigators can measure mRNA levels for thousands of genes simultaneously, rather than one gene at a time. In fact, DNA microarrays are constituted by small glass or filter matrix that contain arrays of DNA sequences (each highly specific to a single gene) and by means hybridization of fluorescent cDNA, they are capable of simultaneously quantifying the expression of thousands of genes in a single experiment (Fig.1).

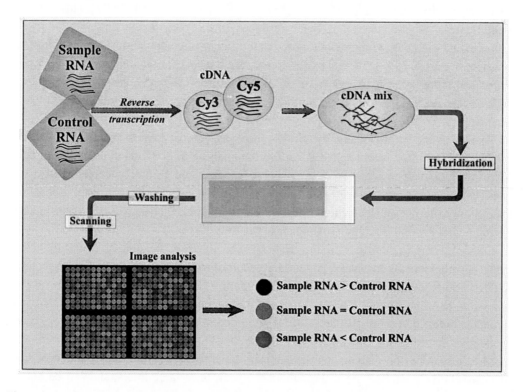

Figure 1. A schematic design of a microarray experiment.

The results of these experiments are spots whose brightness varies from gene to gene corresponding to the transcriptional activity of the examined genes (Fig. 1). In fact, the hybridization signals (brightness) are quantified and the amount of nucleic acid complementary to each spot is calculated with respect to local background. This analysis requires sophisticated bioinformatics tools that are reviewed by Breitling [2006] and Hovatta [2005].

Basically, with this technique it is possible to simultaneously measure the activity of virtually all genes in a genome. Moreover, the vast amount of data obtained by microarrays experiments need to be grouped to define specific information of biochemical pathways or gene function of the differentially expressed genes identified. To this purpose more important are freely available database such as Gene Ontology (GO; http://www.geneonotlogy.org) and the Kyoto Encyclopaedia of Genes and genomes (KEEG; http://www.genome.jp/kegg) (reviewed in the bioinformatics and computational biology paragraphs of this chapter). So, DNA microarrays represent a novel genetic platform which is being widely exploited to bridge the gap between gene sequence and function. This technology, allow researchers to compare large sets of genes in different tissues or cells and thereby identifies pathways and regulatory networks.

Therefore, this technology helps to pinpoint patterns of gene expression associated with disease, assess influences of environmental factors, evaluate effects of therapeutics, and identify molecular markers indicative of disease status and progression risk (Tab. 1).

Table 1. Summary of microarray applications

MICROARRAY	EXPRESSION	TARGET	APPLICATION
SNP	Gene	Single nucleotide variations in a gene sequence	Genotyping
ChIP on chip	Transcription	Protein-DNA complexes	Identification of transcription factors binding sites
DNA (cDNA or oligonucleotides)	mRNA	Steady-state mRNA	Gene expression profile
microRNA	Translation	Small non coding RNA	Gene regulation
Protein	Protein	Protein interaction	Identification of new protein interaction and biochemical activities
Antibody	Protein	Protein expression	Protein profiling, novel biomarker identification
Tissue	Gene	Tissues	Gene expression analysis

In this chapter, we will provide an overview of DNA microarray technology with particular attention to its application to dissect the genetic basis of embryo development.

As previously affirmed, transcriptome is defined as the collection of all RNAs produced in a cell or a tissue at a defined time and is one of many stages that make up a biological system. Analysis of the transcriptome using DNA microarrays has become a standard approach for investigating for example, the molecular basis of human disease in both clinical and experimental settings. In fact, the pattern of the transcriptional deregulation provides insights into the cause of abnormal phenotypes.

The determination of genes expressed in sick tissue compared with the normal counterpart allows understanding disease pathology and identifying potential points for therapeutic intervention. Other studies can focus on different stages of disease, different time points of treatments, different individual cells and the different locations. Moreover, microarray expression-profiling approaches can be applied also to complex diseases or polygenic conditions such as cardiovascular [Ashley, 2007] and renal diseases [Hayden, 2003] [for a review see Gu, 2002; Lyons, 2002; Grant, 2008].

Specific questions related to DNA methylation and microRNA expression (both important for embryonic development) have also been addressed to DNA microarrays.

DNA methylation has an important role in many aspects of biology, including development and disease. However, the genomic distribution of methylated sequences, the methylome, is poorly understood. Cytosine methylation is the most common covalent modification of DNA in eukaryotes. Methylation can be detected using bisulfite conversion, methylation-sensitive restriction enzymes, methyl-binding proteins and anti-methylcytosine antibodies. The combination of these techniques with DNA microarrays and high-throughput sequencing has made the mapping of DNA methylation feasible on a genome-wide scale [Zilberman, 2007; Schumacher, 2006; Wilson, 2006].

Mature microRNAs (miRNAs) are about 18 to 24 nucleotides (nt) in length and are derived from longer transcripts containing primary miRNA (pri-miRNA) sequences [reviewed in Berezikov, 2006] (Fig. 2). MicroRNAs have emerged as important regulators of gene expression [Chen, 2007].

The importance of miRNAs during animal development is clear from several points. First, animals cannot live without miRNAs [Giraldez, 2005]. Second, there is a large repertoire of conserved miRNAs. Third, mRNAs contain conserved miRNA target sites. Fourth, most miRNAs are highly abundant and exhibit striking tissue-specific expression [Kloostermann, 2006]. Together, these studies demonstrate that miRNAs play an essential role in animal development. Currently, about 5,000 miRNA sequences from 58 different species have been annotated in miRBase (release 10.0) [Griffiths-Jones, 2008].

Classical methods for quantification of the expression level of specific miRNAs are northern blotting, RNase protection assay, RT–PCR and microarray [reviewed in Berezikov, 2006]. Recently microarray technology has been implemented with great success to facilitate the analysis of global miRNA expression [Thomson, 2004; Baskerville, 2005]. These microarrays have been able to identify miRNA genes that are specifically expressed or highly enriched in a specific tissue; these data may be important for the implication of these miRNAs in the differentiation and development of such tissues. The conservation in expression of miR-1, for example, in muscles; miR-124 in the central nervous system; and miR-10 in anterior-posterior patterning suggests that the functions of these and possibly many other miRNAs have been conserved. These results indicated that miRNAs are possibly involved in specifying and maintaining tissue identity [Chen, 2006; Cao, 2007; Woltering, 2008]

Recent advances in analyzing the spatial expression of miRNAs have shown that miRNAs are expressed in a very tissue-specific manner during development [Wienholds, 2005; Kloosterman, 2006; Ason, 2006; Aboobaker, 2005].

The miRNA profile obtained by Mineno et al. [2006] provides insights into the embryonic stage-specific miRNA transcriptome, and will facilitate the identification of the primary target for each miRNA and thereby the pathways regulated by embryonic specific mRNAs. These types of studies have shown that the expression of miRNAs, like that of protein-coding genes, is highly regulated according to the cell's developmental lineage and stage. So, the elucidation of the spatial and temporal patterns of miRNAs expression is critical for understanding their precise role in mammalian development.

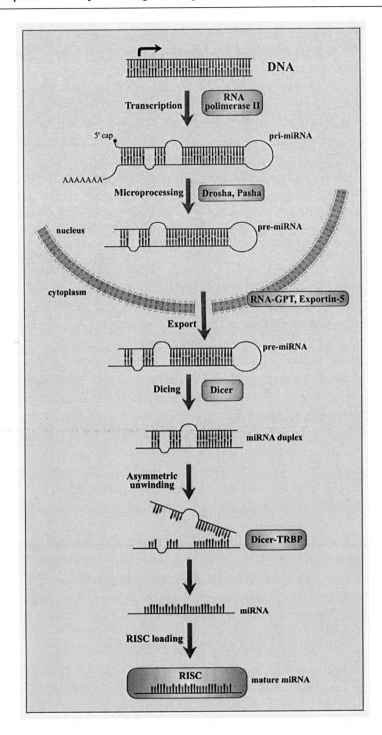

Figure 2. A model for the miRNA biogenesis pathway.

Over the past decade, alternative splicing has increasingly been recognized as a major regulatory process with a critical role in normal development. Furthermore, the importance of alternative splicing in disease development and treatment is starting to be appreciated. Alternative splicing, defined as the generation of multiple RNA transcript species from a common mRNA precursor, is one of the mechanisms for the diversification and expansion of cellular proteins from a smaller set of genes. Current estimates indicate that at least 60% of genes in the human genome exhibit alternative splicing.

Therefore, a recent and very interesting modification of microarrays for analysing this mechanism is the alternative splicing microarrays [Matlin, 2005]. Alternative splicing microarrays contain oligonucleotide probes, typically 25–60 nucleotides in length, that have been designed to hybridize to isoform-specific mRNA regions. Recently, alternative splicing microarrays have been designed with probes that are specific to both exons and exon–exon junctions. Some arrays also contain intron probes to indicate signals from pre-mRNA.

An analysis performed by using an alternative splicing microarray designed with probes for all adjacent exon–exon junctions in 10,000 human genes and hybridized with samples from 52 human tissues and cell lines, revealed cell-type-specific clustering of alternative splicing events, and allowed the discovery of new alternative splicing events [Johnson, 2003]. Pan et al. [2004] analysed 3,126 known cassette-type alternative splicing events in mouse using exon-specific and exon–exon junction probes. Analysis of RNAs in ten tissues showed clustering of alternative splicing events by tissue type, and further revealed that tissue-specific programmes of transcription and alternative splicing operate on different subsets of genes.

An interesting overview of the tools and analysis methods that were developed specifically for alternative splicing microarrays is described in Cuperlovic-Culf [2006].

The recently developed DNA–protein binding assay technique, commonly named ChIP-on-chip (ChIP/chip) is able to produce comprehensive genome-wide location information for specific transcription factors. Transcription factor are one of the best studied gene regulatory mechanisms; however, there are many other layers of gene regulation, including: miRNA, cell signalling; mRNA splicing, polyadenylation and localization; chromatin modifications; and mechanisms of protein localization, modification and degradation.

The ChIP-on-chip technique allows the identification of protein–DNA interactions *in vivo*.

It combines chromatin immunoprecipitation (ChIP) with DNA microarray (chip) detection [Weinmann, 2002]. In fact, DNA fragments that are bound to proteins of interest are isolated by chromatin immunoprecipitation (ChIP) and analysed with DNA microarrays (chips), thereby identifying the partners that are involved in protein–DNA interactions.

Originally developed to study protein–DNA interactions in yeast, this technique was rapidly adopted to eukaryotic cell lines and tissue samples.

An interesting application of this technology to embryonic development permitted the isolation of *cis*-regulatory regions bound by *Pax7* [White, 2008]. This analysis gives greater insight into the direct functional role played by Pax7 during embryonic development. In fact they have identified a suite of genes that are directly targeted by *Pax7* during embryonic development [White, 2008]. Another group used this combined approach to identify gene targets of the chicken transcription factor Hoxa8 [Lei, 2005]. A comprehensive revison of ChIP-on-chip technology to identify transcriptional regulators was in Wang [2005], Blais [2005], Négre [2006], Bock [2008].

In general drawing conclusions from the wealth of data obtained by microarray technology has remained problematic. There have been difficulties with accurate reporting of results, with experimental reproducibility and with identifying and interpreting the biologically relevant information.

Nevertheless microarrays studies are important and fundamental tools for dissecting the genetic basis of embryo development.

Application of Microarrays to Embryogenomics

Fundamental questions in developmental biology are: what genes are expressed, where and when they are expressed, what is the level of expression and how are these programs changed by the functional and/or structural alteration of genes?

These questions have been addressed by studying one gene at a time, but the recent advent of large-scale genomic analysis has changed dramatically this experimental approach. The use of DNA microarray to examine mammalian development is a small but rapidly growing field of study [Smith, 2003]. The application of microarray approach to various embryological samples will lead us to deepen our knowledge of gene expression regulation during embryogenesis [Niemann, 2007].

In this paragraph, we discuss the more recent application of microarray technology to mammalian developmental biology (embryogenomics).

One of the first steps in analysing global gene expression pattern of embryos is the use of the best EST collections. ESTs databases (http://www.ncbi.nlm.nih.gov) provide the precise list of all the genes expressed in a particular cell or tissue; since it is reasonable to expect that there are genes only expressed during embryonic and foetal stages, attention may be put in choosing the right EST collection. This is particularly important for humans because of the cost of obtaining, the limited availability of, and the ethical and legal constraints on studies of human embryos. So developmental studies on animal models are very important to overcome these problems. An interesting tool to circumvent these limitations is the Primate Embryo Gene Expression Resource (PREGER). PREGER resource (http://www.preger.org) provides a valuable resource for basic embryological studies [Latham, 2006]. In fact, this resource contains a set of over 200 samples of rhesus monkey oocytes and embryos that has been converted to cDNA libraries, which are, in turn, used for a variety of molecular analyses [Zheng, 2004]. For example, by using this resource Zheng et al. [2006] provide the first quantitative analysis of WNT and WNT signalling pathway genes in oocytes and embryos of a non-human primate species. In addition, they compare the expression data obtained from mouse embryos and those obtained from non-human primate and concluded that Wnt signalling didn't play a crucial role in the developing rhesus embryo [Zheng, 2006].

To understand the molecular changes that occur during normal pre-implantation development is of fundamental importance to the field of mammalian embryology [Hamatani, 2006].

Large-scale systematic gene expression analyses of early embryos and stem cells provide useful information to identify genes expressed differentially or uniquely in these cells [Carter, 2003; Aiba 2006; Yoshikawa, 2006]. In fact, microarray analysis in mouse oocytes and pre-implantation stage embryos revealed that multiple WNT signaling pathway genes are expressed, and this has led to the suggestion that WNTs may function in early cell fate determination events [Wang, 2004; Zeng, 2004]. Moreover these expression data indicate that WNT signaling pathways are likely not functional in the early embryo until after implantation [Kemler, 2004], raising the possibility that expression of WNT signaling genes in the oocyte and early embryo are instead most likely related to functions in oogenesis (e.g., oocyte-follicle cell interactions, oocyte growth or maturation).

Recent studies on global gene expression and dynamic changes of gene expression in oocytes and pre-implantation stages are reviewed by Hamatani et al. [2008].

Large-scale analysis of gene expression have also been made in surplus human oocytes or embryos discarded from IVF procedures [Chen 2005; Bermudez 2004; Dobson 2004]. These profiling are important also because extracellular factors and embryo manipulation can perturb the pattern of gene expression, success of development in the embryo, and even on gene function in the adult animal.

Large data sets obtained by studies on transcriptome, proteasome and metabolome of embryos as well as biochemical information, provide the basis for systems biology [Hood, 2004]. A major premise of emerging systems biology is that an understanding of the genotype and the biological processes that contribute to the expression of the phenotype should allow reconstruction of the phenotype.

Understanding this network is important because proteins do not function in isolation, but rather interact with one another and with DNA, RNA, and small molecules to form molecular machines. A full and complete revision of state-of-art of studies on biological network analysis is found in Ideker et al. [2008].

Gene Expression Profiling in Mouse Embryos

Early embryogenesis depends on a tightly choreographed succession of gene expression patterns which define normal development. However, little is known of early regulatory mechanisms that operate during development. In fact, the spatial and temporal regulation of gene expression is an important means by which cells respond to physiological and environmental signals.

This section illustrates the power of DNA microarrays for transcriptional profiling during development and for the discovery of functional genes in the mouse.

The advent of array technology has, for the first time, made possible to determine the transcriptional profile of all 20,000 mammalian genes during embryogenesis [reviewed in Niemann, 2007]. Gene expression patterns in mammalian embryos are more complex than those in most somatic cells due to modification of paternal chromatin, the major onset of embryonic transcription (different from species to species), blastocyst formation, expansion and hatching and finally implantation. Moreover embryo development is firstly dependent on maternally stored transcripts which are gradually degraded after a switch to the embryonic expression program [Niemann, 2007]. Nevertheless, a lot of paper analysing gene expression patterns in mammalian embryos and oocytes have been published [reviewed in Sudheer, 2007; Hamatami, 2008; Aiba, 2006]. A DNA microarray approach was also used to identify genes expressed during cytotrophoblast differentiation [Handwerger, 2003].

We applied microarray analysis to evaluate the expression profile of a specific set of genes (68 genes) in mouse embryos at different developmental stages (from 4.5 dpc to 14.5 dpc), corresponding to the pharyngeal development [Amati, 2007]. To analyse the expression changes of individual genes over time we performed a time-course analysis. This approach also makes it possible to evaluate genes that are constantly expressed, i.e., that do not significantly change expression in two subsequent stages while being differentially expressed compared to the reference RNA (pooled 18.5 dpc embryos). Through this statistical analysis, we were able to identify genes expressed during the entire pharyngeal development (Fig. 3). This is an important information because with this methodological approach we may better know the expression pattern of a gene during all the different stages of development.

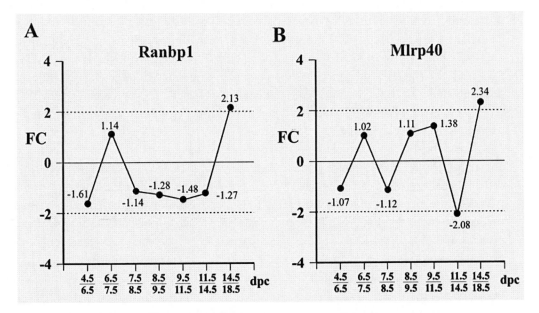

Figure 3. The dynamic expression pattern of two genes (mapping in the commonly deleted region on chromosome 22) constantly present during mouse pharyngeal development. Each FC is obtained by a comparison between the expression value at a developmental stage and the following one (i.e. 4.5 dpc vs 6.5 dpc). Modified from Amati et al., 2007.

Many researchers investigate gene expression pattern during organogenesis. For example, to identify genes regulated by Tbx5 during heart development, a microarray analysis was performed on RNA isolated from cardiac derived 1H cells with ectopic expression of Tbx5. Several Tbx5-induced genes were identified, many of which are expressed and function during heart development [Plageman, 2006].

Various mouse tissues have been used for large-scale transcriptional analysis during normal development or after specific perturbations, such as drug administration or gene targeting or "knockout" experiments [Chisaka and Capecchi, 1991]. A great number of studies used microarray analysis to identify differentially regulated genes in KO mutant mouse embryos [Ko, 2006]. This approach was for example used by Lioubinski et al [2006] to identify novel target genes regulated by FGF signalling in mouse embryos. More recently Beliakoff et al [2008] investigate the biological role of Zimp10 in zimp10-deficient mice. They analyse the yolk sac of mouse embryos deficient for Zimp10, a transcriptional coregulator that may be involved in the modification of chromatin through interactions with the SWI/SNF chromatin-remodeling complexes. The authors demonstrate a crucial role for Zimp10 in vasculogenesis.

The impact of drug administration on gene expression profiling may reveal the drug target, drug metabolism, and disease pathways.

We have developed a mouse model of congenital defects by induction of a triple retinoic acid competitive antagonist (BMS-189453) [Cipollone, 2006]. Pregnant mice treated with 5mg/kg of BMS-189453 at 7.25 and 7.75 dpc have newborn with congenital heart defects (76%), thymic abnormalities (98%) and neural tube defects (20%). TGA is the prevalent cardiac defect obtained (61%) [Cipollone, 2006]. To identify genes or molecular pathways affected in this experimental model we performed a global transcription analysis by microarray. Expression level of two retinoic acid sensitive gene, Rarα and Hoxa1, were analysed by QRT-PCR in 8.5, 9.5 and 11.5 dpc treated embryos. At 8.5 dpc all embryos analysed showed a downregulation of expression of both genes (range: 1% to 18,5%). At 9.5dp 60% of analysed embryos showed a decreased expression of target genes, while at 11.5 dpc only 50% of embryos are downregulated. QRT-PCR results demonstrated that the teratogenic effect of BMS-189453 is diluted during time, and the developmental stage more exposed to its effect is 8.5 dpc.

Pooled treated 8.5 dpc embryos were analysed vs control embryos by using slides containing about 32,000 mouse genes (http://microarray1k.aecom.yu.edu/). A total of 287 differentially expressed genes with a FC= ± 1,5 were identified. 155 genes are decreased while 132 are increased (Fig. 4). According to GO classification, decreased genes are mainly involved in transcription (11%), protein metabolism (16.8%), transport (9.7%) and structural proteins (10.3%) (Fig. 4A). Among them particularly interesting for their function and expression are *Angpt4* (Angiopoietin 4; FC= -1,9), *Hif1α* (hypoxia inducible factor 1, alpha subunit; FC= -1,8), *Tpm1* (Tropomiosin 1, alpha, FC=-1,69) *Tgfb2* (transforming growth factor, beta 2, FC=-1,5).

Increased genes are involved in transcription (9.8%), protein metabolism (12.1%), transport (12.9%), cellular metabolism (9.1%), cell cycle (6.1%) and structural protein (8.3%) (Fig. 4B). Among them *Mela* (Melanoma antigen, FC= +13,4) and *Prps2* (phosphoribosyl pyrophosphate synthetase 2, FC= +3) genes have the highest expression level. A bioinformatics analysis of KEGG pathways database revealed that differentially expressed genes are involved primarily in TGF-beta signalling pathway, Wnt signalling pathway and cell communication.

This study allowed to evidence the modulation at the transcriptional level of a discrete number of genes relevant in biological processes which are altered in this mouse model and to identify genes that might be analysed as candidate genes for congenital heart disease in humans [Amati, 2008].

So, microarrays are an important tool for dissecting gene expression changes in normal physiological processes and disease.

A recent powerful method of identifying novel functional genes involved in embryogenesis is high-throughput whole-mount in situ hybridization [Bell, 2004]. The application of this methods allowed the identification of tissue-specific and temporal patterns of more than 100 newly discovered miRNA genes in the early chick embryo [Ason, 2006; Darnell, 2006].

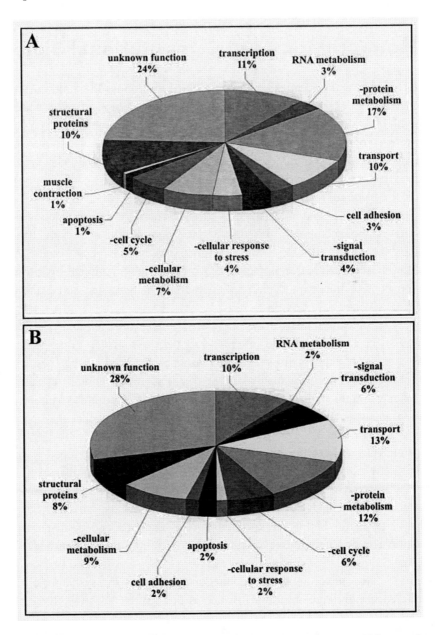

Figure 4. Differentially expressed genes in BMS-treated embryos. A) decresead genes; B) increased genes (Amati et al., 2008)

In general, we may affirm that microarray technology allows co-regulated genes to be identified. In order to identify genes that are controlled by specific regulators, gene expression can be compared in mutant and wild-type cell or tissues. Once co-regulated genes have been identified, protein binding motifs can be identified. By combining these data with a map of promoters, or operons (the operome), the regulatory networks in the cell (the regulome) can start to be built up [Kendall, 2002]. In the next section we discuss deeper the bioinformatics tools for analysing the regulome of a cell.

Bioinformatics and Computational Biology

Bioinformatics and genomics are closely related disciplines that hold great promises for the advancement of research and development in complex biomedical systems, as well as public health, drug design, comparative genomics, personalized medicine and so on.

The genome of an organism fundamentally defines all its characteristics by providing the blueprint for the functional molecules that control life (i.e., genes and proteins). Public accessible databases provide structural and functionally annotated genome sequences, which are critical for genomics research.

Very important for the understanding of the complex data obtained by microarray experiments is the submission of such a data into public accessible database. Gene Expression Omnibus (GEO; http://www.ncbi.nlm.nih.gov/geo/) and ArrayExpress (http://www.ebi.ac.uk/microarray-as/ae/) are two of the main national repositories for high-information content transcript data from microarray analysis and serial analysis of gene expression (SAGE).

The advent of cDNA microarrays for global gene expression analysis [Schena, 1995; Schena, 1996] created the need to interpret vast data sets and to organize genes by their temporal expression patterns. In fact it is reasonable to think that genes with a similar function cluster together, presumably due to common transcriptional regulation, and they usually belong to similar metabolic or regulatory pathways. Bioinformatics will deal with information about gene regulation and developmental pathways that may be modelled on computers.

So, hierarchical clustering and SOM, for example, were developed to visualize and understand gene expression patterns and to group genes with a similar expression profile [Eisen, 1998; Tamayo, 1999; Törönen, 1999]. Beyond these analyses, there has been a parallel demand in methods for the interpretation of the results, which involves translating these data into useful biological knowledge. The methods and strategies used for this interpretation are in continuous evolution and new proposals are constantly arising [Al-sharour, 2008].

One of the most important links for functional genomics and systems biology is the GO database (Gene Ontology Consortium), which is the central repository for functional gene annotations.

The GO project was established as the GO Consortium in 1998 to use the vast volume of accessible biological information for functional annotation of gene products [Lewis, 2005]. The primary mission of the GO project is to facilitate the use of the genome information and enable understanding of the functional units of the genome (genes and proteins). The GO Consortium now includes numerous prokaryotes and eukaryotes. Gene products are classified by molecular function, biological process, and cellular compartment, primarily by using peer-reviewed literature and proteomic data or, if these are unavailable, by homology to orthologous products in other species. Genes and gene products for every organism are functionally annotated by using controlled structured vocabularies (ontologies) [Ashburner and Lewis, 2002; Ashburner, 2000].

Another important database for functional studies is the KEGG pathways database (http://www.genome.ad.jp/kegg/pathway.html). KEGG pathways database is a collection of manually drawn pathway maps representing our knowledge on the molecular interaction and reaction networks for most of the known metabolic pathways and some of the known regulatory pathways [Ogata, 1999]. In addition, biological annotations often provide the better way to obtain functional information on differentially expressed genes identified with microarrays, for which no other data is available.

The application of GO and KEGG pathways databases to a complete set of mouse cDNAs leaded to the construction of the mouse metabolome [Bono, 2003].

The cis-element profile (or cis-profile) of a gene refers to the collection of transcription factor binding sites (TFBS) regulating the transcription of the gene. Underlying the various published studies that attempt to discover cis-elements in the vicinity of co-expressed genes via pattern detection algorithms, there is an implicit assumption that a correlation exists between co-expressed genes and their cis-profiles. This interesting bioinformatic approach is further described in the next paragraph.

In Silico Analysis of Genes Promoters

Comparative approaches may identify regulatory sequences (TFBS and miRNA) that control gene expression. However, predicting regulatory sequences in the genome is much more difficult than gene prediction. Regulatory sequences are *cis*-acting modules (i.e., promoters and enhancers) that regulate the timing and abundance of gene expression.

Genome sequences harbouring these functional elements seem to be conserved among related species. Genome sequences of several related species are aligned to identify evolutionary conserved sequences that contain transcription regulatory elements. This process is called "phylogenetic footprinting" [reviewed in Hooghe, 2008]. Comparative analyses of the human, mouse, rat, and dog genome sequences have identified many common regulatory motifs in promoters and 3'-untranslated regions [Xie, 2005]. So, a lot of bioinformatics tools have been developed for this purpose. Pairwise sequence comparison between evolutionarily distant species represents a powerful method for identification of functional noncoding sequences that regulate gene expression [Ahituv, 2005]. Mulan is another Webbased multiple-sequence local alignment and visualization tool developed for detection of evolutionarily conserved TFBS in multiple-species alignments [Ovcharenko, 2005]. The code MatInspector identifies TBFS using weight matrices and can be invoked from other sequence analysis programs, such as DiAlignTF, FrameWorker or SequenceShaper [Carharius, 2005; Morgenstern, 2007] (Fig. 5).

An original algorithm for the identification of conserved TBFS was recently implemented in the code WeederH [Pavesi, 2007].

A number of important cellular processes, such as transcriptional regulation, recombination, replication, repair, and DNA modification, are performed by DNA binding proteins. Of particular interest are transcription factors (TFs) which, through their sequence-specific interactions with DNA binding sites, modulate gene expression in a manner required for normal cellular growth and differentiation, and also for response to environmental stimuli.

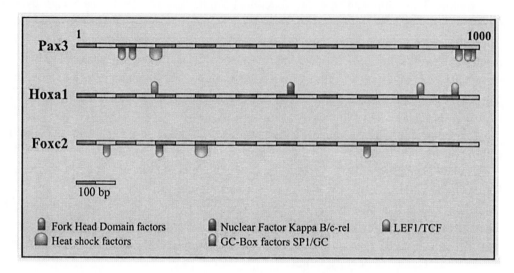

Figure 5. Identication of TFBSs in co-regulated genes in mouse embryos (data from Amati et al. 2007).

Despite their importance, the DNA binding specificities of most DNA binding proteins still remain unknown, since prior technologies aimed at identifying DNA-protein interactions have been laborious, not highly scalable, or have required limiting biological reagents. Recently a new DNA microarray-based technology, termed protein binding microarrays (PBMs), has been developed that allows rapid, high-throughput characterization of the in vitro DNA binding site sequence specificities of TFs, other DNA binding proteins, or synthetic compounds [reviewed in Hudson, 2006]. DNA binding site data from PBMs combined with gene annotation data, comparative sequence analysis, and gene expression profiling, can be used to predict what genes are regulated by a given TF, what the functions are of a given TF and its predicted target genes, and how that TF may fit into the cell's transcriptional regulatory network [Bulyk 2007; Mukherjee, 2004].

This approach was useful to identify developmental PcG target in murine ES cells. The genomic DNA associated with proteins and total (control) DNA were combined and hybridized to microarrays that contained 60-mer oligonucleotide probes covering the region from 28 kb to 2 kb relative to the transcription start sites for 15,742 annotated mouse genes. Genomic sites occupied by PcG proteins were identified as peaks of chromatin immunoprecipitation (ChIP)- enriched DNA using a previously validated algorithm. The results of this paper confirmed the hypothesis that PcG target genes are preferentially upregulated during ES cell differentiation [Boyer, 2006].

Regulation of gene expression in eukaryotic genomes is established through a complex cooperative activity of proximal promoters and distant regulatory elements (REs) such as enhancers, repressors and silencers. Microarray experiments can be used to identify co-regulated sequences, with common motifs being identified through statistical analysis. Mapping of gene and regulatory networks requires high-throughput analysis of transcriptional scans, clustering of co-regulated genes, and computational searches for functional motifs [i.e., *cis*-regulatory elements and TFBSs [Banerjee and Zhang, 2002]. Multiple transcriptional snapshots taken in a time series provide a dynamic dimension of gene expression [Carinci, 2007]. Rigorous interrogation of the *cis*-regulatory regions and GO annotation of genes provides a more comprehensive view of genetic control over metabolic and developmental processes. Large-scale transcriptional profiling and refined bioinformatics analyses have revealed exquisite detail of the biological processes and molecular networks involved in transcriptional regulation of fat cell development [Hackl, 2005].

Even if conservation of a TFBS among several species observed in a multiple alignment is not a proof that it is functional; nevertheless this information is useful for a better understanding of the regulation of gene expression.

Conclusions

High-throughput methods facilitate studies of the cellular transcriptome or proteome and the networks of interactions between the genome, the transcriptome, and the proteome.

Changes in the patterns of gene expression are widely believed to underlie many of the phenotypic differences within and between species. Although much emphasis has been placed on changes in transcriptional regulation, gene expression is regulated at many levels, all of which must ultimately be studied together to obtain a complete picture of the evolution of gene expression.

Very useful to this purpose is a deep determination of the transcriptome of embryos at different developmental stages by means of array technology. However, all the data obtained by these high-throughput experiments need to be condensed and supplemented by other important information such as the regulation of chromatin structure or of transcription process. Recent development in array technology, such as alternative splicing arrays, microRNA arrays and DNA methylation arrays, will help researchers in addressing these questions. However, only an integrated analysis of the enormous number of data obtained from these experiments by means bioinformatic methods for classification and data management will help researcher to obtain a complete picture of the embryonic transcriptome.

In conclusion, functional genomics of embryo development must represent the integration of information from genome sequence and structure, gene and protein expression, and metabolite profiles with knowledge databases by using computational and bioinformatics tools (Figure 6).

Sophisticated genomic and proteomic techniques combined with complex bioinformatics data analyses will be further required to translate these recent basic science discoveries to useful information on regulation of development.

With advances in functional genomics and high-throughput techniques, the future holds much promise for understanding gene expression regulation during embryonic development.

Finally we think that only through concerted effort can developmental researchers make sense of the staggering output of microarrays and the even more staggering complexity of the developing embryo.

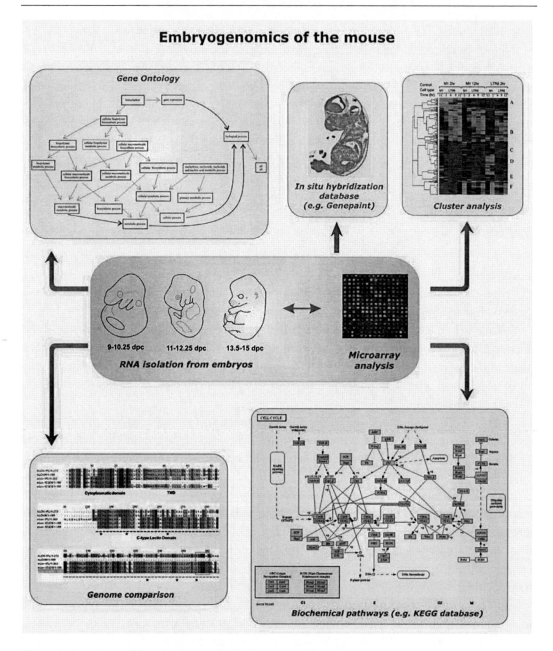

Figure 6. A road map of functional genomics in the mouse embryo.

Acknowledgements

We would like to thank Susana Bueno and Alessandro Desideri for their critical discussions about this manuscript and Graziano Bonelli for preparation of figures.

References

Aboobaker, A.A.; Tomancak, P.; Patel, N.; Rubin, G.M.; Lai, E.C. Drosophila microRNAs exhibit diverse spatial expression patterns during embryonic development. *Proc Natl Acad Sci*, 2005, 102, 18017-18022.

Abril, J.F.; Agarwal, P.; Alexandersson, M.; Antonarakis, S.E.; Baertsch, R.; Berry, E. et al. Initial sequencing and comparative analysis of the mouse genome. *Nature,* 2002, 420, 520-562.

Ahituv, N.; Prabhakar, S.; Poulin, F.; Rubin, E.M.; Couronne, O. Mapping cis-regulatory domains in the human genome using multi-species conservation of synteny. *Hum Mol Genet*, 2005, 14, 3057-3063.

Aiba, K.; Carter, M.G.; Matoba, R.; Ko, M.S. Genomic approaches to early embryogenesis and stem cell biology. *Semin Reprod Med*. 2006, 24, 330-339.

Al-Shahrour, F.; Carbonell, J.; Minguez, P.; Goetz, S.; Conesa, A.; Tárraga, J.; Medina, I.; Alloza, E.; Montaner, D.; Dopazo, J. Babelomics: advanced functional profiling of transcriptomics, proteomics and genomics experiments. *Nucleic Acids Res*, 2008, 36(Web Server issue):W341-6.

Amati, F.; Biancolella, M.; Farcomeni, A.; Giallonardi, S.; Bueno, S.; Minella, D.; Vecchione, L.; Chillemi, G.; Desideri, A.; Novelli, G. Dynamic changes in gene expression profiles of 22q11 and related orthologous genes during mouse development. *Gene*, 2007, 391, 91-102.

Amati, F.; Giallonardi, S.; Cipollone, D.; Bueno, S.; Vecchione, L. Prosperini, G.; Desideri, A.; Chillemi, G.; Marino, B.; Novelli, G. Genome-wide transcriptome profiling of mouse embryos with cardiac and thymic defects induced by an antagonist of retinoic acid. In: BI-ANNUAL MEETING OF THE WORKING GROUP ON DEVELOPMENTAL ANATOMY AND PATHOLOGY,. Alberobello (BA), March 12 - 15, 2008

Aparicio, S.; Chapman, J.; Stupka, E.; Putnam, N.; Chia, J.M.; Dehal, P.; et al. Whole-genome shotgun assembly and analysis of the genome of Fugu rubripes. *Science*, 2002, 297, 1301-1310.

Ashburner, M. & Lewis, S. On ontologies for biologists: the Gene Ontology--untangling the web. *Novartis Found Symp*, 2002, 247, 66-80.

Ashburner, M.; Ball, C.A.; Blake, J.A.; Botstein, D.; Butler, H.; Cherry, J.M.; Davis, A.P.; Dolinski, K.; Dwight, S.S.; Eppig, J.T.; Harris, M.A.; Hill, D.P.; Issel-Tarver, L.; Kasarskis, A.; Lewis, S.; Matese, J.C.; Richardson, J.E.; Ringwald, M.; Rubin, G.M.; Sherlock, G. Gene ontology: tool for the unification of biology. The Gene Ontology Consortium. *Nat Genet*, 2000, 25, 25-29.

Ashley, E.A.; Spin, J.M.; Tabibiazar, R.; Quertermous, T. Frontiers in nephrology: genomic approaches to understanding the molecular basis of atherosclerosis. *Am Soc Nephrol,* 2007, 18, 2853-2862.

Ason, B.; Darnell, D.K.; Wittbrodt, B:; Berezikov, E.; Kloosterman, W.P.; Wittbrodt, J.; Antin, P.B.; Plasterk, R.H.A. Differences in vertebrate microRNA expression. *PNAS*, 2006, 103, 14385-14389.

Banerjee, N. & Zhang, M.Q. Functional genomics as applied to mapping transcription regulatory networks. *Curr Opin Microbiol*. 2002, 5, 313-317.

Baskerville, S. & Bartel, D.P. Microarray profiling of microRNAs reveals frequent coexpression with neighboring miRNAs and host genes. *RNA*, 2005, 11, 241-247.

Beliakoff, J.; Lee, J.; Ueno, H.; Aiyer, A.; Weissman, I.L.; Barsh, G.S.; Cardiff, R.D.; Sun, Z. The PIAS-like protein Zimp10 is essential for embryonic viability and proper vascular development. *Mol Cell Biol*, 2008, 28, 282-292.

Bell, G.W.; Yatskievych, T.A.; Antin, P.B. GEISHA, a whole-mount in situ hybridization gene expression screen in chicken embryos. *Dev Dyn*, 2004, 229, 677-687.

Berezikov, E.; Cuppen, E.; Plasterk, R.H. Approaches to microRNA discovery. *Nat. Genet,* 2006, 38, S2–S7.

Bermudez, M.G.; Wells, D.; Malter, H.; Munne, S.; Cohen, J.; Steuerwald, N.M. Expression profiles of individual human oocytes using microarray technology. *Reprod Biomed Online*, 2004, 8, 325–337.

Biemont, C. & Vieira, C. Junk DNA as evolutionary force. *Nature,* 2006, 443, 521–524.

Blais, A. & Dynlacht, B.D. Constructing transcriptional regulatory networks. *Genes Dev*, 2005, 19, 1499-1511.

Blohm, D.H. & Guiseppi-Elie, A. New developments in microarray technology. *Curr Opin Biotechnol,* 2001, 12, 41-47.

Bock, C. & Lengauer, T. Computational epigenetics. *Bioinformatics*, 2008, 24, 1-10.

Bono, H.; Nikaido, I.; Kasukawa, T.; Hayashizaki, Y.; Okazaki, Y; RIKEN GER Group; GSL Members. Comprehensive analysis of the mouse metabolome based on the transcriptome. *Genome Res*, 2003, 13, 1345-1349.

Boyer, L.A.; Plath, K.; Zeitlinger, J.; Brambrink, T.; Medeiros, L.A.; Lee, T.I.; Levine, S.S.; Wernig, M.; Tajonar, A.; Ray, M.K.; Bell, G.W.; Otte, A.P.; Vidal, M.; Gifford, D.K.; Young, R.A.; Jaenisch, R. Polycomb complexes repress developmental regulators in murine embryonic stem cells. *Nature*, 2006, 441, 349-353.

Brazhnik, P.; de la Fuente, A.; Mendes, P. Gene networks: how to put the function in genomics. *Trends Biotechnol*, 2002, 20, 467-472.

Breitling, R. Biological microarray interpretation: the rules of engagement. *Biochim Biophys Acta*, 2006, 1759, 319-327.

Bulyk, M.L. Protein binding microarrays for the characterization of DNA-protein interactions. *Adv Biochem Eng Biotechnol*, 2007, 104, 65-85.

Byers, R.J.; Hoyland, J.A.; Dixon, J.; Freemont, A.J. Subtractive hybridization--genetic takeaways and the search for meaning. *Int J Exp Pathol,* 2000, 81, 391-404.

Cao, X.; Pfaff, S.L.; Gage, F.H. A functional study of miR-124 in the developing neural tube. *Genes Dev*. 2007, 21, 531-536.

Carninci, P. Constructing the landscape of the mammalian transcriptome. *J Exp Biol*. 2007, 210, 1497-1506.

Carter GW. Inferring network interactions within a cell. *Brief Bioinform*, 2005, 6, 380-389.

Carter, M.G.; Hamatani, T.; Sharov, A.A.; Carmack, C.E.; Qian, Y.; Aiba, K.; Ko, N.T.; Dudekula, D.B.; Brzoska, P.M.; Hwang, S.S.; Ko, MS. In situ-synthesized novel microarray optimized for mouse stem cell and early developmental expression profiling. *Genome Res*, 2003, 13, 1011-1021.

Cartharius, K.; Frech, K.; Grote, K.; Klocke, B.; Haltmeier, M.; Klingenhoff, A.; Frisch, M.; Bayerlein, M.; Werner, T. MatInspector and beyond: promoter analysis based on transcription factor binding sites. *Bioinformatics*, 2005, 21, 2933-2942.

Carulli, J.P.; Artinger, M.; Swain, P.M.; Root, C.D.; Chee, L., Tulig, C.; Guerin, J.; Osborne, M.; Stein, G.; Lian, J.; Lomedico, P.T. High throughput analysis of differential gene expression. *J Cell Biochem,* 1998, 30-31, 286-296.

Chen, H.W.; Chen, J.J.; Yu, S.L.; Li, H.N.; Yang, P.C.; Su, C.M. et al. Transcriptome analysis in blastocyst hatching by cDNA microarray. *Hum Reprod*, 2005, 20, 2492–2501.

Chen, J.F.; Mandel, E.M.; Thomson, J.M.; Wu, Q.; Callis, T.E.; Hammond, S.M.; Conlon, F.L.; Wang, D.Z. The role of microRNA-1 and microRNA-133 in skeletal muscle proliferation and differentiation. *Nat Genet*. 2006, 38, 228-233.

Chen, K. & Rajewsky, N. The evolution of gene regulation by transcription factors and microRNAs. *Nature Rev Genetics,* 2007, 8, 93-103.

Chisaka, O. & Capecchi, M.R. Regionally restricted developmental defects resulting from targeted disruption of the mouse homeobox gene hox-1.5. *Nature*. 1991, 350, 473-479.

Cipollone, D.; Amati, F.; Carsetti, R.; Placidi, S.; Biancolella, M.; D'Amati, G.; Novelli, G.; Siracusa, G.; Marino, B. A multiple retinoic acid antagonist induces transposition of the great arteries in the mouse. *Cardiovasc Pathol,* 2006, 15, 194-202.

Collins, F.S.; Morgan, M.; Patrinos, A. The Human Genome Project: lessons from large-scale biology. *Science*, 2003, 300, 286-290.

Cuperlovic-Culf, M.; Belacel, N.; Culf, A.S.; Ouellette, R.J. Microarray analysis of alternative splicing. *OMICS*, 2006, 10, 344-357.

Darnell, D.K.; Kaur, S.; Stanislaw, S.; Konieczka, J.H.; Yatskievych, T.A.; Antin, P.B. MicroRNA expression during chick embryo development. *Dev Dyn*, 2006, 235, 3156-3165.

Davidson, E.H.; Rast, J.P.; Oliveri, P.; Ransick, A.; Calestani, C.; Yuh, C.H.; Minokawa, T.; Amore, G.; Hinman, V.; Arenas-Mena, C.; Otim, O.; Brown, C.T.; Livi, C.B.; Lee, P.Y.; Revilla, R.; Rust, A.G.; Pan, Z.; Schilstra, M.J.; Clarke, P.J.; Arnone, M.I.; Rowen, L.; Cameron, R.A.; McClay, D.R.; Hood, L.; Bolouri, H. A genomic regulatory network for development. *Science*, 2002, 295, 1669-1678.

de la Fuente, A. & Mendel P. Quantifying gene networks with regulatory strengths. *Mol Biol Rep*, 2002, 29, 73-77.

Dobson, A.T.; Raja, R.; Abeyta, M.J.; Taylor, T.; Shen, S.; Haqq, C. et al. The unique transcriptome through day 3 of human preimplantation development. *Hum Mol Genet*, 2004, 13, 1461–1470.

Eisen, M.B.; Spellman, P.T.; Brown, P.O.; Botstein D. Cluster analysis and display of genome-wide expression patterns. *Proc Natl Acad Sci*, 1998, 95, 14863-14868.

Fadiel, A.; Anidi, I.; Eichenbaum, K.D. Farm animal genomics and informatics: an update. *Nucleic Acids Res*, 2005, 33, 6308–6318.

Giraldez, A.J.; Cinalli, R.M.; Glasner, M.E.; Enright, A.J.; Thomson, J.M.; Baskerville, S.; Hammond, S.M.; Bartel, D.P.; Schier, A.F. MicroRNAs regulate brain morphogenesis in zebrafish. *Science,* 2005, 308, 833-888.

Goodstadt, L. & Ponting, CP. Phylogenetic reconstruction of orthology, paralogy, and conserved synteny for dog and human. *PLoS Comput Biol,* 2006, 2, e133.

Grant, S.F. & Hakonarson, H. Microarray technology and applications in the arena of genome-wide association. *Clin Chem*, 2008, 54, 1116-1124.

Gregory, S.G.; Sekhon, M.; Schein, J.; Zhao, S.; Osoegawa, K.; Scott, C.E.; et al. A physical map of the mouse genome. *Nature*, 2002, 418, 743–750.

Griffiths-Jones, S.; Saini, H.K.; van Dongen, S.; Enright, A.J. miRBase: tools for microRNA genomics. *Nucleic Acids Res*. 2008, 36(Database issue), D154-158.

Gu, C.C.; Rao, D.C.; Stormo, G.; Hicks, C.; Province, MA. Role of gene expression microarray analysis in finding complex disease genes. *Genet Epidemiol*, 2002, 23, 37-56.

Hackl, H.; Burkard, T.R.; Sturn, A.; Rubio, R.; Schleiffer, A.; Tian, S.; Quackenbush, J.; Eisenhaber, F.; Trajanoski, Z. Molecular processes during fat cell development revealed by gene expression profiling and functional annotation. *Genome Biol*. 2005, 6, R108.

Hamatani, T.; Ko, M.S.; Yamada, M.; Kuji, N.;, Mizusawa, Y.; Shoji, M.; Hada, T.; Asada, H.; Maruyama, T.; Yoshimura, Y. Global gene expression profiling of preimplantation embryos. *Hum Cell*, 2006, 19, 98-117.

Hamatani, T.; Yamada, M.; Akutsu, H.; Kuji, N.; Mochimaru, Y.; Takano, M.; Toyoda, M.; Miyado, K.; Umezawa, A.; Yoshimura, Y. What can we learn from gene expression profiling of mouse oocytes? *Reproduction*, 2008, 135: 581-592.

Handwerger, S. & Aronow, B. Dynamic changes in gene expression during human trophoblast differentiation. *Recent Prog Horm Res*, 2003, 58, 263-281.

Hayashizaki, Y. & Kanamori, M. Dynamic transcriptome of mice. *Trends Biotechnol*, 2004, 22, 161–167.

Hayden, P.S.; El-Meanawy, A.; Schelling, J.R.; Sedor, J.R. DNA expression analysis: serial analysis of gene expression, microarrays and kidney disease. *Curr Opin Nephrol Hypertens*, 2003, 12, 407-414.

Heintzman, N.D. & Ren, B. The gateway to transcription: identifying, characterizing and understanding promoters in the eukaryotic genome. *Cell Mol Life Sci*, 2007, 64, 386–400.

Hood, L.; Heath, J.R.; Phelps, M.E.; Lin, B. Systems biology and new technologies enable predictive and preventative medicine. *Science*, 2004, 306, 640–643.

Hooghe, B.; Hulpiau, P.; van Roy, F.; De Bleser, P. ConTra: a promoter alignment analysis tool for identification of transcription factor binding sites across species. *Nucleic Acids Res*, 2008, 36(Web Server issue), W128-132.

Hovatta, I.; Kimppa, K.; Lehmussola, A.; Pasanen, T.; Saarela, J.; Saarikko, I.; Saharinen, J.; Tiikkainen, P.; Toivanen, T. et al. DNA Microarrays Data Analysis, second edition. Picaset Oy, Helsinki, The authors and CSC Scientific Computing Ltd, 2005

Hudson, M.E. & Snyder, M. High-throughput methods of regulatory element discovery. *Biotechniques*, 2006, 41, 673, 675, 677 passim.

Ideker, T. & Sharan, R. Protein networks in disease. *Genome Res*, 2008, 18, 644-52.

Jirtle, R.L. & Skinner, M.K. Environmental epigenomics and disease susceptibility. *Nat Rev Genet*, 2007, 8, 253–256.

Johnson, J.M.; Castle, J.; Garrett-Engele, P.; Kan, Z.; Loerch, P.M.; Armour, C.D.; Santos, R.; Schadt, E.E.; Stoughton, R.; Shoemaker, D.D. Genome-wide survey of human alternative pre-mRNA splicing with exon junction microarrays. *Science*, 2003, 302, 2141-2144.

Kemler, R.; Hierholzer, A.; Kanzler, B.; Kuppig, S.; Hansen, K.; Taketo, M.M.; de Vries, W.N.; Knowles, B.B.; Solter, D. Stabilization of beta-catenin in the mouse zygote leads to premature epithelial-mesenchymal transition in the epiblast. *Development*. 2004, 131, 5817-5824.

Kendall, S.L.; Movahedzadeh, F.; Wietzorrek, A., Stoker, N.G. Microarray analysis of bacterial gene expression: towards the regulome. *Comp Funct Genomics*, 2002, 3, 352-354.

Kloosterman, W.P. & Plasterk R.H.A. The Diverse Functions of MicroRNAs Review in Animal Development and Disease Developmental. *Cell*, 2006, 11, 441–450.

Ko, M.S. Embryogenomics: developmental biology meets genomics. *Trends Biotechnol*, 2001, 19, 511-518.

Ko, M.S. Expression profiling of the mouse early embryo: reflections and perspectives. *Dev Dyn*. 2006, 235, 2437-2448.

Latham, KE. The Primate Embryo Gene Expression Resource in embryology and stem cell biology. *Reprod Fertil Dev*, 2006, 18, 807-810.

Lei, H.; Wang, H.; Juan, A.H.; Ruddle, F.H. The identification of Hoxc8 target genes. *Proc Natl Acad Sci*, 2005, 102, 2420-2424.

Lewis, SE. Gene Ontology: looking backwards and forwards. *Genome Biol*, 2005, 6, 103.

Lindblad-Toh, K.; Wade, C.M.; Mikkelsen, T.S.; Karlsson, E.K.; Jaffe, D.B.; Kamal, M. et al. Genome sequence, comparative analysis and haplotype of the domestic dog. *Nature*, 2005, 438, 803–819.

Lioubinski, O.; Alonso, M.T.; Alvarez, Y.; Vendrell, V.; Garrosa, M.; Murphy, P.; Schimmang, T. FGF signalling controls expression of vomeronasal receptors during embryogenesis. *Mech Dev*, 2006, 123, 17-23.

Lyons, PA. Gene-expression profiling and the genetic dissection of complex disease. *Curr Opin Immunol*, 2002, 14, 627-630.

Matlin, A.J.; Clark, F.; Smith, C.W.J. Understanding alternative splicing: towards a cellular code. *Nat. Rev. Mol. Cell Biol*. 2005, 6, 386-398.

McClelland, M.; Mathieu-Daude, F.; Welsh, J. RNA fingerprinting and differential display using arbitrarily primed PCR. *Trends Genet*, 1995, 11, 242-246.

Mineno, J.; Okamoto, S.; Ando, T.; Sato, M.; Chono, H.; Izu, H.; Takayama, M.; Asada, K.; Mirochnitchenko, O.; Inouye, M.; Kato, I. The expression profile of microRNAs in mouse embryos. *Nucleic Acids Research*, 2006, 34, 1765–1771.

Morgan, T.H. (1926). *The Theory of the Gene*. Yale University Press, New Haven.

Morgenstern, B. Alignment of genomic sequences using DIALIGN. *Methods Mol Biol*, 2007, 395, 195-204.

Mukherjee, S.; Berger, M.F.; Jona, G.; Wang, X.S.; Muzzey, D.; Snyder, M.; Young, R.A.; Bulyk, M.L. Rapid analysis of the DNA-binding specificities of transcription factors with DNA microarrays. *Nat Genet*, 2004, 36, 1331-1339.

Nègre, N.; Lavrov, S.; Hennetin, J.; Bellis, M.; Cavalli, G. Mapping the distribution of chromatin proteins by ChIP on chip. *Methods Enzymol*, 2006, 410, 316-341.

Niemann, H.; Carnwath, J.W.; Kues, W. Application of DNA array technology to mammalian embryos. *Theriogenology*, 2007, 68S, S165–S177.

Ogata, H.; Goto, S.; Sato, K.; Fujibuchi, W.; Bono, H.; Kanehisa, M. KEGG: Kyoto Encyclopedia of Genes and Genomes. *Nucleic Acids Res*, 1999, 27, 29-34.

Okazaki, Y.; Furuno, M.; Kasukawa, T.; Adachi, J.; Bono, H.; Kondo, S.; et al. Analysis of the mouse transcriptome based on functional annotation of 60,770 full-length cDNAs. *Nature*, 2002, 420, 563–573.

Oliveri, P.; Tu, Q.; Eric H. Davidson† Global regulatory logic for specification of an embryonic cell lineage 2008, 105, 5955–5962

Ooi, L. & Wood, I.C. Regulation of gene expression in the nervous system. *Biochem J*, 2008, 414, 327-341.

Ovcharenko, I. & Nobrega, M.A. Identifying synonymous regulatory elements in vertebrate genomes. *Nucleic Acids Res*, 2005, 33(Web Server issue), W403-407.

Pan, Q. et al. Revealing global regulatory features of mammalian alternative splicing using a quantitative microarray platform. *Mol. Cell*, 2004, 16, 929–941.

Pavesi, G.; Zambelli, F.; Pesole, G. WeederH: an algorithm for finding conserved regulatory motifs and regions in homologous sequences. *BMC Bioinformatics*, 2007, 8, 46.

Plageman, T.F. & Yutzey, K.E. Microarray analysis of Tbx5-induced genes expressed in the developing heart. *Dev. Dynam*, 2006, 235:2868–2880.

Powell J. SAGE. The serial analysis of gene expression. *Methods Mol Biol*, 2000, 99, 297-319.

Rassoulzadegan, M.; Grandjean, V.; Gounon, P.; Vincent, S.; Gillot, I.; Cuzin, F. RNA-mediated non-mendelian inheritance of an epigenetic change in the mouse. *Nature*, 2006, 441, 469–474.

Schena, M. Genome analysis with gene expression microarrays. *Bioessays*, 1996, 18, 427-431.

Schena, M.; Shalon, D.; Davis, R.W.; Brown, P.O. Quantitative monitoring of gene expression patterns with a complementary DNA microarray. *Science*, 1995, 270, 467-470.

Schumacher, A.; Kapranov, P.; Kaminsky, Z.; Flanagan, J.; Assadzadeh, A.; Yau, P.; Virtanen, C.; Winegarden, N.; Cheng, J.; Gingeras, T.; Petronis, A. Microarray-based DNA methylation profiling: technology and applications. *Nucleic Acids Res*, 2006, 34, 528-542.

Segal, E.; Fondufe-Mittendorf, Y.; Chen, L.; Thastrom, A.; Field, Y.; Moore, I.K. et al. A genomic code for nucleosome positioning. *Nature*, 2006, 442, 772–778.

Siddiqui, A.S.; Khattra, J.; Delaney, A.D.; Zhao, Y.; Astell, C.; Asano, J.; Babakaiff, R.; Barber, S. et al. A mouse atlas of gene expression: large-scale digital gene-expression profiles from precisely defined developing C57BL/6J mouse tissues and cells. *Proc Natl Acad Sci*, 2005, 102, 18485-18490.

Smith, J.; Theodoris, C.; Davidson, EH. A gene regulatory network subcircuit drives a dynamic pattern of gene expression. *Science*, 2007, 318, 794-797.

Smith, L. & Greenfield A. DNA microarrays and development. *Hum. Mol. Genet.*, 2003, 12, 1-8.

Sudheer, S. & Adjaye, J. Functional genomics of human pre-implantation development. *Brief Funct Genomic Proteomic*, 2007, 6, 120-132.

Tamayo, P.; Slonim, D.; Mesirov, J.; Zhu, Q.; Kitareewan, S.; Dmitrovsky, E.; Lander, E.S.; Golub, T.R. Interpreting patterns of gene expression with self-organizing maps: methods and application to hematopoietic differentiation. *Proc Natl Acad Sci,* 1999, 96, 2907-2912.

Thomson, J.M.; Parker, J.; Perou, C.M.; Hammond, S.M. A custom microarray platform for analysis of microRNA gene expression. *Nat Methods*, 2004, 1, 47-53.

Törönen, P.; Kolehmainen, M.; Wong, G.; Castrén, E. Analysis of gene expression data using self-organizing maps. *FEBS Lett*, 1999, 451, 142-146.

van Someren, E.P.; Wessels, L.F.; Backer, E.; Reinders, M.J. Genetic network modeling. *Pharmacogenomics,* 2002, 3, 507-525.

Venkatesh, B.; Kirkness, E.F.; Loh, Y.H.; Halpern, A.L.; Lee, A.P.; Johnson, J. et al. Ancient noncoding elements conserved in the human genome. *Science,* 2006, 314, 765.

Venter, JC, Adams, MD, Myers, EW, Li, PW, Mural, RJ, Sutton, GG, et al. The sequence of the human genome. *Science,* 2001, 291, 1304-1351.

Visel, A.; Thaller, C.; Eichele, G. GenePaint.org: an atlas of gene expression patterns in the mouse embryo. *Nucleic Acids Res*, 2004, 32 (Database issue), D552-556.

Wagner, A. How to reconstruct a large genetic network from n gene perturbations in fewer than n(2) easy steps. *Bioinformatics.* 2001, 17, 1183-1197.

Wallis, J.W.; Aerts, J.; Groenen, M.A.; Crooijmans, R.P.; Layman, D.; Graves T.A. et al. A physical map of the chicken genome. *Nature,* 2004, 432 761–764.

Wang, JC. Finding primary targets of transcriptional regulators. *Cell Cycle,* 2005, 4: 356-358.

Wang, Q.T.; Piotrowska, K.; Ciemerych, M.A.; Milenkovic, L.; Scott, M.P.; Davis, R.W. et al. A genome wide study of gene activity reveals developmental signalling pathways in the preimplantation mouse embryo. *Dev Cell,* 2004, 6, 133–44.

Weinmann, A.S.; Yan, P.S.; Oberley, M.J.; Huang, T.H.; Farnham, P.J. Isolating human transcription factor targets by coupling chromatin immunoprecipitation and CpG island microarray analysis. *Genes Dev,* 2002, 16, 235–244.

Weiss, K. M. The phenogenetic logic of life. *Nat rev genet,* 2005, 6, 36-45.

Werner A. Natural antisense transcripts. *RNA Biol,* 2005, 2, 53–62.

White R.B. & Ziman, M.R. Genome-wide discovery of Pax7 target genes during development. *Physiol Genomics,* 2008, 33, 41–49.

Wienholds, E. & Plasterk, R.H.A. MicroRNA function in animal development. *FEBS Letters,* 2005, 579, 5911–5922.

Wilson, I.M; Davies, J.J.; Weber, M.; Brown, C.J.; Alvarez, C.E.; MacAulay, C:; Schübeler, D.; Lam, W.L. Mapping the Methylome. *Cell Cycle*, 2006, 2, 155-158.

Woltering, J.M. & Durston, A.J. MiR-10 represses HoxB1a and HoxB3a in zebrafish. *PLoS ONE.* 2008, 3, :e1396.

Xie, X.; Lu, J.; Kulbokas, E.J.; Golub, T.R.; Mootha, V.; Lindblad-Toh, K.; Lander, E.S.; Kellis, M. Systematic discovery of regulatory motifs in human promoters and 3' UTRs by comparison of several mammals. *Nature*, 2005, 434, 338-345.

Yoshikawa, T.; Piao, Y.; Zhong, J.; Matoba, R., Carter, M.G.; Wang, Y.; Goldbergb, I.; Ko, M.S.H. High-throughput screen for genes predominantly expressed in the ICM of mouse blastocysts by whole mount in situ hybridization. *Gene Exp. Patt*, 2006, 6, 213–224.

Zeng, F.; Baldwin, D.A.; Schultz, R.M. Transcript profiling during preimplantation mouse development. *Dev Biol,* 2004, 272, 483–496.

Zhang DY, Sabla G, Shivakumar P, Tiao G, Sokol RJ, Mack C, Shneider BL, Aronow B, Bezerra JA. Coordinate expression of regulatory genes differentiates embryonic and perinatal forms of biliary atresia. *Hepatology.* 2004, 39, 954-962.

Zheng, P.; Patel, B.; McMenamin, M.; Reddy, S.E.; Paprocki, A.M.; Schramm, R.D.; Latham, K.E. The Primate Embryo Gene Expression Resource: a novel resource to facilitate rapid analysis of gene expression patterns in non-human primate oocytes and preimplantation stage embryos. *Biol. Reprod.* 2004, 70, 1411–1418.

Zheng, P.; Vassena, R.; Latham, K. Expression and downregulation of WNT signaling pathway genes in rhesus monkey oocytes and embryos. *Mol Reprod Dev,* 2006, 73, 667-677.

Zhu, F.; Yan, W.; Zhao, Z.L.; Chai, Y.B.; Lu, F.; Wang, Q.; Peng, W.D.; Yang, A.G.; Wang, C.J. Improved PCR-based subtractive hybridization strategy for cloning differentially expressed genes. *Biotechniques,* 2000, 29, 310-313.

Zilberman, D. & Henikoff, S. Genome-wide analysis of DNA methylation patterns. *Development,* 2007, 134, 3959-3965.

In: Development Gene Expresión Regulation
Editor: Nathan C. Kurzfield

ISBN: 978-60692-794-6
©2009 Nova Science Publishers, Inc.

Chapter IX

Gene Expression Regulation in the Developing Brain

Ching-Lin Tsai[1], and Li-Hsueh Wang[2]*

[1]Department of Marine Biotechnology and Resources,
National Sun Yat-sen University, Kaohsiung 804, Taiwan
[2]National Museum of Marine Biology and Aquarium,
2 Houwan Road, Checheng, Pingtung 944, Taiwan

Abstract

The developing central neural circuits are genetically controlled and initiated by developmental signals. Recent progress in molecular and cellular developmental biology provides evidence of how the brain is feminized or masculinized during the critical developmental period. Research into the development of brain architecture requires experimenting with animals, specifically, interfering with normal development and with environmental conditions. Drosophilae, sea urchins, and metazoans are simple invertebrates used for standard research models. Recently, the teleosts, bony fish with biological and genomic complexity found in the higher vertebrates, have become important models for developmental and molecular neurobiology studies. As in mammals, sexual dimorphic genetic expression is found in the developing brain of teleosts. The cellular and synaptic organization of brain architecture is determined by the genomic program and triggered by environmental cues such as photoperiod and temperature. This review highlights some of the methodological issues related to current findings about the gene expression regulation involved in the complex process of neural development, particularly in brain-sex differentiation.

Introduction

The development of neural architecture in the brain is crucial to the physiological functions and behaviors of an animal. Sexual dimorphism in brain structure has been recognized since the pioneering studies by Raisman and Field (1973). The different brain neural circuits are initiated by sex-biased gene expression and are hormone-controlled (Davies and Wilkinson, 2006). The cellular and synaptic organization of the central nervous system is determined not only by genetic regulation, but also by extrinsic, environmental, and epigenetic influences that operate during development (Carrer and Cambiasso, 2002). The generation of neurons and glial cells in the developing brain is mediated by estrogen and neurotransmitters (Carrer and Cambiasso, 2002; Nguyen et al., 2001). The rodent brain has been a prominent model, and non-mammalian species have become leading models for studies of development and genetics. We devote a major discussion to gene expression regulation in the developing brain, to investigate brain-sex differentiation and highlights in a non-mammalian model.

Sexual Dimorphism of the Brain

Brain architecture and functions are developed with sex-specific patterns. The most widely known animal model of neural sexual dimorphism is found in the rat's medial preoptic area (mPOA). Within the mPOA of rats, the region of the sexually dimorphic nucleus of the mPOA (SDN-POA) shows between 2.5 and 5.0 times more neurons in males than in females (Morris et al., 2004). The function of sexual dimorphism in the SDN-POA is not clear.

Sexual dimorphism of the SDN-POA in humans also shows more neurons in males than in females (Swaab et al., 2004a,b; Swaab, 2007). In another famous animal model, the vocal control area of the songbird brain, the vocal control area in zebra finches is about 6 times larger in males than in females (Gurney and Konishi, 1980). The sexually dimorphic distribution of neurotransmitters in the brain is widely investigated in mammals. In tilapia, teleosts, the central neurotransmitters such as serotonin (5-HT), norepinephrine (NE) in different brain areas (telencephalon, optic lobe, and hypothalamus) show sexual dimorphic distribution (Tsai et al., 1995). Furthermore, in response to different environmental cues (higher- or lower-than-normal water temperatures) or to chemical pollutants (mercury), tilapia showed a sex-difference response (Tsai et al., 1995; Tsai and Wang, 1997a). The 5-HT concentration, measured using a high-performance liquid chromatography system with an electrochemical detector, was significantly different between each region: the hypothalamus had a higher concentration than did the telencephalon and optic lobe. The 5-HT concentration in the female hypothalamus was significantly lower than in males. However, 5-HT concentration in the telencephalon and optic lobe was not different between males and females (Tsai et al., 1995). After they had been exposed to mercuric chloride ($HgCl_2$) for 6 months, male fish showed a significantly dose-dependent decrease in 5-HT concentration in the hypothalamus, but not in other regions of the brain. These data provide evidence that the influence of $HgCl_2$ on the central serotonergic system is sex-specific and brain-region-specific (Tsai et al., 1995). In sexually mature males and females exposed to elevated water temperatures, 26°C, 29°C, or 32°C, for 3 weeks, hypothalamic 5-HT concentrations decreased. Similar results were found in the hypothalamic NE system. In the optic lobe, acclimation to elevated temperature resulted in higher 5-HT concentrations in both males and females; however, NE concentrations increased in females but were not altered in males. In the telencephalon, elevated temperature had no affect on 5-HT concentrations in males or females, but they did result in lower NE concentrations in both. These finding show that neurotransmitter activity is influenced by thermal acclimation in a sexually and regionally dependent pattern. Sexually and regionally specific responses of central neurotransmitter systems were also found in tilapia that had been exposed to lower-than-normal temperatures (Tsai and Wang, 1997a). Changes in 5-HT and NE concentrations in the central nervous system may be involved in the physiological and biochemical responses that occur during thermal acclimation. This segment of brain architecture seems to have developed to produce the necessary neural circuitry for integrating information from the external environment with the internal physiological states of a male or female animal. The sex-based difference in brain circuits is based on the sexual dimorphism of neurogenesis. The type of cells, the number of cells, neurogenesis, and the organization of the neural connections are crucial to the development of brain architecture.

Brain-Sex Differentiation in the Development of Brain Neurotransmitter Systems

In addition to the volumetric difference, synaptic connectivity patterns are an important indicator for the structural sexual dimorphism of the brain. Brain neural circuits genetically initiated and regulated by estrogen and neurotransmitters (Toran-Allerand et al., 1999; Schwarz et al., 2008). While the sexual differences in the architecture and function of central neurotransmitter systems is widely known, the development of central neurotransmitter systems has scarcely been studied *in vivo* in mammals. Brain-sex differences in the development of neurotransmitter systems have been well studied in tilapia, *Oreochromis mossambicus*, by interfering with their normal development or with their environmental conditions. Zero-day-old (the hatching day) tilapia were kept at four different temperatures: 20 (lower), 24 (control), and 28 and 32°C (elevated), respectively. On the 5th day, brain 5-HT, NE, γ-aminobutyric acid (GABA), and glutamate (Glu) contents were quantified. Similar experiments on days 5, 10, 15, 20, and 25 posthatching showed that before day 30 posthatching (day 33 postfertilization) is a developing period of brain neurotransmitter systems in tilapia. During this period, the neurotransmitter content consistently increased with age. Subsequently, the influence of both lower and elevated temperatures on the neurotransmitter content differed according to the stage of development. This is evidence that the development of central neurotransmitter systems is differentially influenced by aquatic temperature, according to the stage of development, during its specific effective period (Wang and Tsai, 2000b).

Combined with an understanding of the critical period of brain-sex differentiation, whether a neurotransmitter system is involved in brain-sex differentiation can be predicted. Being exposed to lower temperature before day 10 posthatching induced a high proportion of females, whereas being exposed to elevated temperature after day 10 posthatching induced a high proportion of males (Wang and Tsai, 2000a). Before day 10 posthatching, a critical period for low temperature to induce a female tilapia, both elevated and lower temperatures downregulated brain 5-HT and brain NE, but only lower temperature downregulated GABA and Glu. It is possible that the suppression of brain Glu and GABA, but neither 5-HT nor the NE system, is involved in low-temperature-induced brain feminization, because neural excitation by GABA and Glu during the developmental period is thought to be a potential mechanism that mediates the masculinization of brain neural circuits in mammals (McCarthy et al., 1997; Todd et al., 2007).

The influence of photoperiod (light/dark cycle) on the development of central neurotransmitter systems has also been investigated. During the developing period of central neurotransmitter systems, brain neurotransmitter content is consistently increased with age. Zero-day-old tilapia were raised in three different photoperiods: 12/12, 24/0, and 0/24 h, respectively. On the 5th day, brain 5-HT, NE, GABA, and Glu contents were quantified. Similar experiments on days 5, 10, 15, 20, and 25 posthatching showed that, before day 10 posthatching, the photoperiod altered both brain NE and GABA content. Brain 5-HT content was differentially influenced, either up- or downregulated, according to the developing stage, but brain Glu content was not altered by being exposed to different photoperiods (Huang et al., 2004; Wang and Tsai, 2004). A serial study showed that development of the central 5-HT, NE, and GABA systems was regulated by both environmental temperature and photoperiod, but that the development of the Glu system was modified by the environmental temperature and not by the photoperiod. These facts show that the development of central neurotransmitter systems is age-specifically and neurotransmitter-system-specifically influenced by environmental cues (Wang and Tsai, 2000b; Huang et al., 2004; Wang and Tsai, 2004). The effects of sex steroids, *viz.* estrogen and androgen, on the development of brain neurotransmitter systems in the early developing tilapia brain have been well investigated (Tsai et al., 2001a; Tsai and Wang, 1997b; Tsai and Wang, 1998; Tsai and Wang, 1999; Wang and Tsai, 1999; Huang et al., 2004). Before day 30 posthatching, the brain's NE, 5-HT, GABA, and Glu content significantly increased with age, which showed that before 30 days old is a developing period for the NE, 5-HT, GABA, and Glu systems in the tilapia brain. During this period, both *in vivo* treatment of 17β-estradiol (E_2) and methyltestosterone (MT) upregulate the GABA and Glu systems during a restricted effective period: before day 20 posthatching. Treatment with E_2 *in vivo* upregulates, but does not inhibit, the development of the brain NE system during a specific period. These serial *in vivo* studies provide evidence that the development of brain neurotransmitter systems are differentially altered, both developmental-stage- and neurotransmitter-system-specifically, by being exposed to the sex steroids androgen and estrogen during the restricted developing period.

5-HT is an important signal for neural development. In tilapia, before day 10 posthatching, *in vivo* E_2 treatment induces a significantly higher proportion of females, as happens when the development of the 5-HT system is inhibited. These effects can be mimicked by treating tilapia with *para*-chlorophenyalanine, a 5-HT synthesis blocker (Tsai et al., 2000). Suppression of the central 5-HT system, therefore, may be an indicator of *in vivo* E_2-treatment-induced brain-sex differentiation, *viz.* the formation of female brain neural circuits. During the critical period for *in vivo* E_2-treatment-induced brain feminization, being exposed to low temperature also induces a higher proportion of females (Wang and Tsai, 2000a). However, being exposed to both low and elevated temperatures during this period significantly decreased the development of the brain 5-HT system (Wang and Tsai, 2000b). The effect of *in vivo* E_2 treatment on sexual differentiation may be mediated by the 5-HT system, which is consistent with what occurs in mammals. The development of brain 5-HT is influenced by water temperature in tilapia; however, the alteration of both 5-HT content and the gene expression of 5-HT receptors (5-HT$_{1A}$ and 5-HT$_{1D}$) does not coincide with the brain-sex differentiation in tilapia. Serial studies indicate that low-temperature-induced feminization is mediated neither by 5-HT$_{1A}$ or 5-HT$_{1D}$ receptors nor by altering brain 5-HT content. Therefore, neither brain neurotransmitter 5-HT content nor the gene expression of brain 5-HT receptors (5-HT$_{1A}$ and 5-HT$_{1D}$) is critical for temperature-induced sexual differentiation (Wang and Tsai, 2006).

In summary, the mechanism of sex steroid-induced brain-sex differentiation is not consistent with temperature-induced brain-sex differentiation. One thing that should be mentioned is that in the *in vivo* study of tilapia, the estrogen concentration in the brain is lower in the E_2-treated group than that in the untreated group during the critical period of feminization (unpublished data). This result is consistent with the notion that brain estrogen acts on neurons to feminize brain neural circuits. How to define the brain feminization/masculinization process is still an open question not easily resolved by the mammalian model.

Brain-Sex-Biased Gene Expression in the Sex-Differentiating Brain

Brain-sex differentiation is regulated by estrogen, a product of androgen converted by aromatase, and neurotransmitters. Serotonin, a differentiation signal in neural development, induces a dimorphic sexual structure (Azmitia, 2001). The brain 5-HT concentration in perinatal rats is correlated with the process of brain-sex differentiation (Hardin, 1973). It is not clear whether the 5-HT receptor is activated when the brain is feminized. Tilapia were used to resolve this question. cDNA sequences of 5-HT 1A and 1D receptors were cloned from the brain of the tilapia, *O. mossambicus*. Quantitative real-time polymerase chain reaction (PCR) showed that the ontogenetic expression of neither 5-HT$_{1A}$ nor 5-HT$_{1D}$ was altered by *in vivo* E_2 treatment during the critical period of brain-sex differentiation. No correlation between the ontogenetic expression of the 5-HT receptors 5-HT$_{1A}$ and 5-HT$_{1D}$ and temperature-induced brain feminization/masculinization has been found in *in vivo* studies. Neither 5-HT$_{1A}$ nor 5-HT$_{1D}$ gene expression was associated with either temperature-induced or sex-steroid-induced brain-sex differentiation (Wang and Tsai, 2006).

Brain aromatase and brain estrogen receptors (ERs) are thought to be involved in brain differentiation in teleosts as they are in mammals. In the latter, estrogen-forming (aromatase) and estrogen-sensitive (ER-containing) networks of neurons developing peri- and postnatally are crucial in brain differentiation (Naftolin, 1994; Beyer, 1999). Aromatase, a key enzyme for converting androgen to estrogen (Naftolin et al., 1975; Balthazart et al., 1998), is involved in neural differentiation and maturation in the brain (Hutchison et al., 1997; Horvath et al., 1999). There is a remarkable sex difference in both aromatase (Hutchison et al., 1997; Lauder et al., 1997; Jeyasuria et al., 1998; Karolczak et al., 1998) and ER expression (Kuhnemann et al., 1994; Karolczak et al., 1998; Kuppers and Beyer, 1999; Ivanova and Beyer, 2000) in the developing brain. There is greater aromatase activity and gene expression in neurons developing in the embryonic male brain than in the female brain (Hutchison et al., 1997; Lauder et al., 1997; Jeyasuria et al., 1998; Karolczak et al., 1998). The estrogen-forming capacity of the male hypothalamus may affect brain differentiation at specific sex-steroid-sensitive stages in the ontogeny of mammals (Hutchison et al., 1997). In zebra fish, the timely and appropriate expression of aromatase should be important in development, and the expression of aromatase b (the 'extragonadal' form) may be associated with sexual differentiation if not sexual determination (Trant et al., 2001). In turn, many of the effects of estrogens during development are mediated through neuronal ERs that produce changes in the expression of estrogen-responsive genes, which ultimately influence important developmental processes, such as neuronal proliferation, migration, synapse formation, and apoptosis (Beyer, 1999). Estrogen-receptor concentration is an important component of the mechanism of brain-sex differentiation (Beyer, 1999; MacLusky et al, 1997; Simerly et al., 1997). Numerous recent studies (Toran-Allerand et al., 1999; Kuhnemann et al., 1994; Kuppers and Beyer, 1999; Ivanova and Beyer, 2000; Simerly et al., 1997) have shown that the simultaneous expression of ERs and aromatase during pre- and postnatal ontogeny are potentially involved in mammalian brain differentiation. As in mammals, the ontogeny of brain aromatase and brain ER expression are involved in the process of temperature-induced sex differentiation in tilapia. The feminizing thermosensitive period, before day 10 posthatching, is the same as the estrogen-sensitive period in tilapia (Tsai et al., 2000; Wang and Tsai, 2000a). Similar to the low-temperature-induced effects, exogenous estrogen has a feminizing effect on the gonad before day 10 posthatching when the expression of brain aromatase and brain estrogen receptor α (ERα) is downregulated (Tsai et al., 2001a, 2003). It seems that the downregulation of brain ERα and aromatase gene expression is the common pathway for the development of sex-specific brain neural circuits.

Similar to *in vivo* E$_2$ treatment, water temperature regulates neural development in the tilapia brain. In an *in vivo* study on tilapia (Tsai et al., 2000; Wang and Tsai, 2000a; Tsai et al., 2001a, 2003), exogenous estrogen and low temperature during the critical period separately induced brain feminization, while brain 5-HT content was decreased in the E$_2$-treated group but not altered in the low-temperature-treated group. However, both ERα and aromatase mRNA expression were downregulated by both exogenous estrogen and low temperature. The gene expression of both aromatase-containing and ER-containing systems in the developing brain might be an indicator of both estrogen- and temperature-induced brain-sex differentiation in tilapia. Research on tilapia provides evidence that the gene expression of brain-sex differentiation-related genes is controlled by sex-steroids and temperature, and subsequently determines whether brain neural circuits are feminized or masculinized. In mammals, the ontogenetic expression of both brain aromatase and ERs during the peri- and postnatal periods is crucial in brain differentiation (Toran-Allerand et al., 1999; Kuhnemann et al., 1994; Kuppers and Beyer, 1999; Ivanova and Beyer, 2000; Simerly et al., 1997). Aromatase expression is higher in neurons developing in the embryonic male brain than in the female brain (Hutchison et al., 1999). Both ERα and ERβ are expressed in the developing brain of mammals. There are significant sex differences in ER mRNA expression in the developing brain (Kuhnemann et al., 1994; Karolczak and Beyer, 1998; Kuppers and Beyer, 1999; Ivanova and Beyer, 2000). Estrogen treatment during either the peri- or the postnatal period decreases ER expression. ER mRNA downregulation may be an important estrogen-regulated event in the process of brain-sex differentiation (Kuhnemann et al., 1994; Karolczak and Beyer, 1998; Kuppers and Beyer, 1999; DonCarlos et al., 1995), and changes in ER concentrations may be one of the hallmarks of this process (Kuhnemann et al., 1994). In songbirds, brain aromatase activity and mRNA expression may serve a sexually dimorphic function (Saldanha et al., 2000). Estrogen plays a major role in masculinizing the song function (Ramachandran et al., 1999). The ontogenetic investigation of tilapia during the critical period of brain-sex differentiation provides evidence that the sex-biased gene expression of brain aromatase and ERα is one of the hallmarks of brain-sex differentiation processing.

Estrogen acts directly on neurons to mediate the sex differentiation of brain neural circuits. The generation of neurons and glial cells mediated by estrogen in the developing brain during this specific period is expected to lead to broad changes in the structure and functions of the brain. A serial review in mammals (Carrer and Cambiasso, 2002) reported that male-type brain circuitry results from exposure to androgens during a "critical period" of brain development, whereas female-type brain circuitry develops in the absence of testicular secretion, irrespective of chromosomal sex. This is consistent with the finding in tilapia that *in vivo* E_2 treatment during the restricted developing period induced brain feminization. Brain alpha-fetoprotein (AFP)-binding estrogen is thought to be critical in the developing brain. The inhibiting estrogen rescues the brain masculinization found in female mice lacking this gene, which suggests that α-fetoprotein inhibits estrogen activity in females. In females, the AFP binding of estrogens blocks them from the brain and keeps them circulating long enough to be metabolized into inactive steroids. Another hypothesis is that AFP-escorted estrogen may result in very selective and specific estrogen delivery to particular sets of neurons (Bakker et al., 2006), which may contribute to feminine brain development (Puts et al., 2006). The masculinizing actions of androgen are mediated by estrogen, the product of aromatization by aromatase. Serial studies in tilapia provide evidence that the downregulation of brain ERα and brain aromatase gene expression during the brain-sex differentiation period is an indicator of the processing of sex-steroid- and temperature-induced brain feminization. Based on this understanding of the critical period for exogenous sex-steroid- and temperature-induced brain-sex differentiation, the genetic and brain-sexual differentiation in tilapia may become a powerful animal model providing insights into questions that have remained unanswered by studies on other vertebrate systems.

The Development of Brain Neural Circuitry

For the past 50 years, sexually differentiated development has been thought to be the effect of estrogen on neurons in vertebrates. However, growing evidence suggests that there are probably direct genetic effects that induce sex-specific neural circuits in the brain. A different pattern of synaptic connections in the mPOA of male and female rats was reported 35 years ago (Raisman and Field, 1973). Histological studies have revealed the sex-specific brain neural circuits in mammals and non-mammals. Recently, microarray analysis of gene expression has become a powerful tool for discovering the sex-specific expression of sex-biased genes involved in the development of brain neural circuitry. A comprehensive microarray analysis of gene expression in mouse brains (Yang et al., 2006) reported 355 female-biased genes and 257 male-biased genes. The expression of several X escapee genes is indeed higher in female than in male brain tissue. Some of the sex differences are found only in the adult brain but not in other types of tissue (Yang et al., 2006). The sex-linked genes, which are asymmetrically inherited between males and females, may directly influence sexually dimorphic neurobiology (Arnold et al., 2004; Davies and Wilkinson, 2006). However, because the brain is comprised of highly heterogeneous tissue, sex differences in gene expression within individual regions of the brain may be masked when the whole brain is studied, owing the limited sensitivity of microarrays for detecting genes expressed at low levels (Isensee and Ruiz Noppinger, 2007). The transitional gene-by-gene approach is necessary to study brain-sex-differentiation-related genes. Many sexual dimorphisms in brain neural circuitry and gene expression have been proved to be due to sex steroids and metabolites that act in the developing brain and permanently write to the brain in a sex-specific brain architecture (McCarthy and Konkle, 2005; Becker et al., 2005; Morris et al., 2004). Genetic and physiological studies in teleosts with the biological and genomic complexity found in the higher vertebrate have been especially important in providing insights into the answers to questions that remained unanswered by studying other vertebrates. We are fortunate that tilapia are an almost ideal animal model for investigating the progressing alteration at the molecular, genetic, cellular, and circuitry levels in the brain in parallel with the critical period of brain-sex differentiation.

In vertebrates, the accumulated evidence suggests that genetic mechanisms controlling gender-specific neural characteristics precede or are concomitant with hormonal effects (Carrer and Cambiasso, 2002). Sexual dimorphism of gene expression in the mouse brain, for example, occurs before gonadal differentiation (Dewing et al., 2003). As in mammals, sexual dimorphic gene expression in the teleost brain precedes gonadal differentiation. As with the feminization effect of estrogen treatment *in vivo*, being exposed to lower temperature during the critical period of feminization in tilapia (*O. mossambicus*) before posthatching day 10 induces a high proportion of females when the mRNA expression of brain aromatase and brain ERα are downregulated (Tsai et al., 2000; Wang and Tsai, 2000a; Tsai et al., 2001a, 2003). The genetic expression of the developing brain, therefore, is critical for brain-sex differentiation. However, the molecular mechanism for this differentiation is not clear.

Expressed sequence tags (ESTs) vary with species, tissue, age, sex, and physiological conditions. Adult brain ESTs have been derived from many species, mammalian and non-mammalian. EST cataloging and profiling provides a basis for functional genomic research. Developing brain ESTs will be useful for studying the cellular and molecular mechanisms of the development of the brain. A few ESTs have been cloned from the developing animal, particularly from the brain. The analysis of ESTs provides significant addition functional structure and evolutionary information (Quackenbush et al., 2000). However, there is no transcriptome analysis concentrated specifically on the developing brain during the critical period of brain-sex differentiation. Identifying genes expressed in the cells of the developing brain, particularly during the critical period of brain-sex differentiation, is important for studying both the molecular mechanism of sexual differentiation in brain neural circuits and the physiological functions during the developmental stages. A list of transcripts expressed in the developing tilapia brain using ESTs has been derived from the developing brain during the critical period for the formation of sex-specific brain neural circuits (Tsai et al., 2007). Based on available protein domain information and GO annotation, 14 genes have been classified as neural-development-related genes: discs large homolog 5, dishevelled-1 isoform (DVL-1), endothelial differentiation-related factor 1, inhibitor of differentiation protein 2 (Id2), midkine-related growth factor 2 (Mdk2), mitogen-activated protein kinase 14b (mitogen-activated protein kinase p38b), myelin expression factor 2, nuclear protein NAP (beta-catenin-like isoform 1), odd Oz/ten-m homolog 1 (tenascin M), p53 tumor suppressor protein, plasticity-related protein 2 (PRG-2), pleiotropic factor b (heparin-binding neurite-promoting factor), tsc2 gene product, and ubiquitin-activating enzyme E1. All of these neural development-related genes are expressed in the early developing brain. Their ontogenetic expression of the remaining neural development-related genes varies with the stage of development. These neural development-related genes have been classified into four types based on real-time-quantification-reverse transcriptase PCR analysis of their responses to different temperatures.

- Type 1: The ontogenetic expression was not altered by different temperatures.
- Type 2: The ontogenetic expression was particularly influenced by elevated temperature (32°C).
- Type 3: The ontogenetic expression was particularly influenced by lower temperature (20°C).
- Type 4: The ontogenetic expression was differentially influenced by elevated and lower temperatures according to the stage of development.

Discs large homolog 5, myelin expression factor 2, plasticity-related protein-2, tsc2 gene-product-related genes, and an inhibitor of differentiation protein 2 (Id2) were differentially temperature-influenced according to their developmental stages. Endothelial-differentiation-related factor 1, midkine-related growth factor b, and mitogen-activated protein kinase 14b are specifically influenced by elevated temperature, and beta-catenin-like isoform 1 by lower temperature. Neural-development-related genes, expressed in the sex-differentiating brain, are related to the development of sex-specific brain neural circuits. These neural development genes with thermosensitive ontogenetic expression should be involved in temperature-induced brain-sex differentiation. In order to screen the brain-sexual-differentiation-inducing gene, a gene-by-gene approach in neural/glial culture is a powerful tool. Indicators of brain-feminized neural circuits and brain-masculinized neural circuits are required for the research.

The completion of the genome sequence of *Drosophila melanogaster* (Adams and Sekelsky, 2000; Rubin, 2000; Yoshihara et al., 2001) provides an important foundation for an important animal model that will allow us to apply genetic screens to identify mutants that interrupt specific neural functions. The annotation of approximately 14,000 genes contained within the 120-megabase euchromatic genome allows the fly-research community to make homology-based comparisons to comprehensively identify gene families and homologs of known prokaryotic and eukaryotic proteins. Neurobiology in *Drosophila* has covered a wide range of experimental questions ranging from the specification of the nervous system during early development to the molecules involved in learning and memory. Behavioral mutants initiated a wave of genetic studies into the function of the nervous system and led to the characterization of mutants such as Shibire (Ikeda et al., 1976) and Shaker (Kaplan and Trout 1969). These studies provided the foundation for a new generation of fly neurobiologists that employed systematic genetic screens for specific neurological phenotypes. The fru gene of *Drosophila* functions at the head of one of the branches of the sex-determination pathway, and acts specifically in the central nervous system (CNS) to govern sexual orientation and male courtship behavior (Ryner et al., 1996; Taylor et al., 1994; Goodwin, 1999). The synaptic formation-related gene and protein are found at *Drosophila* neuromuscular junctions. Using microarrays has allowed us to identify the sex-enriched transcripts expressed during three different stages of the development of *C. elegans* larvae. The TRA-1 Zinc finger protein is a sex-determination controller in *C. elegans*, a Ci/GLI homolog that determines the fate of female cells throughout the body. The sex specific neuron and, consequently, the underlying sex-specific neural circuitry are important for producing sex-specific behaviors (Goodwin, 1999; Thoemke et al., 2005). Though sex dimorphism in the neural circuitry is ubiquitous, the expression and regulation of the gene to induce brain-architecture sex-differentiation remain poorly understood.

Estrogen acts directly on neurons to mediate the sex differentiation of brain neural circuits. The generation of neurons and glial cells mediated by estrogen in the developing brain is expected to lead to broad change in the structure and functions of brain. A serial review in mammals (Carrer and Cambiasso, 2002) reported that male-type brain circuitry is the result of the brain's having been exposed to androgens during a "critical period" of brain development, whereas female-type brain circuitry is the result of the absence of testicular secretion, irrespective of the animal's chromosomal sex. In our *in vivo* study of tilapia (unpublished data), the estrogen concentration in the brain was lower in the E_2-treated group than in the untreated group during the critical period of brain feminization. Alpha-fetoprotein (AFP) in the developing brain is thought to inhibit estrogen activity in females. Selective estrogen delivery to a specific neural circuit may contribute to the development of a female brain (Puts, 2006). Estrogen directs the formation of sexually dimorphic circuits by influencing axonal guidance and synaptogenesis. *In vivo* and *in vitro* studies show that the development of glial cells and neurons in the brain is regulated by estrogen, which controls the formation of sex-specific brain neural circuits. Estrogen is thought to be involved in the formation of the sexually dimorphic distribution of central serotonergic innervation via the ER-containing 5-HT cells in rats (Lu et al., 2004). However, little is known about the mechanism that induces the development of sex-specific brain neural circuits at the genetic level.

The genetic and epigenetic mechanisms involved in determining the development of brain neural circuits have been explored in cultures of neurons, cultures of glial cells, and co-cultures. The genomic determinants expressed in the brain, those that shape the sex-specific characteristics of neurons and glial cells, include these sex-, region-, and time-specific responses to estrogen. The effects of estrogen and neurotransmitters on the proliferation of brain cells have been investigated in the primary neuronal culture of the tilapia brain (Fig. 1). In an *in vitro* study (Tsai et al., 2001b), E_2 significantly increased the proliferation of neurons in the primary neural culture. *In vitro* E_2 treatment significantly and dose-dependently increased the serotonergic cells of the primary brain neural cell (Fig. 2). 5-HT treatment also induced the 5-HT-induced proliferation of neurons in the neural culture. The antagonist of 5-HT_{1A} receptor, WAY-100635, inhibited this proliferation (Fig. 3). These results showed that E_2 increased the number of neural cells, including 5-HT-containing neurons, which is evidence that estrogen acts on neurons to induce a masculinized brain neural circuit, in part, by increasing the number of 5-HT-containing neurons associated with 5-HT_{1A} receptor. This is consistent with the *in vivo* E_2-treatment-induced brain feminization while brain 5-HT content is decreased. An *in vivo* study (Tsai et al., 2001a, 2003) showed that the gene expression of neither brain 5-HT_{1A} nor 5-HT_{1D} is associated with the formation of sex-specific brain neural circuits, which is induced by *in vivo* sex-steroid treatment or temperature-induced brain-sex differentiation. *In vitro* studies, however, show that 5-HT_{1A} receptor deals with the sex-steroid-induced mechanism. 5-HT may be mediated in part by both 5-HT_{1A} and 5-HT_{1D} receptors to induce the proliferation of neurons in mammals (Barnes and Sharp, 1999). 5-HT_{1A} receptor may play a neurotrophic role in both the developing brain and the adult brain. The cell-specific role for the formation of sex-specific brain architecture may be a reason for the different results between the *in vivo* and *in vitro* studies. Cell-specific gene expression will be marked when the whole brain is studied. Therefore, cell-by-cell-specific gene expression should be investigated, using cell culture, for the formation of sex-specific neural circuits. Combined with the gene-by-gene approach, the knockdown and silencing of genes in neural and glial cell cultures is a powerful tool for studying the regulation of the expression of feminization-/masculinization-related genes in brain neural circuits. Finally, *in vitro* results should be confirmed by *in vivo* studies.

(a)   (b)

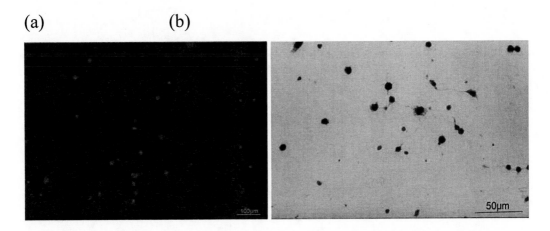

Figure 1. Cell cultures derived from the adult hypothalamus of tilapia, *Oreochromis mossambicus*. Approximately 48 h after plating, the cells started to differentiate. (a) Neurons were stained with MAP2 (2a+2b) (red) and nuclei were stained with Hoechst 33342 (blue). (b) 5-HT-containing neurons (black).

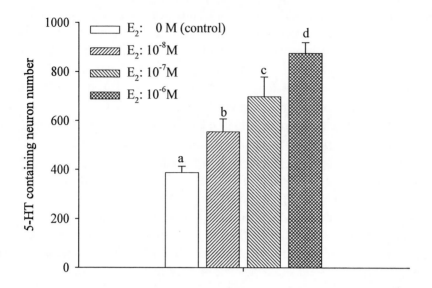

Figure 2. Effects of E_2 on the 5-HT-containing proliferated neurons cultured from the tilapia hypothalamus. Statistical data are means ± SD. Different letters indicate significant differences between groups at the same stage (one-way ANOVA, and then Duncan's multiple-range test).

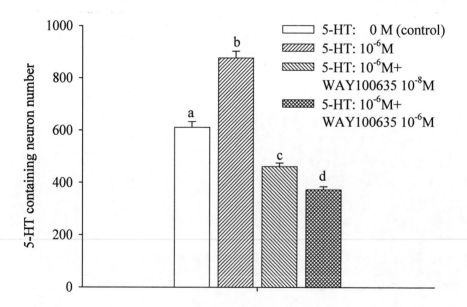

Figure 3. Effects of 5-HT on the 5-HT-containing proliferated neurons cultured from the tilapia hypothalamus. Statistical data are means ± SD. Different letters indicate significant differences between groups at the same stage (one-way ANOVA, and then Duncan's multiple-range test).

Conclusion

Sex-specific genes control the formation of brain architecture. Environmental cues regulate sex-biased gene expression in the brain. The regulation of the expression of feminization-/masculinization-related genes in brain neural circuits is an open question. The mechanisms for sex-steroid- and temperature-induced brain-sex differentiation as well as for the development of central neurotransmitter systems have been well investigated in tilapia, *O. mossambicus*. Their critical periods for forming male and female brain neural circuits are well defined. The development of the central neurotransmitter system and the up- and downregulation of the number of males and females caused by *in vivo* sex-steroid treatment during the critical period of brain feminization/masculinization, respectively, may be an important indicator for the progression of brain-sex differentiation. Based on this information, the regulation of environmental cues affecting sex-biased gene expression during a critical period of brain-sex differentiation can be studied. Gene expression of brain aromatase and brain ERα is associated with brain-sex differentiation in both temperature-induced and sex-steroid-induced brain-sex differentiation. It will be interesting to determine the role of the brain's 5-HT system in both. *In vivo* studies on genetic and brain sexual differentiation induced by sex steroids and temperature in tilapia, have been especially important in providing insights into questions that remain unanswered from other vertebrate models. The *in vivo* genomic approach using tilapia appears to be a good model for this investigation. Downregulation of the gene expression of brain ERα and brain aromatase induced by estrogen or low-temperature treatments in *in vivo* studies during the critical developing period indicates the feminization of brain neural circuits. Transcriptome analysis is useful for functional genomic research on development, comparative genomic studies, and genomic evolution. The functional analyses of the ESTs derived from the developing brain cDNA library of tilapia will provide more information for the molecular mechanism of the physiological functions as well as the cellular and synaptic organization of the brain during its developmental stages. The genes identified as neural-development-related, and cloned from the developing tilapia brain, should play a role in the cellular and synaptic organization of the central neural circuits. Based on this information, knockdown-gene expression, using a gene-by-gene approach *in vitro* combined with microarray analysis, will be an animal model for resolving the open questions about the regulation of gene-expression regulation in sex-differentiation brain neural circuits at the cellular level. The participation of glial cells and neurons in the formation of brain-sex-specific neural circuits is an emerging field of great interest. Regulation of sex-specific gene expression at the level of glial cells and neurons requires more study, and all *in vitro* findings should be confirmed by *in vivo* studies.

References

Adams, M.D., & Sekelsky, J.J. (2002). From sequence to phenotype: reverse genetics in *Drosophila melanogaster*. *Nat. Rev. Genet.*, *3*, 189-98.

Arnold, A.P., Xu, J., Grisham, W., Chen, X., Kim, Y.H., & Itoh, Y. (2004). Sex chromosomes and brain sexual differentiation. *Endocrinology*, *145*, 1057-1062.

Azmitia, E.C. (2001). Modern views on an ancient chemical: serotonin effects on cell proliferation, maturation, and apoptosis. *Brain Res. Bull.*, *56*, 413-424.

Bakker, J., De Mees, C., Douhard, Q., Balthazart, J., Gabant, P., Szpirer, J., & Szpirer, C. (2006). Alpha-fetoprotein protects the developing female mouse brain from masculinization and defeminization by estrogens. *Nature Neuroscience, 9*, 220-226.

Balthazart, J., & Ball, G.F. (1998). New insights into the regulation and function of brain estrogen synthase (aromatase). *Trends Neurosci., 21*, 243-249.

Barnes, N.M., & Sharp, T. (1999). A review of central 5-HT receptors and their function. *Neuropharmacology, 38*, 1083-1152.

Becker, J.B., Arnold, A.P., Berkley, K.J., Blaustein, J.D., Eckel, L.A., Hampson, E., Herman, J.P., Marts, S., Sadee, W., Steiner, M., Taylor, J., & Young, E. (2005). Strategies and methods for research on sex differences in brain and behavior. *Endocrinology, 146*, 1650-73.

Beyer, C. (1999). Estrogen and the developing mammalian brain. *Anat. Embryol. (Berlin), 199*, 379-390.

Carrer, H.F., & Cambiasso, M.J. (2002). Sexual differentiation of the brain: genes, estrogen, and neurotrophic factors. *Cell Mol Neurobiol, 22*, 479-500.

Davies, W., & Wilkinson, L.S. (2006). It is not all hormones: alternative explanations for sexual differentiation of the brain. *Brain Res., 1126*, 31-45.

Dewing, P., Shi, T., Horvath, S., & Vilain, E. (2003). Sexual dimorphic gene expression in mouse brain precedes gonadal differentiation. *Mol. Brain Res., 118*, 82-90.

DonCarlos, L.L., McAbee, M., Ramer-Quinn, D.S., & Stancik, D.M. (1995). Estrogen receptor mRNA levels in the preoptic area of neonatal rats are responsive to hormone manipulation. *Dev. Brain Res., 84*, 253-260.

Goodwin, S.F. (1999). Molecular neurogenetics of sexual differentiation and behaviour. *Curr. Opinion Neurobiol., 9*, 759-765.

Gurney, M.E., Konishi, M. (1980). Hormone-Induced Sexual Differentiation of Brain and Behavior in Zebra Finches. *Science, 208*, 1380-1383.

Hardin, C.M., (1973). Sex differences in serotonin synthesis from 5-hydroxytryptophan in neonatal rat brain. *Brain Res., 59*, 437-439.

Horvath, T.L., & Wikler, K.C. (1999). Aromatase in developing sensory systems of the rat brain. *J. Neuroendocrinol., 11*, 77-84.

Huang, Y.S., Wang, L.H., & Tsai, C.L. (2004). Photoperiod a.ects the development of central neurotransmitter systems of tilapia, *Oreochromis mossambicus. Neurosci. Lett., 355*, 201-204.

Hutchison, J.B., Beyer, C., Hutchison, R.E., & Wozniak, A. (1997). Sex differences in the regulation of embryonic brain aromatase. *J. Steroid Biochem. Mol. Biol., 61*, 315-322.

Ikeda, K., Ozawa, S., & Hagiwara, S. (1976). Synaptic transmission reversibly conditioned by single-gene mutation in *Drosophila melanogaster. Nature, 259*, 489-91.

Isensee, J., & Ruiz Noppinger, P. (2007). Sexually dimorphic gene expression in mammalian somatic tissue. *Genet Med., 4*, S75-95.

Ivanova, T., & Beyer, C. (2000). Ontogenetic expression and sex differences of aromatase and estrogen receptor-alpha/beta mRNA in the mouse hippocampus. *Cell Tissue Res., 300*, 231-237.

Jeyasuria, P., & Place, A.R. (1998). Embryonic brain-gonadal axis in temperature dependent sex determination of reptiles: a role for P450 aromatase (CYP19). *J. Exp. Zool., 281,* 428-449.

Kaplan, W.D., & Trout, W.E. 3rd. (1969). The behavior of four neurological mutants of *Drosophila. Genetics, 61,* 399-409.

Karolczak, M., & Beyer, C. (1998). Developmental sex differences in estrogen receptor-beta mRNA expression in the mouse hypothalamus/preoptic region. *Neuroendocrinology, 68,* 229-234.

Karolczak, M., Kuppers, E., & Beyer, C. (1998). Developmental expression and regulation of aromatase- and 5alpha-reductase type I mRNA in the male and female mouse hypothalamus. *J. Neuroendocrinol., 10,* 267-274.

Kuhnemann, S., Brown, T.J., Hochberg, R.B., & MacLusky, N.J. (1994). Sex differences in the development of estrogen receptors in the rat brain. *Horm. Behav., 28,* 483-491.

Kuppers, E., & Beyer, C. (1999). Expression of estrogen receptor-alpha and beta mRNA in the developing and adult mouse striatum. *Neurosci. Lett., 276,* 95-98.

Lauber, M.E., Sarasin, A., & Lichtensteiger, W. (1997). Sex differences and androgendependent regulation of aromatase (CYP19) mRNA expression in the developing and adult rat brain. *J. Steroid Biochem. Mol. Biol., 61,* 359-364.

Lu, H., Nishi, M., Matsuda, K., & Kawata, M. (2004). Estrogen reduces the neurite growth of serotonergic cells expressing estrogen receptors. *Neurosci. Res., 50,* 23-28.

MacLusky, N.J., Bowlby, D.A., Brown, T.J., Peterson, R.E., & Hochberg, R.B. (1997). Sex and the developing brain: suppression of neuronal estrogen sensitivity by developmental androgen exposure. *Neurochem. Res., 22,* 1395-1414.

McCarthy, M.M., Davis, A.M., & Mong, J.A. (1997). Excitatory neurotransmission and sexual differentiation of the brain. *Brain Res., 44,* 487-495.

McCarthy, M.M., & Konkle, A.T. (2005). When is a sex difference not a sex difference? *Front. Neuroendocrinol., 26,* 85-102.

Morris, J.A., Jordan, C.L., & Breedlove, S.M. (2004). Sexual differentiation of the vertebrate nervous system. *Nature: Neuroscience, 7,* 1034-1039.

Naftolin, F. (1994). Brain aromatization of androgens. *J. Reprod. Med., 39,* 257-261.

Naftolin, F., Ryan, K.J., Davies, I.J., Reddy, V.V., Flores, F., Kuhn, M., White, R.J., Takaoka, Y., Wolin, L. (1975). The formation of estrogens by central neuroendocrine tissues. *Recent Prog. Horm. Res., 31,* 295-319.

Nguyen, L., Rigo, J.M., Rocher, V., Belachew, S., Malgrange, B., Rogister, B., Leprince, P., & Moonen, G. (2001). Neurotransmitters as early signals for central nervous system development. *Cell Tissue Res., 305,* 187-202.

Puts, D.A., Jordan, C.L., & Breedlove, S.M. (2006). Defending the brain from estrogen. *Nature Neuroscience, 9,* 155-156.

Quackenbush, J., Liang, F., Holt, I., Pertea, G., & Upton, J. (2000). The TIGR gene indices: reconstruction and representation of expressed gene sequences. *Nucleic Acids Res., 28,* 41-45.

Raisman, G., Field, P.M. (1973). Sexual dimorphism in the neuropil of the preoptic area of the rat and its dependence on neonatal androgen. *Brain Res., 54,* 1-29.

Ramachandran, B., Schlinger, B.A., Arnold, A.P., & Campagnoni, A.T. (1999). Zebra finch aromatase expression is regulated in the brain though an alternate promoter. *Gene, 240,* 209-216.

Rubin, G.M. (2000). Biological annotation of the *Drosophila* genome sequence. *Novartis Found Symp., 229,* 79-82.

Ryner, L.C., Goodwin, S.F., Castrillon, D.H., Anand, A., Villella, A., Baker, B.S., Hall, J.C., Taylor, B.J., & Wasserman, S.A. (1996). Control of male sexual behavior and sexual orientation in *Drosophila* by the fruitless gene. *Cell, 87,* 1079-1089.

Saldanha, C.J., Tuerk, M.J., Kim, Y.H., Fernandes, A.O., Arnold, A.P., & Schlinger, B.A. (2000). Distribution and regulation of telencephalic aromatase expression in the zebra finch revealed with a specific antibody. *J.Comp. Neurol., 423,* 619–630.

Schwarz, J.M., Liang, S.L., Thompson, S.M., & McCarthy, M.M. (2008). Estradiol induces hypothalamic dendritic spines by enhancing glutamate release: a mechanism for organizational sex differences. *Neuron, 58,* 584-598.

Simerly, R.B., Zee, M.C., Pendleton, J.W., Lubahn, D.B., & Korach, K.S. (1997). Estrogen receptor-dependent sexual differentiation of dopaminergic neurons in the preoptic region of the mouse. *Proc. Nat. Acad. Sci. USA, 94,* 14077-14082.

Swaab, D.F. (2007). Sexual differentiation of the brain and behavior. *Best Pract. Res. Clin. Endocrinol. Metabol., 21,* 431-444.

Swaab, D.F. (2004a). The Human Hypothalamus. Basic and Clinical Aspects. Part II: Neuropathology of the Hypothalamus and Adjacent Brain Structures. In Aminoff MJ, Boller F & Swaab DF (eds.). Handbook of Clinical Neurology. (pp. 596) Amsterdam: Elsevier, 2004.

Swaab, D.F. (2004b). The Human Hypothalamus. Basic and Clinical Aspects. Part I: Nuclei of the Hypothalamus. In Aminoff MJ, Boller F & Swaab DF (eds.). Handbook of Clinical Neurology. (pp. 476) Amsterdam: Elsevier.

Taylor, B.J., Villella, A., Ryner, L.C., Baker, B.S., & Hall, J.C. (1994). Behavioral and neurobiological implications of sex-determining factors in *Drosophila. Dev. Genet., 15,* 275-296.

Thoemke, K., Yi, W., Ross, J.M., Shinseog, K. S, Reinke, V., & Zarkower, D. (2005). Genome-wide analysis of sex-enriched gene expression during *C. elegans* larval development. *Dev. Biol., 284,* 500-508.

Todd, B.J., Schwarz, J.M., Mong, J.A., & McCarthy, M.M. (2007). Glutamate AMPA/kainate receptors, not GABA(A) receptors, mediate estradiol- induced sex differences in the hypothalamus. *Dev. Neurobiol., 15,* 304-15.

Toran-Allerand, C.D., Singh, M., Setalo Jr., G. (1999). Novel mechanisms of estrogen action in the brain: new players in an old story. *Front. Neuroendocrinol., 20,* 97-121.

Trant, J.M., Gavasso, S., Ackers, J., Chung, B.C., & Place, A.R. (2001). Developmental expression of cytochrome P450 aromatase genes (CYP19a and CYP19b) in zebrafish fry (*Danio rerio*). *J. Exp. Zool., 290,* 475-483.

Tsai, C.L., Jang, T.H., & Wang, L.H. (1995). Effects of mercury on serotonin concentration in the brain of tilapia, *Oreochromis mossambicus. Neurosci, Lett., 184,* 208-211.

Tsai, C.L., & Wang, L.H. (1997a). Effects of thermal acclimation on the neurotransmitters, serotonin and norepinephrine in the discrete brain of male and female tilapia, *Oreochromis mossambicus. Neurosci. Lett., 233,* 77-80.

Tsai, C.L., & Wang, L.H. (1997b). Effects of estradiol and testosterone on the serotonin content and turnover in the brain of tilapia embryo. *Biogenic Amines, 13,* 19-28.

Tsai, C.L., & Wang, L.H. (1998). Effects of gonadal steroids on the noradrenergic activity in the early developing tilapia brain. *Biogenic Amines, 14,* 591-598.

Tsai, C.L., & Wang, L.H. (1999). Effects of gonadal steroids on the serotonin synthesis and metabolism in the early developing tilapia brain. *Neurosci. Lett., 264,* 45–48.

Tsai, C.L., Wang, L.H., Chang, C.F., & Kao, C.C. (2000). Effects of gonadal steroids on brain serotonergic and aromatase activity during the critical period of sexual differentiation in tilapia, *Oreochromis mossambicus. J. Neuroendocrinol., 12,* 894–898.

Tsai, C.L., Wang, L.H., & Fang, L.S. (2001a). Estradiol and *para-* Chlorophenylalanine down-regulate the expression of brain aromatase and estrogen receptor α mRNA during the critical period of feminization in tilapia, *Oreochromis mossambicus. Neuroendocrinology, 74,* 325-334.

Tsai, C.L., Wang, L.H., & Lin, Y.H., (2001b). Effects of estrogen and neurotransmitters on the primary cultures of tilapia brain from the different ages. *Dev. Brain Res., 129,* 111-113.

Tsai, C.L., Chang, S.L., Wang, L.H., & Chao, T.Y. (2003). Temperature influences the ontogenetic expression of aromatase and estrogen receptor mRNA in the developing tilapia brain. *J. Neuroendocrinol., 15,* 97-102.

Tsai, C.L., Wang, L.H., Shiue, Y.L., & Chao, T.Y. (2007). The influence of temperature on the ontogenetic expression of neural development-related genes from developing tilapia brain expressed sequence tags. *Marine Biotechnol., 9,* 243-267.

Wang, L.H., & Tsai, C.L. (1999). Effects of gonadal steroids on the GABA and glutamate contents of the early developing tilapia brain. *Dev. Brain Res., 114,* 273-276.

Wang, L.H., & Tsai, C.L. (2000a). Effects of temperature on the deformity and sex differentiation of tilapia, *Oreochromis mossambicus. J. Exp. Zool., 286,* 534-537.

Wang, L.H., & Tsai, C.L. (2000b). Temperature affects the development of the central nervous system in tilapia, *Oreochromis mossambicus. Neursci. Lett., 285,* 95-98.

Wang, L.H., and Tsai, C.L. (2004). Effects of photoperiod on the development of central glutamate system in tilapia, *Oreochromis mossambicus. Dev. Brain Res., 152,* 79-82.

Wang, L.H., & Tsai, C.L. (2006). Cloning and characterization of tilapia serotonin 1A and 1D receptor cDNAs: Influence of temperature and gonadal steroids on the ontogenetic expression of brain serotonin 1A and 1D receptors during the critical of sexual differentiation in tilapia, *Oreochromis mossambicus. Comp. Biochem. Physiol. B, 143,* 117-126.

Yang, X., Schadt, E.E., Wang, S., Wang, H., Arnold, A.P., Ingram-Drake, L., Thomas, A., Drake, T.A. & Lusis, A.J., (2006). Tissue-specific expression and regulation of sexually dimorphic genes in mice. *Genome Res., 16,* 995-1004.

Yoshihara, M., Ensminger, A.W., Littleton, J.T. (2001). Neurobiology and the *Drosophila* genome. *Funct. Integr. Genomics, 1,* 235-40.

In: Development Gene Expresión Regulation
Editor: Nathan C. Kurzfield

ISBN: 978-60692-794-6
©2009 Nova Science Publishers, Inc.

Chapter X

Expression and Action of SRY during Gonadal Sex Differentiation in the Mouse

Teruko Taketo and Chung-Hae Lee

Departments of Surgery and Biology, McGill University,
Royal Victoria Hospital, 687 Pine Avenue West, Montreal, Quebec H3A 1A1

Abstract

SRY/Sry, a single-copy gene on the Y-chromosome, was identified to play the critical role in initiating testicular differentiation during gonadal development in humans and mice two decades ago. Nonetheless, neither the regulation of *Sry* expression nor the mode of SRY action during gonadal differentiation is well understood. The B6.Y^{TIR} mouse carries a Y-chromosome originally from a *Mus musculus domesticus* mouse caught in Tirano, Italy (Y^{TIR}) and the X-chromosome and autosomes from the C57BL/6J (B6) inbred mouse strain, which belongs to *Mus musculus molossinus*. It has been demonstrated that the SRY protein is expressed normally both in pattern and onset, yet, B6.Y^{TIR} mice develop only ovaries or ovotestes. Therefore, this mouse model provides an opportunity to study the mechanism of SRY action during gonadal sex determination. We hypothesize that the testis determining pathway in the B6.Y^{TIR} gonad is impaired by at least two mechanisms that act synergistically. First, *Sry* transcript levels from the Y^{TIR}-chromosome are reduced on the B6 genetic background. Second, polymorphisms of *Sry* sequences lead to inefficient biological activity of the SRY protein encoded on the Y^{TIR}-chromosome. Both dysfunctions are requisite to impairing testicular differentiation.

Introduction

In normal eutherian mammalian development, the presence or absence of a Y-chromosome determines the differentiation of a gonadal primordium into a testis or an ovary.

Consequently, the hormones produced by a testis induce the development of male reproductive organs as well as sexual behaviors. On the other hand, the absence of testicular hormones results in female development as a default pathway (Jost et al., 1973). The gene on the Y-chromosome responsible for primary testis determination (*TDF* or *Tdy*) has been identified and named *SRY/Sry* (Sex-determining Region on the Y-chromosome) (Berta et al., 1990; Gubbay et al., 1990; Sinclair et al., 1990; Koopman et al., 1991). The *SRY/Sry* gene encodes a protein with an HMG-box type of DNA-binding domain, which probably acts as a transcription factor (Nasrin et al., 1991). It has been demonstrated that in vitro-translated SRY proteins bind to a specific DNA sequence, resulting in a sharp bend of its flanking DNA (Ferrari et al., 1992; Giese et al., 1994). This structural change in the target DNA may be important for its regulation of down-stream genes essential for testicular differentiation (Pontiggia et al., 1994; Ukiyama et al., 2001). A role for the SRY protein in pre-mRNA splicing has also been suggested (Ohe et al., 2002). Despite the progress in our understanding of the chemical nature of *Sry* gene products in vitro, the mode of SRY action during gonadal sex differentiation in vivo remains unknown.

Gonadal sex differentiation is characterized by differentiation of several distinct cell types and their organization. Testicular differentiation is initiated by *Sry* transcription in the gonadal primordium at embryonic day (ED) 11 in the mouse. However, morphological characteristics of testicular differentiation are not discernible until ED12, at which stage several testis cords are formed with tunica albuginia under the surface epithelium. Sertoli cells and germ cells are enclosed in the testis cords whereas Leydig cells differentiate in the interstitium. By contrast, ovarian differentiation is recognized by the entry of germ cells into meiosis at ED13-14. Primordial follicles are formed around individual oocytes, which are arrested at the end of meiotic prophase, only at or after birth (e.g., ED19-20). Granulosa cells, the somatic cells in follicles, become a distinct cell type at this stage.

In this review, we will discuss our hypothesis regarding the mechanism of SRY action in testis determination based on its expression patterns in the B6.Y[TIR] mouse gonad, which undergoes complete or partial sex reversal. We do not extend our discussions to include other genes involved in gonadal sex differentiation downstream of *SRY/Sry* unless necessary since many excellent reviews are available on such topics (Brennan & Capel, 2004; Polanco & Koopman, 2007; Wilhelm et al., 2007).

B6.Y[TIR] (B6.Y[DOM]) Sex Reversal

In 1982, Eicher et al. reported an interesting case of sex reversal in the mouse. They found that the transfer of the Y-chromosome from local variants of *Mus musclus* (*M. m.*) *domesticus* to the genetic background of the C57BL/6J (B6) inbred strain causes disruption of the normal testis determining process and results in complete or partial sex reversal of XY individuals. These mice were named B6.Y[DOM] or B6.Y[POS] because the original mouse carrying the Y[DOM]-chromosome was caught in Poschiavinus Valley, Switzerland. Soon, similar sex reversal was confirmed and extended to include Y-chromosomes of more, but not all, variants of *M. m. domesticus* (Nagamine et al., 1987a; Biddle & Nishioka, 1988). For the sake of convenience, we use B6.Y[DOM] to represent the mouse strains which undergo sex reversal although many strains carrying Y^{DOM}-chromosomes on the B6 background do not undergo sex reversal. For example, the mouse strain, named B6.Y^{TIR}, carrying the Y^{DOM}-chromosome from a male mouse caught in Tirano, Italy, display sex reversal as strongly as the B6.Y^{POS} strain whereas the B6.Y^{AKR} strain only partial and transient sex reversal. On the other hand, B6.Y^{FBV}, B6.Y^{SJL} and many other strains develop only normal testes[2]. These studies indicate that the Y-chromosome does not act alone and instead requires coordination of autosomal (or X-encoded) genes to initiate normal testicular differentiation. It is remarkable that two *Tdy* alleles, DOM- and B6-types, were predicted before the discovery of *Sry*. It is now evident that *Sry* is the gene on the Y-chromosome involved in the sex reversal in the B6.Y^{DOM} mouse strains (Eicher et al., 1995). On the other hand, the autosomal testis-determining genes responsible for the sex reversal have been found only in the B6 strain, so far (Nagamine et al., 1987b). They have been characterized to be recessive and strongly linked to some areas of Chromosomes 2 and 4, but remain unidentified (Eicher et al., 1996).

Despite the pivotal role of SRY in mammalian sex determination, SRY proteins from different species share homology within but diverge considerably outside of their DNA binding (HMG) domains (Whitfield et al., 1993). In some rodents, including the laboratory rat, the *SRY* sequence exists in 2-13 copies and even on the X-chromosome (Nagamine et al., 1994; Marchal et al., 2008). A high degree of polymorphism in the *Sry* variable region has been found within *M. m.* subspecies. Restriction fragment-length polymorphisms detectable by Y-specific DNA probes divide the *M. m.* Y-chromosomes into two, *domesticus* and *molossinus*, groups (Bishop et al., 1985; Nagamine et al., 1992; Nishioka & Lamothe, 1986). A striking difference between the two groups is a C to T substitution that turns a CAG (Gln) codon into a TAG stop codon, truncating the open-reading-frame prematurely, in *M. m. domesticus* (Coward et al., 1994). In addition, the mouse *Sry* gene contains unique Gln repeats at its C-terminus and its repeat number varies among *M. m.* subspecies. It was proposed that the Gln repeat number correlates with the severity of sex reversal (Coward et al., 1994). However, further studies did not support this hypothesis (Carlisle et al., 1996; Albrecht & Eicher, 1997).

Development of Ovotestes and Ovaries in the B6.Y^{TIR} Gonad

Sex reversal in B6.Y^{DOM} mouse strains is not 100% penetrant even after many generations of backcrossing (Eicher et al., 1982; Nagamine et al., 1987a; Taketo-Hosotani et al., 1989). About half or more of the adult B6.Y^{TIR} mice are completely sex reversed, carrying bilateral ovaries and the female phenotype. The rest develop uni- or bi-lateral testes, which are always smaller than the testes in the B6.XY male mouse. We examined gonadal sex differentiation in B6.Y^{TIR} fetuses and found that none has complete testicular structures; all contain only ovarian structures (= ovary) or both testicular and ovarian structures (= ovotestis) (Taketo et al., 1991). Furthermore, we observed that the onset of testicular differentiation is substantially delayed in the ovotestis. We used immunohistochemical

---

2AKR, FBV, and SJL are inbred mouse strains.

staining of Mullerian Inhibiting Substance (MIS), a specific marker of Sertoli cells, to demonstrate testicular differentiation. In control B6.XY fetuses, testis cords are morphologically conspicuous and fetal Sertoli cells are intensely stained for MIS in all gonads at ED12. We confirmed the central-to-peripheral progression of testis cord formation transiently at ED12. On the other hand, only weak MIS staining can be seen in a small number of B6.Y^{TIR} gonads at the same gestation age. Some other B6.Y^{TIR} gonads show early signs of testicular differentiation, i.e., vascularization near the surface epithelium (named tunica albuginia), but testis cords are hardly recognizable and remain unstained for MIS. At ED13, MIS staining was clearly seen in clusters of cells forming the testis cords in the central region of some B6.Y^{TIR} gonads. At ED14, testis cords are well organized, but only in the central region, and MIS staining is intense in Sertoli cells forming the testis cords, while the cranial and caudal regions remained unstained. The percentage of B6.Y^{TIR} gonads which possess testis cords with MIS staining is low at ED12, increases at ED13, reaches the peak at ED14, and remains constant afterwards. Based on these observations, we speculate a critical period between ED13 and ED14, before which testicular differentiation must be initiated, otherwise the ovarian differentiation pathway autonomously ensues (Fig. 1).

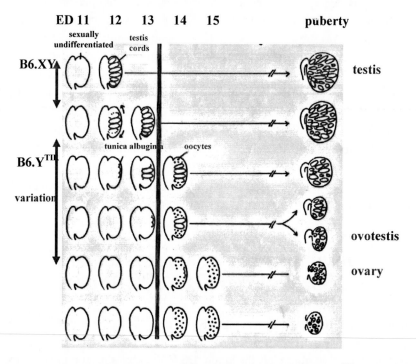

Figure 1. Schematic presentation of gonadal sex differentiation in B6.XY and B6.Y^{TIR} mouse strains. In control B6.XY fetuses, testis cords develop in a central-to-peripheral pattern and reach the entire region of gonad within ED12. By contrast, the onset of testicular differentiation is substantially delayed in B6.Y^{TIR} fetuses. Testis cords can be seen only in the central regions of some gonads at ED13 and in some more gonads at ED14, but no further increase is observed afterwards. The remaining gonads never develop testis cords and instead develop ovarian structures containing oocytes in meiosis. The ovotestis in fetal life appears to develop into a full testis, which is always smaller than the testis in the B6.XY strain, by puberty. On the other hand, the ovary in fetal life continues to develop as an ovary. Occasionally, both ovarian and testicular structures remain, recognized as an ovotestis, in puberty.

To explain sex reversal in B6.Y^{DOM} mouse strains, Eicher's group proposed a hypothesis that the *Tdy* allele of *M. m. domesticus* (*Tdy^{DOM}*) initiates the testis-determining pathway later in development than does the *Tdy* of B6 (*Tdy^{B6}*) (Eicher & Washburn, 1986). Although such difference in timing should not affect testicular differentiation in *M. m. domesticus* if the onset of the ovary-determining (*Od*) pathway is proportionally delayed, it may become critical when the *Tdy^{DOM}* allele is combined with the *Od* allele of B6 since the *Od^{B6}* pathway can be switched on while the *Tdy^{DOM}* allele remains silent. This hypothesis is consistent with our results and supported by Palmer and Burgoyne (1991). However, it was an open question whether the delay in testicular differentiation is intrinsic to the *Tdy^{TIR}* allele or secondary to the sex reversal. To assess the effect of the *Tdy^{TIR}* allele on testicular differentiation in normal males, we crossed B6.Y^{TIR} males with SJL females, producing (SJL.B6)F1.Y^{TIR} genetic construct. In such F1 mice, normal XY males and XX females develop as previously reported (Nagamine et al., 1987b). We did not find any delay in testis cord organization or MIS staining in the F1.Y^{TIR} gonad. We conclude that the delay of testicular differentiation in the B6.Y^{TIR} mouse is not due to the *Tdy^{TIR}* allele itself, but to the impaired interaction of the *Tdy^{TIR}* allele with the B6 background.

The Molecular Mechanism of Sex Reversal in the B6.Y^{TIR} Mouse Gonad (Hypothesis)

The molecular mechanism of sex reversal in the B6.Y^{DOM} mouse remains controversial. In particular, none of a simple hypothesis, proposed so far, can satisfy the mechanism of sex reversal. We have proposed a hypothesis that at least two mechanisms, expression levels and efficiency of SRY proteins, are involved in the sex reversal in the B6.Y^{TIR} mouse gonad (Fig. 2). We predict that the longer form of SRY^{B6} is more efficient than the shorter form of SRY^{DOM} in initiating testis differentiation. Therefore, the low levels of SRY^{B6} are sufficient whereas greater levels of SRY^{DOM} are required for effective initiation of testis differentiation. However, this hypothesis alone is not sufficient because many B6.Y^{DOM} mouse strains which carry the short form of SRY do not result in sex reversal. On the other hand, we found that *Sry* transcript levels differ among B6.Y^{DOM} mouse strains and correlates with the extent of sex reversal. Again, this hypothesis alone is also not sufficient because *Sry* transcript levels in the B6.XY gonad are as low as in the B6.Y^{TIR} gonad. When these two conditions work synergistically, the low levels and inefficiency of the *Sry^{TIR}* gene products may delay both the onset and progression of testicular differentiation. Consequently, either both poles or the entire region of the B6.Y^{TIR} gonad misses the critical time for testis determination and initiates ovarian differentiation. In the following sections, we will present the evidence in more details to support our hypothesis.

Mechanism of Sex Reversal (Hypothesis)

B6 background

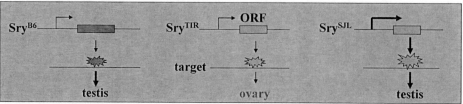

(SJLxB6)F1 background

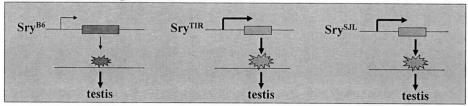

Figure 2. Molecular mechanism of sex reversal in the B6.Y^{TIR} gonad (hypothesis). On the B6 background, the levels of *Sry* transcripts from both Sry^{B6} and Sry^{TIR} alleles are low. However, the higher efficiency of SRY^{B6} protein can initiate normal testicular differentiation whereas the lower efficiency of SRY^{TIR} protein results in a failure in testicular differentiation. The SRY^{SJL} protein is anticipated to have low efficiency as the SRY^{TIR} protein. However, its high levels of Sry transcripts secure normal testicular differentiation. On the (SJLxB6)F1 background, the levels of *Sry* transcripts from the Sry^{TIR} allele increases and support normal testicular differentiation.

Sry Transcript Levels during Gonadal Sex Differentiation

We examined the ontogeny of *Sry* transcription in the B6.Y^{TIR} gonad in comparison to that in the B6.XY gonad. We found that the onset of *Sry* transcription is comparable in both strains, while *Sry* down-regulation is delayed in the B6.Y^{TIR} gonad (Lee & Taketo, 1994). For comparison, the transcription of *MIS*, which is expressed as a consequence of Sertoli cell differentiation but not essential for testis differentiation, is delayed. The genes required for testosterone synthesis, such as 3β-hydroxysteroid dehydrogenase and 17α-hydroxylase, which are expressed in Leydig cells, are also delayed in the B6.Y^{TIR} ovotestis and absent in the B6.Y^{TIR} ovary. These results agree with our morphological observations and further support our hypothesis that the onset of *Sry* is not altered but the onset of *Sry* downstream genes is delayed in the B6.Y^{TIR} gonad.

Effects of Genetic Backgrounds on *Sry* Transcript Levels

A widely accepted hypothesis is that low levels of *Sry* transcripts may account for the aberrant testis differentiation in the B6.Y^{TIR} mouse (Nagamine et al., 1999). These authors demonstrated that the levels of *Sry* transcripts are the highest in B6.Y^{FVB}, intermediate in B6.Y^{AKR}, and the lowest in B6.Y^{TIR} strains. This order corresponds to the extent of sex reversal. We confirmed their findings by using the SJL strain in place of FVB (Lee & Taketo, 2001). However, we found the lowest levels in the B6.Y^{B6} strain, which develops normal testes. This result suggests that the low levels of *Sry^{B6}* transcripts are sufficient to initiate testicular differentiation in the B6 background. We conclude that low levels of *Sry* transcripts cannot be the sole cause of sex reversal in the B6.Y^{TIR} gonad. Furthermore, we found that when the three Y-chromosomes are placed on the (SJL.B6)F1 background, the levels of *Sry* transcripts increase in the F1.Y^{TIR} strain and only normal testes develop. In contrast, transcript levels remain low in the F1.Y^{B6} strain and high in the F1.Y^{CD1} strain. These results indicate that the promoter/enhance region of *Sry* is polymorphic among B6, TIR, and SJL alleles and regulated differently in various genetic backgrounds. Similar findings have been reported by Albrecht et al. (2005).

Expression of the SRY Protein during Gonadal Sex Differentiation

Most studies on *Sry* expression have focused on analyses of its transcripts. Some of these studies have demonstrated the existence of several types of *Sry* transcripts, including a circular form and linear forms with multiple transcription initiation and polyadenylation sites (Capel et al., 1993; Hacker et al., 1995; Jeske et al., 1995). Therefore, analyses of *Sry* transcripts alone are not sufficient to clarify the mechanism or site of SRY action. All previous studies concerning the mechanism of sex reversal in B6.Y^{DOM} mouse strains were also limited to *Sry* transcript levels. To clarify the molecualr mechanism of testis determination and sex reversal, it is essential to detect and characterize the endogenous SRY protein, the ultimate mediator of *Sry* functions, during gonadal sex differentiation.

We developed monoclonal antibodies against the SRY recombinant protein, which detected the endogenous SRY protein in normal XY fetal mouse gonads by Western blotting and immunohistochemistry (Taketo et al., 2005). The tissue-specificity and ontogeny of the detected protein are consistent with those of *Sry* transcripts. Immunofluorescent double labeling reveals that the SRY protein is detectable in the Serotli cell lineage and swiftly down-regulated concurrently with testis cord organization (Fig. 3).

The SRY protein is also detectable in the entire region of all B6.Y^{TIR} gonadal primordial at ED11. SRY down-regulation is considerably delayed, compared to control B6.XY gonads, and is not associated with testis cord organization. Furthermore, the SRY protein is detectable in all B6.Y^{TIR} gonads until later developmental stages and down-regulated simultaneously. In normal XY gonads, by contrast, its presence or absence mirrors the pattern of testis cord formation that initiates centrally and spreads to the polar regions. We conclude that the SRY protein is expressed in the B6.Y^{TIR} gonad or region that fails to develop testicular structures. All these results suggest that SRY is expressed, but its action is impaired in the entire region of the B6.Y^{TIR} gonad. We further speculate that testis cords develop only when other factors that favor testicular organization overcome the inefficient action of the SRY^{TIR} protein.

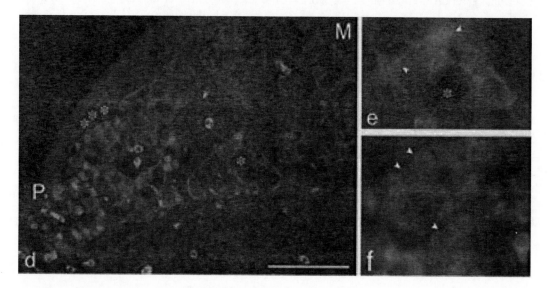

Figure 3. Double immunofluorescence labeling of SRY (green) and MIS (red) in CD1.XY fetal gonads. **d**: Medial (M) – posterior (P) region of an XY gonad at ED12. Intense SRY labeling is seen in the nuclei of abundant cells in the posterior pole, whereas much reduced labeling is seen in the Steroli cells within the testis cords near the medial region. On the other hand, intense MIS labeling is seen in the cytoplasm of Sertoli cells within the testis cords near the medial region, although it is less frequently seen near the posterior pole. Both labeling are seen in the cells in an areas indicated with asterisks and also in a few cells within the testis cords. Yellow staining indicates nonspecific binding of both secondary antibodies to the blood cells. **e**: An area (asterisk) of **d** at a higher magnification. Intense SRY labeling is seen in the nuclei, and MIS labeling is seen in the cytoplasm of the same cells (arrowheads). **f**: The area marked with three asterisks in **d** is shown here at a higher magnification. Most cells have intense SRY labeling in their nuclei. A few cells have both SRY labeling in their nuclei and MIS labeling in their cytoplasm (arrowheads). Scale bar = 80 μm in **d** and 32 μm in **e** and **f**. (Taketo et al., 2005)

Endogenous SRY protein levels in B6.Y^{TIR} gonads were not less than those in normal B6.XY gonads, in agreement with *Sry* transcript levels (Lee & Taketo, 2001; Albrecht et al., 2003). These results further support our hypothesis that *Sry* expression levels alone, at either transcript or protein levels, cannot explain sex reversal in the B6.Y^{TIR} mouse. We observed weaker SRY staining in B6.XY and B6.Y^{TIR} gonads compared with CD1.XY gonads even at their peak expression time. Although almost all somatic cells in the genital ridge are positive for SRY in CD1.XY gonads, a considerably fewer cells are positive in B6.XY and B6.Y^{TIR} gonads. By Western blotting, the SRY protein in CD1.XY gonads is much easier to be detected than those in B6.XY and B6.Y^{TIR} gonads. Although these quantitative comparisons are not accurate, they are in agreement with the levels of SRY transcripts in different mouse strains (Lee & Taketo, 1994). The genetic background appears to influence the levels of both SRY transcription and translation.

Efficiency of SRY Proteins in Transactivation

SRY^{TIR} and SRY^{B6} proteins are not only different in sizes but also polymorphic at multiple sites. Therefore, differences in their biochemical properties or three-dimensional structures may be sufficient to impair the ability of the SRY^{TIR} protein to interact with cofactors or target genes of the B6 strain. The SRY protein is anticipated to interact with cofactors to form a functional transcriptome in regulation of downstream target genes (Dubin et al., 1995; Poulat et al., 1997: Lau and Zhang, 1998; Oh et al., 2005). However, direct evidence to support the differences in the efficiency of SRY action during testicular differentiation is lacking. Nonetheless, a few lines of indirect evidence are available to support such assumptions. First, Eicher et al. have demonstrated that introduction of an *Sry* transgene, either from *molossinus* or *domesticus*, into B6.Y^{POS} embryos secures the development of normal testes (Eicher et al., 1995; Albrecht et al., 2003). These observations suggest that overexpression of SRY proteins overcome the inefficient action of the SRY^{DOM} protein. Second, Dubin et al. (1995) have shown that the SRY^{MOL} protein activates transcription of a CAT reporter gene containing multiple copies of a putative SRY-binding site AACAAT in HeLa cells, whereas SRY^{DOM} proteins, originally from TIR, AKR, FVB, or SJL strains, do not. This difference results from the truncation of SRY^{DOM} proteins by a premature stop codon within the C-terminal activation domain. Although these results clearly show different transactivation efficiency between the two types of SRY proteins, one must note that in vitro conditions may generate artificial consequences, particularly in consideration of the fact that the endogenous SRY^{DOM} proteins are fully functional on proper genetic backgrounds during testicular differentiation.

Considerations on the Mechanism of Testicular Differentiation

The molecular mechanism of SRY action cannot be clarified until its endogenous target sequence and gene are identified. Nonetheless, the observations in the B6.Y^{DOM} sex reversal provide valuable information about the mechanism of testicular differentiation. In B6.Y^{DOM} mouse strains, the bilateral gonads often develop into an ovary and a contra lateral ovotestis. This observation alone convincingly indicates that gonadal sex is determined not only by the genetic background but also by the developmental factors. In addition, our observations in both normal and sex-reversed gonads show that SRY expression does not directly influence the spatiotemporal pattern of testis cord formation. We predict a mechanism that coordinates with SRY and facilitates testis cord organization in a central-to-polar direction.

SOX9, which shares the DNA binding motif with SRY, is up-regulated by SRY during testicular differentiation. It has been shown that SOX9 is necessary and sufficient for testis determination (Cameron et al., 1996; Bishop et al., 2000; Vidal et al., 2001; Chaboissier et al., 2004; Sekido & Lovell-Badge, 2008). Furthermore, SOX9 may up-regulate MIS in Sertoli cells (de Santa-Barbara et al., 1998) Therefore, it is particularly interesting to address the relationship between SRY and SOX9 in B6.Y^{DOM} sex reversal. It has been reported that

SOX9 is up-regulated only in the testis cords, similarly to MIS, in the B6.Y^{TIR} ovotestis (Moreno-Mendoza et al., 2004). However, our preliminary results demonstrate more complex association between SRY and SOX9 or between SOX9 and MIS; SOX9 is expressed in all B6.Y^{TIR} gonadal primordia, most likely including future ovaries, but become limited to the central region afterwards (Fig. 4). MIS expression is further limited to the region which harbors relatively well-organized testis cords. These results suggest that regulation of SOX9 by SRY is impaired but the SRY action is also impaired further downstream in B6.Y^{TIR} gonads. Cell migration from the adjacent mesonephros into developing gonads is required for testicular organization (Buehr et al., 1993; Merchant-Larios et al., 1993; Tilmann & Capel, 1999), and may contribute as a developmental factor. The addition of single recombinant growth factors, known to be expressed by Sertoli cells, can induce cell migration from the mesonephros into XX gonads in culture although evidence for their roles in vivo has yet to be provided (Brennan & Capel, 2004). SRY may regulate not only SOX9 up-regulation but also other events critical for testicular differentiation.

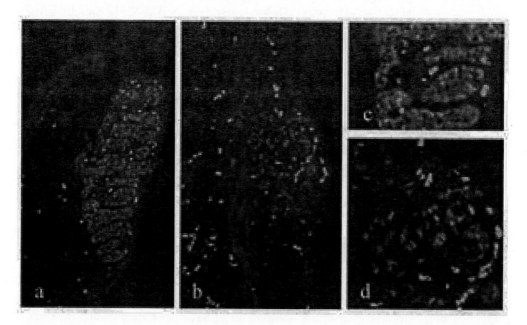

Figure 4. Double immunofluorescence labeling of SOX9 (green) and MIS (red) in B6.XY and B6.Y^{TIR} fetal gonads at ED13. **a.** A B6.XY gonad. Testis cords are formed in the entire gonad, aligned with Sertoli cells (right). Intense SOX9 labeling is seen in the nucleus whereas MIS labeling is seen in the cytoplasm of Sertoli cells. Less intense SOX9 labeling is seen in the epithelium of the mesonephric duct (left). **b.** B6.Y^{TIR} gonad without apparent testis cord formation. Intense SOX9 labeling is seen in scattered cells in the medial region of the gonad. MIS labeling is seen around some, but not all, SXO9-positive cells. **c.** The medial region of the gonad shown in **a** at a higher magnification. SOX9-positive nuclei of Sertoli cells form palisade-like structures along the basement membrane of testis cords. MIS-positive cytoplasm of Sertoli cells fill the space between the germ cells. **d.** The medial region of the gonad shown in **b** at a higher magnification. SOX9-positive cells occasionally form ring-like structures, but not organized as normal testis cords. MIS-labeling is seen around these structures but not associated with more scattered SOX9-positive cells.

Considerations on the Mechanism
of Ovarian Differentiation

That ovarian differentiation is a default pathway is a dogma. Yet, comparison of ovarian vs. testicular differentiation is not simple as they are not symmetrical in time and pattern. Ovarian differentiation is recognized by the entry of germ cells into meiosis and organization of somatic cells into follicles around individual oocytes[3] at much later developmental stages (Merchant-Larios & Centeno, 1981). On the other hand, testicular differentiation is characterized by the differentiation and organization of somatic cells into testis cords. Germ cells are not essential for this process and their sexual differentiation into prospermatogonia is imposed by the testicular environment (Merchant-Larios, 1975). Therefore, the critical questions are, first, what regulates the onset of meiosis and, second, at which stage does the differentiation of ovarian somatic cells occurs. Although some genes are expressed in the somatic cells of XX gonads as early as ED12, null mutantation of these genes do not result in aberrant phenotype until follicles are formed in postnatal life (Schmidt et al., 2004; Pangas et al., 2006). In other words, the differentiation of germ cells into oocytes remains the only criterion for ovarian differentiation in fetal life. It has long been believed that the entry of germ cells into meiosis is a default pathway (McLaren & Southee, 1997; Upadhyay & Zamboni, 1982). However, it has recently been proposed that retinoic acid (RA) synthesized in the adjacent mesonephros is responsible for the induction of meiosis in the ovary whereas Cry26b1, which is expressed in Sertoli cells and metabolizes RA, suppresses meiosis in the germ cells within the testis cords (Bowles et al., 2006; Koubova et al., 2006; MacLean et al., 2007). These hypotheses remain controversial since neither induction nor inhibition of meiosis has yet been convincingly demonstrated. Furthermore, Best et al. (2008) could not reproduce the role of RA or Cry26b1 in germ cell sex differentiation. Instead, they found that *Sdmg1* is required for the induction of meiosis in the germ cells in fetal testes.

The coexistence of both testicular and ovarian structures in the B6.Y^{TIR} ovotestis provides a unique opportunity to examine the regulation of germ cell sex differentiation. We have observed that most germ cells surrounded by MIS-positive cells are arrested at the stage of prospermatogonia (Taketo et al., 1991). By contrast, many germ cells have entered meiosis in the cranial and caudal regions which are negative for MIS. Between the two immunocytochemically distinct regions within the epithelial cords, meiotic germ cells are usually surrounded by MIS-negative cells, but also occasionally surrounded by MIS-positive cells. Prospermatogonia are exclusively surrounded by MIS-positive cells. Despite the exceptional meiotic germ cells, the striking consistency between MIS expression and prospermatogonia differentiation suggest that both gonadal sex and germ cell sex are determined during a short period of development. These observations support the hypothesis that direct contact with Sertoli cells is required for prospermatogonia differentiation. Our observations also cast a doubt on the hypothesis that germ cells in meiosis suppress testis cord formation (Yao et al., 2003). Since germ and somatic cells were dissociated and aggregated before cultured in these studies, the timing of meiotic entry and the resistance to testis cord formation might have been shifted. It would be interesting to examine the

---

[3] Germ cells that have entered meiosis in the ovary are termed oocytes.

relationship between the germ cell sex and the expression of Cry26b1 or Sdmg1 in the adjacent epithelial cells in the B6.Y^{TIR} ovotestis.

References

Albrecht, K.H. and Eicher, E.M. (1997). DNA sequence analysis of *Sry* alleles (subgenus *Mus*) implicates misregulation as the cause of C57BL/6J-Y^{pos} sex reversal and defines the *SRY* functional unit. Genetics *147*, 1267-1277.

Albrecht, K.H., Young, M., Wahsburn, L.L., and Eicher, E.M. (2003). *Sry* Expression level and protein isoform differences play a role in abnormal testis development in C57BL/6J mice carrying certain *Sry* alleles. Genetics. *164*, 277-288.

Berta, P., Hawkins, J.R., Sinclair, A.H., Taylor, A., Griffiths, B.L., Goodfellow, P.N., and Fellous, M. (1990). Genetic evidence equating *SRY* and the testis-determining factor. Nature *348*, 448-450.

Best, D., Salender. D.A, Walther, N., Peden, A.A., and Adams, I.R. (2008). Sdmg1 is a conserved transmembrane protein associated with germ cell sex determination and germline-soma interactions in mise. Development *135*, 1415-1425.

Biddle, F.G. and Nishioka, Y. (1988). Assays of testis development in the mouse distinguish three classes of domesticus-type Y chromosome. Genome *30*, 870-878.

Bishop, C.E., Boursot, P., Baron, B., Bonhomme, F., and Hatat, D. (1985). Most classical *Mus musculus domesticus* laboratory mouse strains carry a *Mus musculus musculus* Y chromosome. Nature *315*, 70-72.

Bishop, C.E., Whitworth, D.J., Qin, Y., Agoulnik, A.I., Agoulnik, I.U., Harrison, W.R., Behringer, R.R., and Overbeek, P.A. (2000). A transgenic insertion upstream of *Sox9* is associated with dominant XX sex reversal in the mouse. Nature Genet. *26*, 490-494.

Bowles, J., Knight, D., Smith, C., Wilhelm, D., Richman, J., Mamiya, S., Yashiro, K., Chawengsaksophak, K., Wilson, M.J., Rossant, J., Hamada, H., and Koopman, P. (2006). Retinoid signaling determines gem cell fate in mice. Science *312*, 596-600.

Brennan, J.B. and Capel, B. (2004). One tissue, two fates: Molecular genetics events that underlie testis versus ovary development. Nature Rev. Genet. *5*, 509-521.

Buehr, M., Gu, S., and McLaren, A. (1993). Mesonephric contribution to testis differentiation in the fetal mouse. Development *117*, 273-281.

Cameron, F.J., Hageman, R.M., Cooke-Yarborough, C., Kwok, C., Goodwin, L.L.S.D.O., and Sinclair, A.H. (1996). A novel germ line mutation in *SOX9* causes familial campomelic dysplasia and sex reversal. Hum. Mol. Genet. *5*, 1625-1630.

Capel, B., Swain, A., Nicolis, S., Hacker, A., Walter, M., Koopman, P., Goodfellow, P., and Lovell-Badge, R. (1993). Circular transcipts of the testis-determining gene *Sry* in adult mouse testis. Cell *73*, 1019-1030.

Carlisle, C., Winking, H., Weichenhan, D., and Nagamine, C.M. (1996). Absence of correlation between *Sry* polymorphisms and XY sex reversal caused by the *M. m. domesticus* Y chromosome. Genomics *33*, 32-45.

Chaboissier, M.-C., Kobayashi, A., Vidal, V.I.P., Lutzkendorf, S., van de Kant, H.J.G., Wegner, M., de Rooij, D.G., Behringer, R.R., and Schedl, A. (2004). Functional analysis of *Sox8* and *Sox9* during sex determination in the mouse. Development *131*, 1891-1901.

Coward, P., Nagai, K., Chen, D., Thomas, H.D., Nagamine, C.M., and Lau, Y.C. (1994). Polymorphism of a CAG trinucleotide repeat within *Sry* correlates with B6.Y^{DOM} sex reversal. Nature Genet. *6*, 245-250.

de Santa Barbara, P., Bonneaud, N., Boizet, B., Desclozeaux, M., Moniot, B., Sudbeck, P., Scherer, G., Poulat, F., and Berta, P. (1998). Direct interaction of SRY-related protein SOX9 and steroidogenic factor 1 regulates transcription of the human Anti-Mullerian Hormone Gene. Mol. Cell. Biol. *18*, 6653-6665.

Dubin, R.A., Coward, P., Lau, Y.-F.C., and Oster, H. (1995). Functional comparison of the *Mus musculus molessinus* and *Mus musculus domesticus Sry* genes. Mol. Endocrinol. *9*, 1645-1654.

Eicher, E.M., Shown, E.P., and Washburn, L.L. (1995). Sex reversal in C57BL/6J-Y^{POS} mice corrected by a *Sry* transgene. Phil. Trans. R. Soc. Lond. *B350*, 263-269.

Eicher, E.M. and Washburn, L.L. (1986). Genetic control of primary sex determination in mice. Ann. Rev. Genet. *20*, 327-360.

Eicher, E.M., Washburn, L.L., Schork, N.J., Lee, B.K., Shown, E.P., Xu, X., Dredge, R.D., Pringle, M.J., and Page, D.C. (1996). Sex-determining genes on mouse autosomes identified by linkage analysis of C57BL/6J-Y^{POS} sex reversal. Nature Genet. *14*, 206-209.

Eicher, E.M., Washburn, L.L., Whitney, J.B.I., and Morrow, K.E. (1982). *Mus poshiavinus* Y chromosome in the C57BL/6J murine genome causes sex reversal. Science *217*, 535-537.

Ferrari, S., Harley, V.R., Pontiggia, A., Goodfellow, P.N., Lovell-Badge, R., and Bianchi, M.E. (1992). SRY, like HMG1, recognizes sharp angles in DNA. EMBO J. *11*, 4497-4506.

Foster, H.A., Abeydeera, L.R., Griffin, D.K., and Bridger, J.M. (2005). Non-random chromosome positioning in mammalian sperm nuclei, with migration of the sex chromosomes during late spermatogenesis. J. Cell Sci. *118*, 1811-1820.

Giese, K., Pagel, J., and Grosschedl, R. (1994). Distinct DNA-binding properties of the high mobility group domain of murine and human SRY sex-determining factors. Proc. Natl. Acad. Sci. USA *91*, 3368-3372.

Gubbay, J., Collignon, J., Koopman, P., Capel, B., Economou, A., Münsterberg, A., Vivian, N., Goodfellow, P., and Lovell-Badge, R. (1990). A gene mapping to the sex-determining region of the mouse Y chromosome is a member of a novel family of embryonically expressed genes. Nature *346*, 245-250.

Hacker, A., Capel, B., Goodfellow, P., and Lovell-Badge, R. (1995). Expression of *Sry*, the mouse sex determining gene. Development *121*, 1603-1614.

Jeske, Y.W.A., Bowles, J., Greenfield, A., and Koopman, P. (1995). Expression of a linear *Sry* transcript in the mouse genital ridge. Nature Genet. *10*, 480-482.

Jost, A., Vigier, B., Prépin, J., and Perchellet, J.P. (1973). Studies on sex differentiation in mammals. Rec. Prog. Horm. Res. *29*, 1-35.

Koopman, P., Gubbary, J., Vivian, N., Goodfellow, P., and Lovell-Badge, R. (1991). Male development of chromosomally female mice transgenic for *Sry*. Nature *351*, 117-121.

Koubova, J., Menke, D.B., Zhou, Q., Capel, B., Griswold, M.D., and Page, D.C. (2006). Retinoic acid regulates sex-specific timing of meiotic initiation in mice. Proc. Natl. Acad. Sci. USA *103*, 1474-2479.

Lau, Y.-F.C. and Zhang, J. (1998). Sry interactive proteins: implication for the mechanisms of sex determination. Cytogenet. Cell Genet. *80*, 128-132.

Lee, C.-H. and Taketo, T. (1994). Normal onset, but prolonged expression, of *Sry* gene in the B6.Y^{DOM} sex-reversed mouse gonad. Dev. Biol. *165*, 442-452.

Lee, C.-H. and Taketo, T. (2001). Low levels of *Sry* transcripts cannot be the sole cause of B6.Y^{TIR} sex reversal. Genesis *30*, 7-11.

MacLean, G., Li, H., Metzger, D., Chambon, P., and Petkovich, M. (2007). Apoptotic extinction of germ cells in testes of Cry26b1 knockout mice. Endocrinology *148*, 4560-4567.

Marchal, J.A., Acosta, M.J., Bullejos, M., de la Guardia, R.D., and Sanchez, A. (2008). Origin and spread of the *SRY* gene on the X and Y chromosomes of the rodent *Microtus cabrerae*: Role of L1 elements. Genomics *91*, 142-151.

Merchant, H. (1975). Rat gonadal and ovarian organogenesis with and without germ cells. An ultrastructural study. Dev. Biol. *44*, 1-21.

Merchant-Larios, H. and Centeno, B. (1981). Morphogenesis of the ovary from the sterile W/W^v mouse. Advances in the Morphology of the Cells and Tissues 383-392.

Merchant-Larios, H., Moreno-Mendoza, N., and Buehr, M. (1993). The role of the mesonephros in cell differentiation and morphogenesis of the mouse fetal testis. Int. J. Dev. Biol. *37*, 407-415.

Nagamine, C.M. (1994). The testis-determining gene, *SRY*, exists in multiple copies in Old World rodents. Genet. Res. *64*, 15-159.

Nagamine, C.M., Morohashi, K., Carlisle, C., and Chang, D.K. (1999). Sex reversal caused by *Mus musclus domesticus* Y chromosomes linked to variant expression of the testis-determining gene *Sry*. Dev. Biol. *216*, 182-194.

Nagamine, C.M., Nishioka, Y., Moriwaki, K., Boursot, P., Bonhomme, F., and Lau, Y.F.C. (1992). The *musculus*-type Y chromosome of the laboratory mouse is of asian origin. Mammal. Genome *3*, 84-91.

Nagamine, C.M., Taketo, T., and Koo, G.C. (1987a). Morphological development of the mouse gonad in *tda-1* XY sex reversal. Differentiation *33*, 214-222.

Nagamine, C.M., Taketo, T., and Koo, G.C. (1987b). Studies on the genetics of *tda-1* XY sex reversal in the mouse. Differentiation *33*, 223-231.

Nasrin, N., Buggs, C., Kong, X.F., Carnazza, J., Goebl, M., and Alexander-Bridges, M. (1991). DNA-binding properties of the product of the testis-determining gene and a related protein. Nature *354*, 317-320.

Nishioka, Y. and Lamothe, E. (1986). Isolation and characterization of a mouse Y chromosomal reptitive Sequence. Genetics *113*, 417-432.

Oh, H., Li, Y., and Lau, Y.F.C. (2005). Sry associates with the heterochromatin protein 1 complex by interacting with a KRAB domain protein. Biol. Reprod. *72*, 407-415.

Ohe, K., Lalli, E., and Sassone-Corsi, P. (2002). A direct role of SRY and SOX proteins in pre-mRNA splicing. Proc. Natl. Acad. Sci. USA *99*, 1146-1151.

Palmer, S.J. and Burgoyne, P.S. (1991). The *Mus musculus domesticus Tdy* allele acts later than the *Mus musculus musculus Tdy* allele: a basis for XY sex-reversal in C57BL/6-Y^{POS} mice. Development *113*, 709-714.

Pangas, S.A., Choi, Y., Ballow, D.J., Zhao, Y., Westphal, H., Matzuk, M.M., and Rajkovic, A. (2006). Oogenesis requires germ cell-specific transcriptional regulators *Sohlh1* and *Lhx8*. Proc. Natl. Acad. Sci. USA *103*, 8090-8095.

Polanco, J.C. and Koopman, P. (2007). *Sry* and the hesitant beginnings of male development. Dev. Biol. *302*, 13-24.

Pontiggia, A., Rimini, R., Harley, V.R., Goodfellow, P.N., Lovell-Badge, R., and Bianchi, M.E. (1994). Sex-reversing mutations affect the architecture of SRY-DNA complexes. EMBO J. *13*, 5115-6124.

Poulat, F., de Santa Barbara, P., Desclozeaux, M., Soullier, S., Moniot, B., Bonneaud, N., Boizet, B., and Berta, P. (1997). The human testis determining factor SRY binds a nuclear factor containing PDZ protein interaction domains. J. Biol. Chem. *272*, 7167-7172.

Sekido, R. and Lovell-Badge, R. (2008). Sex determination involves synergistic action of SRY and SF1 on a specific Sox9 enhancer. Nature *453*, 930-934..

Sinclair, A.H., Berta, P., Palmer, M.S., Hawkins, J.R., Griffiths, B.L., Smith, M.J., Foster, J.W., Frischauf, A.-M., Lovell-Badge, R., and Goodfellow, P.N. (1990). A gene from the human sex-determining region encodes a protein with homology to a conserved DNA-binding motif. Nature *346*, 240-244.

Taketo-Hosotani, T., Nishioka, Y., Nagamine, C., Villalpando, I., and Merchant-Larios, H. (1989). Development and fertility of ovaries in the B6.Y^{DOM} sex-reversed female mouse. Development *197*, 95-105.

Taketo, T., Lee, C.-H., Zhang, J., Li, Y., Lee, C.-Y.G., and Lau, Y.-F.C. (2005). Expression of SRY proteins in both normal and sex-reversed XY fetal mouse gonads. Dev. Dyn. *233*, 612-622.

Taketo, T., Saeed, J., Nishioka, Y., and Donahoe, P.K. (1991). Delay of testicular differentiation in the B6.Y^{DOM} ovotestis demonsrated by immunocytochemical staining for Müllerian inhibiting substance. Dev. Biol. *146*, 386-395.

Tilmann, C. and Capel, B. (1999). Mesonephric cell migration induced testis cord formation and Sertoli cell differentiation in the mammalian gonad. Development *126*, 2883-2890.

Tucker, P.K. and Lundrigan, B.L. (1993). Rapid evolution of the sex determining locus in old world mice and rats. Nature *364*, 715-717.

Ukiyama, E., Jancso-Radek, A., Li, B., Milos, L., Zhang, W., Phillips, N.B., Morikawa, N., King, C.-Y., Chan, G., Happ, C.M., Rodek, J.T., Poulat, F., Donahoe, P.K., and Weiss, M.A. (2001). SRY and architectural gene regulation: The kinetic stability of a bent protein-DNA complex can regulate its transcriptional potency. Mol. Endocrinol. *15*, 363-377.

Upadhyay, S. and Zamboni, L. (1982). Ectopic germ cells: Natural model for the study of germ cell sexual differentiation. Proc. Natl. Acad. Sci. USA *79*, 6584-6588.

Vidal, V.P.I., Chaboissier, M.-C., de Rooij, D.G., and Schedl, A. (2001). *Sox9* induces testis development in XX trasngenic mice. Nature Genet. *28*, 216-217.

Whitfield, L.S., Lovell-Badge, R., and Goodfellow, P.N. (1993). Rapid sequence evolution of the mammalian sex-determining gene *SRY*. Nature *364*, 713-715.

Wilhelm, D., Palmer, S., and Koopman, P. (2007). Sex determination and gonadal development in mammals. Physiol. Rev. *87*, 1-28.

Yao, H.H.C., DiNapoli. L., and Capel, B. (2003). Meiotic germ cells antagonize mesonephric cell migration and testis cord formation in mouse gonads. Development *130*, 5895-5902.

In: Development Gene Expresión Regulation
Editor: Nathan C. Kurzfield

ISBN: 978-60692-794-6
©2009 Nova Science Publishers, Inc.

Chapter XI

The Fibroblast Growth Factor (FGF) Gene Families of Japanese Medaka (*Oryzias latipes*)

Asok K. Dasmahapatra and Ikhlas A. Khan

National Center for Natural Product Research
Department of Pharmacology, School of Pharmacy
University of Mississippi, University, MS38677 USA

Abstract

Fibroblast growth factors (FGFs) constitute a large family of signaling polypeptides that play critical roles in development. During morphogenesis, FGFs are involved in cell proliferation, differentiation and migration; however, in adults these proteins function as homeostatic factors. FGFs mediate their functions through a cell surface receptor, the fibroblast growth factor receptors (FGFRs), which are a member of the tyrosine kinase superfamily. Both the *FGF* and *FGFR* gene families are identified in multicellular organisms but not in unicellular ones and have expanded greatly during evolution. FGF gene organization is highly conserved among vertebrates. In human and mouse, the FGF gene family consists of 22 members; however, in zebrafish (*Danio rerio*) there are 27 identified *fgf* members. Japanese medaka *(Oryzias latipes)*, like zebrafish, is a small aquarium fish used as a model organism in vertebrate development. During evolution, these two fish species (zebrafish and Japanese medaka) were separated from their last common ancestor about 110 million years ago. The medaka genome is only half (800 Mb) of the zebrafish genome (1700 Mb). We have searched medaka genome data bases and identified 28 *fgf* genes in this species of which nine are paralogs. We have done a phylogenetic and conserved gene location (synteny) analysis of the identified *fgf* genes of medaka and analyzed the evolutionary relationships of these genes with human *FGF* gene families.

Introduction

Fibroblast growth factors (FGFs) are multifunctional peptide growth factors identified in many organisms. There are at least 28 distinct members of this family (Cotton et al. 2008). FGFs were originally recognized as growth factors for fibroblast; however, they are now known to be polypeptide growth factors expressed in various cell types from early embryos to adults with diverse biological activities (Xu et al., 1999; Ford-Perriss et al., 2001; Chen and Deng, 2005; Dvorak and Hampl, 2005; Chen and Forough, 2006; Cotton et al., 2008). Affinity for heparin or heparin sulfate proteoglycans is a hallmark of all the memebers of the FGF family. Moreover, the FGF signalling system is not identified in unicellular organisms like *Escherichia coli* or *Saccharomyces cerevisiae*. In multicellular organisms the expression of these genes are essential both in embryonic and adult stages (Itoth, 2007). During embryogenesis, FGFs regulate cellular proliferation, differentiation and migration; however, in adults these proteins are responsible for tissue repair and wound healing, tumor angiogenesis and nervous system functions (Cotton et al., 2008). FGFs play a significant role in neural induction and caudalisation of the developing neuroectoderm (Kengaku and Okamoto, 1993, 1995; Cox and Hemmati-Brivanlou, 1995; Lamb and Harland, 1995), and are involved in separating the midbrain and hindbrain with the development of isthmus (midbrain hind brain boundaries, MHB). In late stages of development, FGFs are essential for the regionalization and patterning of the brain (Bally-Cuif and Wassef, 1995; Wassef and Joyner, 1997), and axon outgrowth and guidance (McFarlane et al., 1996: Saffelle et al., 1997). The *Fgf 8* and *Fgf10* knockout mice died shortly after birth (Itoh, 2007); however, *Fgf4* and *Fgf8* knockout mice died in early embryonic stages (Ithoh and Ornitz, 2008). The *FGF* gene family expanded in two phases during evolution. The first phase is at the time of early metazoan evolution and the second phase is in the early stages of vertebrate evolution (Itoh and Ornitz, 2008). Extracellular FGF mediates their biological responses by binding to and activating cell surface tyrosine kinase FGF receptors (FGFRs). In vertebrates, the *fgfr* gene family consists of four highly related genes, *fgfr1*, *fgfr2*, *fgfr3* and *fgfr4*, encoding polypeptides that are 55 to 72% identical in their amino acid sequences (Cotton et al., 2008). A typical FGFR molecule consists of a signal peptide with three extracellular immunoglobulin-like domains, a single transmembrane domain and a split intracellular tyrosine kinase domain (Robinson, 2006; Itoh, 2007). Transcripts from all four *fgfr* genes are able to undergo alternative splicing and therefore generating numerous receptor isoforms (Johnson and Williams, 1993; Wuechner et al., 1996; Takaishi et al., 2000).

Fish models, particularly zebrafish (*Danio rerio*) and Japanese medaka (*Oryzias latipes*), are currently emerging as an alternative to mammalian models because of their easy availability, low maintenance cost, short life cycle and also their accessibility to the study of gene function (Furutani-Seiki and Wittbrodt, 2004). The *fgf* gene family in zebrafish that comprises 27 members, 6 of which are paralogs, has recently been reviewed (Itoh, 2007; Itoh and Konishi, 2007). Mutation of *fgf8*, *fgf24*, *fgf10* and *fgf20a* in zebrafish produced specific phenotypes which were also utilized in understanding the importance of *fgf* genes during development (Itoh, 2007). Experiments on morpholino knockdown of several *fgf* genes in zebrafish has also been done (Itoh, 2007). In medaka the expression of *fgf* and *fgfr* genes during embryogenesis and adult stages are reported (Emori et al., 1992; Watanabe et al.,

1997; 1998: Carl and Wittbrodt, 1999; Terasaki et al., 2006; Shimada et al., 2008); however, their phylogenetic analysis has not yet been done. In this review, we have collected all the available sequence information of *fgf* genes of medaka reported in different data bases and made a phylogenetic analysis with regard to human *FGF* gene families.

Methods

To identify *fgf* genes of medaka we have searched the databases in pub med (http://www.ncbi.nlm.nih.gov/sites/entrez), medakafish homepage (http://biol1.bio.nagoya-u.ac.jp:8000/), and Ensembl (http://www.ensembl.org/Oryzias_latipes/index.html). In medaka fish homepage we have used genome browser (http://medaka, utgenome.urg) key and searched the genes with the key words "fgf" and "fibroblast growth factor" ; these searches generates 96 hits. This database is also linked to Ensembl Genome Browser (http:// ensemble.org) from which we obtained the information about the structure, function and location of the *fgf* genes on medaka chromosomes. The amino acid sequences obtained through these databases were further used for a homology-based search, TBLASTN (search translated nucleotide database using a protein query).

Results

We have identified 28 members of the *fgf* gene family of medaka. The nomenclature used in this review is either in the database or given as per amino acid identity with human or other species. Moreover, we were unable to identify *fgf9, fgf15, fgf17, fgf 21* and *fgf24* genes of medaka through database searching. *FGF 9* was identified in human but not in zebrafish and it was assumed that *fgf9* in zebrafish genome might have been lost during evolution (Itoh and Konishi, 2007). *Fgf15* is the mouse ortholog of human *FGF19*. We found *fgf19* in medaka. The other *fgf* genes (*fgf17/21/24*) which are found in other fish species such as zebrafish, takifugu (*Takifugu rubripes*) might have been lost from the medaka genome during evolution. The medaka have a smaller genome (800 Mb) than zebrafish (1700 Mb) or human (3000 Mb). We therefore have the reason to believe that these genes (*fgf9/17/21/24*) might have been lost from the medaka genome due to evolution.

3.1. Phylogenetic Analysis of the Medaka *fgf* Genes

One of the common methods used to establish evolutionary relationships among the genes are the phylogenetic analysis. To analyze the phylogenetic relationships among the identified *fgf* genes of medaka we have used vector NTI analysis software (Invitrogen, Carlsbad, CA) which is based on the neighbor-joining method described by Saitou and Nei (1986). From the analysis we found that medaka *fgf* gene (*mfgf*) family like human or zebrafish can be divided into seven subfamilies. They are, *fgf1* [*fgf1a, fgf1b, fgf2*], *fgf4* [*fgf4, fgf5a, fgf5b, fgf6a, fgf6b*], *fgf7* [*fgf3, fgf7, fgf10a, fgf10b, fgf22*], *fgf8* [*fgf8a, fgf8b, fgf18*], *fgf*9 [*fgf16, fgf20a, fgf20b*], *fgf11* [*fgf11a, fgf11b, fgf12, fgf13a, fgf13b, fgf14a, fgf14b*], and *fgf*19 [*fgf19, fgf23*]. This analysis also identified the consistency of medaka *fgf* gene subfamilies with human, mouse or zebrafish (Itoh and Konishi, 2007).

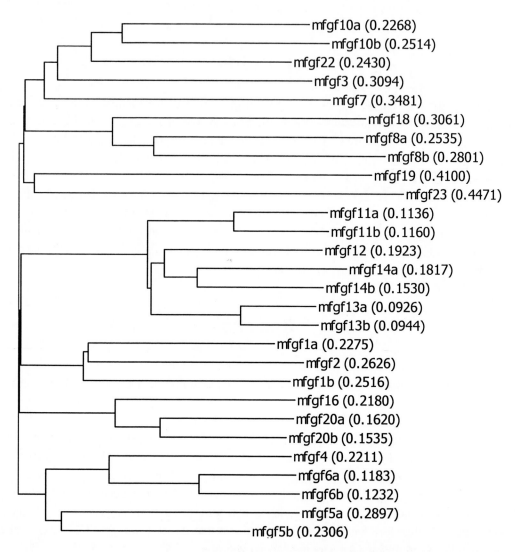

Figure 1. The phylogenetic tree of Japanese medaka *fgf* gene family. Twenty eight *fgf* genes have been identified in Japanese medaka. The tree was constructed by the neighbor-joining method (Saitou and Nei, 1986) using vector NTI analysis software. The number in parenthesis indicates the calculated distance values that are related to the degree of divergence between the sequences. Phylogenetic analysis suggests that *fgf* genes of Japanese medaka can be divided into seven subfamilies, each one containing two to seven members.

3.2. Conserved Synteny Analysis

Although phylogenetic analysis establishes potential evolutionary relationships among the gene families (which is basically a structural similarity of the genes), we have used gene location analysis on chromosomes (Itoh, 2007) as an alternative approach for the confirmation of the results we have obtained through phylogenetic analysis. At the beginning we have identified the *fgf* gene loci on specific chromosomes of medaka and then analyzed the synteny (conserved gene order) of these loci with other linked genes common to both human and medaka.

3.2.1. Chromosomal Loci of fgf Genes in Medaka

The *fgf* gene loci on medaka chromosomes indicate that only 14 out of 24 chromosomes have the *fgf* genes; however, the chromosomal loci of *fgf2* (ultracontig 117), *fgf5b* (ultracontig 256) and *fgf20a* (ultracontig 23) are yet to be identified. Among these 14 chromosomes, nine chromosomes (chromosomes 1, 2, 4, 6, 12, 15, 17, 18, 21), and three ultracontigs (23, 117 and 256) have a single *fgf* locus, two chromosomes (chromosome 9 and chromosome 23) have two *fgf* loci (*fgf5a* and *fgf10a* are on chromosome 9, and *fgf6b* and *fgf23* are on chromosome 23). Moreover, chromosome 14 has three *fgfs* (*fgf1b, fgf11b, fgf13b*), chromosome 3 has four (*fgf3, fgf4, fgf19* and *fgf7*) and chromosome 10 has five *fgfs* gene loci (*fgf13a, fgf16, fgf18, fgf20b* and *fgf1a*). In human, the distributions of twenty two *FGF* genes on 23 chromosomes are slightly different from medaka. Here the *FGF* genes are found only on 13 chromosomes (chromosomes 3, 4, 5, 7, 8, 10, 11, 12, 13, 15, 17 19 and X) of which 5 chromosomes (chromosome 3, 7, 10, 15, and 17) have a single *FGF* locus (*FGF7, FGF8, FGF11, FGF12,* and *FGF15*), six have two (chromosome 1: *FGF2* and *FGF5*; chromosome 8: *FGF17* and *FGF20*; chromosome 12: *FGF6* and *FGF23*; chromosome 13: *FGF9* and *FGF14*; chromosome 19: *FGF21* and *FGF22*; X chromosome: *FGF13* and *FGf16*), and only two chromosomes have three *FGF* loci (chromosome 5: *FGF1, FGF10* and *FGF18*; chromosome 11: *FGF3, FGF4* and *FGF19*). Therefore, the evolution of FGF genes in fish and mammals are not identical. The mammalian line of evolution probably suggests that there is no addition of new FGF genes after the first separation of the ancestral fish from the ancestral mammals; however, in the fish line there was an additional genome duplication and translocation immediately after the separation from mammalian line, which might have added more *fgf* genes to the genome (Itoh and Konishi, 2007). Moreover, the gene order of *FGF19/FGF4/FGF3* in chromosome 11 of human and chromosome 3 of medaka and the gene order of *FGF23/FGF6* in chromosome 12 of human and chromosome 23 (*fgf6b/fgf23*) of medaka are closely linked and identical between these two species. These findings suggest that the evolution of these *FGF* genes (*FGF19/4/3* and FGF23/6) is highly conserved between human and medaka.

3.2.1. 2. Conserved Gene Order Analysis

Subfamily *fgf1*

Phylogenetically, medaka has three members in this subfamily. The amino acid residues range from 152-155 with no signal peptides at the N-terminals. These three genes showed 35.4% amino acid identity with 56 amino acid residues are positionally conserved. The coding regions of *mfgf1a* is separated by three introns, however, *mfgf1b* and *mfgf2* are separated by two introns. Conserved synteny analysis of *mfgf1a* and *mfgf1b* with human *FGF1* indicate that all of them have some commonly linked genes and the gene order is more or less in the same direction (Table 1); however, the amino acid identity of *mfgf1b* (51.6%)

with human *FGF1* is more than *mfgf1a* (45.8%).Therefore, *mfgf1b* is considered to be the medaka ortholog of human *FGF1* and *mfgf1a* is the paralog of *mfgf1a* (Table 2). The chromosomal locus of *mfgf2* is yet to be determined. However, conserved synteny analysis indicates that *mfgf2* locus shows some commonly linked genes with human *FGF2* locus in an identical order. The amino acid identity of these two genes is 38.9%. Therefore, *mfgf2* both from phylogenetic as well from conserved gene location analysis is identified as the medaka ortholog of human *FGF2* gene.

Table 1. Conserved synteny of *fgf* and *FGF* loci of Japanese medaka and human chromosomes.

FGF gene subfamily of medaka	FGF gene	Species (medaka/ human)	Chromosome number	Conserved synteny
fgf1	*fgf1a*	medaka	14	*ndfip1a/fgf1b* (#)
(fgf1a,fg1b,fgf2)	*fgf1b*	medaka	10	*ndfip1b/spry4/fgf1a/arhgap26* (#)
	FGF1	human	5	*NDFIP1/SPRY4/FGF1/ARHGAP26*
	fgf2	medaka	Ultra contig 117	*adad1/bbs12/fgf2/nudt6/spata5* (#)
	FGF2	human	4	*ADAD1/BBS12/FGF2/NUDT6/ SPATA5*
fgf4	*fgf4*	medaka	3	*fgf19/fgf4/fgf3* (#)
(fgf4,fgf5a,fgf5b,	*FGF4*	human	11	*FGF19/FGF4/FGF3*
fgf6a,fgf6b)	*fgf5a*	medaka	9	*fgf5/prdm8/antxr2* (*)
	fgf5b	medaka	Ultra contig 256	*En1/fgf5b/ccdc93/ndufv3* (N)
	FGF5	human	4	*ANTXR2/PRDM8/FGF5*
	fgf6a	medaka	6	*c12orf4/fgf6a/c12orf5/ccnd2a* (*)
	fgf6b	medaka	23	*rad51/fgf6b/fgf23/ccnd2b* (*)
	FGF6	human	6	*CCND2/C12ORF5/FGF23/FGF6/ C12ORF4/RAD51*
fgf7	*fgf3*	medaka	3	Linked with *fgf4* (#)
(fgf3,fgf7,fgf10a				
fgf10b,fgf22)				
	FGF3	human	11	Linked with *FGF4*
	fgf7	medaka	3	*cops2/fgf7/dtwd1* (#)
	FGF7	human	15	*COPS2/FGF7/DTWD1*
	fgf10a	medaka	9	*snag1a/arl15/fgf10a* (*)
	fgf10b	medaka	12	*fgf10b/snag1b* (#)
	FGF10	human	5	*FGF10/ARL15/SNAG1*
	fgf22	medaka	4	*polrmt/fgf22* (#)
	FGF22	human	19	*POLRMT/FGF22*
fgf8	*fgf8a*	medaka	15	*fgf8a/pitx3/kcnip2a/ldb1/hps6/pprc1/m gea5/npm3* (N)
(fgf8a,fgf8b,fgf18)	*fgf8b*	medaka	1	*ldba1/c10orf76/kcnip2b/fgf8b/poll* (*)

Table 1. Continued

FGF gene subfamily of medaka	FGF gene	Species (medaka/ human)	Chromo some number	Conserved synteny
	FGF8	human	10	*POLL/**FGF8**/NPM3/MGEA5/KCNIP2/ C10ORF76/HPS6/LDB1/PPRC1/PITX 3/CNNM2*
	fgf18	medaka	10	*tlx3/**fgf18**/fbxw11* (#)
	FGF18	human	5	*TLX3/**FGF18**/FBXW11*
fgf9 (*fgf16,fgf20a, fgf20b*)	*fgf16*	medaka	10	*artx/**fgf16**/uprt/rnf12* (*)
	FGF16	human	X	*RNF12/UPRT/**FGF16**/ATRX*
	fgf20a	medaka	Ultra contig 23	*efha2a/**fgf20a*** (*)
	fgf20b	medaka	10	***fgf20b**/efha2c/efha2b/mtmr7/pdgfrl* (#)
	FGF20	human	8	***FGF20**/EFHA2/MTMR7/ PDGFRL*
fgf11 (*fgf11a,fgf11b, fgf12, fgf13a, fgf13b, fgf14a,fgf14b*)	*fgf11a*	medaka	18	*chrnb1a/**fgf11a**/c17orf61/nlgn2b* (*)
	fgf11b	medaka	14	*Cldn7/**fgf11b**/nlgn2a* (N)
	FGF11	human	17	*CLDN7/NLGN2/C17ORF61/ **FGF11**/CHRNB1*
	Fgf12	medaka	17	*Ephb1a/**fgf12*** (#)
	FGF12	human	3	*EPHB1/**FGF12***
	fgf13a	medaka	10	*atp11c/mcf2a/**fgf13a**/zic3/arhgef6* (*)
	fgf13b	medaka	14	*mcf2b/**fgf13b*** (*)
	FGF13	human	X	*ARHGEF6/ZIC3/**FGF13**/MCF2/ ATP11C*
	fgf14a	medaka	21	*upf3a/cdc16/**fgf14a**/itgbl1/nalcn* (*)
	fgf14b	medaka	2	*mbnl2/rap2a/**fgf14b**/tubgcp3/atp11a* (#)
	FGF14	human	13	*MBNL2/RAP2A/NALCN/ITGBL1 /**FGF14**/TUBGCP3/ATP11A /CDC16/UPF3A*
fgf19 (*fgf19, fgf23*)	*fgf19*	medaka	3	Linked with *fgf4* (#)
	FGF19	human	11	Linked with **FGF4**
	fgf23	medaka	23	*Rad51ap1/**fgf6b**/**fgf23**/ccnd2b* (*)
	FGF23	human	12	*CCND2/**FGF23**/**FGF6**/ RAD51AP1*

FGF Gene loci of human and medaka were identified by searching Ensembl Genome Browser (Http:// ensemble. org) data base. The gene order was constructed on commonly linked genes at *fgf* loci of Japanese medaka and *FGF* loci of human. During analysis only the direction of gene loci, not the proportional distances between the genes with regard to *FGF/fgf* loci, were considered. The pound (#), asterisk (*) or (N) marks indicate that the gene orders in medaka are in the same (#) or in inverse (*) or in inconclusive (N) direction with regard to human *FGF* loci.

Subfamily *fgf4*

The *fgf4* subfamily members of medaka are *mfgf4*, *mfgf5a*, *mfgf5b*, *mfgf6a* and *mfgf6b*. The amino acid identity of these subfamily members is only 13.7% with 37 amino acid residues positionally conserved. The number of amino acid residues is raged from 123-213. The exons of all these genes are separated by two introns. Except for *mfgf4* and *mfgf5b,* all have the signal peptides at the N-terminal end. Moreover, *mfgf6a* and *mfgf6b* have transmembrane domains. The amino acid identity between *mfgf4* and human *FGF4* is 53.1%. The conserved gene order analysis of *mfgf4* showed an identical order (*fgf19/fgf4/fgf3*) with human *FGF4* locus. Therefore, both from phylogenetic analysis and conserved gene order analysis *mfgf4* has been identified as the medaka ortholog of human *FGF4*. The homologs *mfgf5a* and *mfgf5b* is 28.2% identical with regard to their amino acid identity; however, *mfgf5a* showed 45% and *mfgf5b* shows 21.6% amino acid identity with human *FGF5*. The conserved synteny of *mfgf5a* locus showed some commonly linked genes arranged in an inverse order with human *FGF5* locus; however, the conserved genes of *mfgf5b* are not common with human *FGF5* locus. Therefore, *mfgf5a* is considered to be the medaka ortholog of human *FGF5* due to phylogenetic identity and *mfgf5b* is the paralog of *mfgf5a*. The homologs of *fgf6* show 74.8% amino acid identity with each other; however, *mfgf6a* shows 64.1% and *mfgf6b* shows 65.1% amino acid identity with human *FGF6*. Although the conserved gene order of *mfgf6a* and *mfgf6b* has shown several common genes with human *FGF6* locus, the orders of genes are in inverse positions with human *FGF6*. Due to the closed linkage of *mfgf6b* and *mfgf23* and also a higher amino acid identity of *mfgf6b* with human *FGF6*, *mfgf6b* is considered as the medaka ortholog of human *FGF6* and *mfgf6a* is the paralog of *mfgf6b* potentially generated from *mfgf6b* by genome duplication and gene translocation. Similarly, *mfgf23* is the medaka ortholog of human *FGF23*.

Subfamily *fgf7*

Phylogenetic analysis identified five medaka genes (*fgf3, fgf7, fgf10a, fgf10b and fgf22*) as the members of *fgf7* subfamily, which have only 10.4% amino acid identity with 30 amino acid residues positionally conserved. The number of amino acid residues is ranged from 184-241. The exons of these genes are separated by 2 (*mfgf3* and *mfgf7*), 3 (*mfgf22*) and 4 (*mfgf3* and *mfgf7*) introns. All the members except *mfgf22* have the signal peptides at the N-terminal ends and *mfgf3* and *mfgf7* have transmembrane domains. Moreover, conserved synteny analysis as well as amino acid identity with human *FGF3* (56.8%) identified *mfgf3* as the medaka ortholog of human *FGF3*. Similarly, the amino acid identity of *mfgf7* with human *FGF7* is 50% and the conserved synteny is also identical with human *FGF7* locus. Therefore, *mfgf7* is considered to be the medaka ortholog of human *FGF7*. The *mfgf10* has two homologs, *mfgf10a* and *mfgf10b* which show 42.7% amino acid identity among them. Human *FGF10* shows 51.9% amino acid identity with *mfgf10a* and 41.8% identity with *mfgf10b*. The conserved synteny of *mfgf10a* is in inverse order with human *FGF10* and in identical order with *mfgf10b* (Table1). Therefore, *mfgf10b* despite less amino acid identity with human FGF10 is considered to be the medaka ortholog of *human FGF10* and *mfgf10a* is the paralog of *mfgf10a* potentially generated from *mfgf10b* by gene duplication and translocation (Table 2). The other member of *fgf7* gene subfamily is *mfgf22*, which is located on chromosome 4. The amino acid identity between *mfgf22* and human *FGF22* is 50.3% and the conserved gene order is also identical between these two species. Therefore *mfgf22* is the medaka ortholog of human *FGF22*.

Table2. Human *FGF* and medaka *fgf* ortholog and paralog genes.

Human	Medaka	
	ortholog	paralog
FGF1	*fgf1b* (51.6)	*fgf1a* (45.8)
FGF2	*fgf2* (38.9)	
FGF3	*fgf3* (56.8)	
FGF4	*fgf4* (53.1)	
FGF5	*fgf5a* (45.0)	*Fgf5b* (21.6)
FGf6	*fgf6b* (65.1)	*fgf6a* (64.1)
FGF7	*fgf7* (50.0)	
FGF8	*fgf8a* (59.3)	*fgf8b* (49.1)
FGF9	-	
FGF10	*fgf10b* (41.8)	*fgf10a* (51.9)
FGF11	*fgf11b* (69.2)	*fgf11a* (66.1)
FGF12	*fgf12* (87.7)	
FGf13	*fgf13a* (95.0)	*fgf13b* (79.5)
FGf14	*fgf14b* (89.6)	*fgf14a* (61.5)
FGF16	*fgf16* (42.2)	
FGF17	-	
FGF18	*fgf18* (60.2)	
FGF19	*fgf19* (30.3)	
FGF20	*fgf20b* (69.2)	*fgf20a* (66.1)
FGF21	-	
FGF22	*fgf22* (50.3)	
Fgf23	*fgf23* (27.3)	

The number in parenthesis indicates per cent (%) amino acid identity with human FGF genes.

Subfamily *fgf8*

The *fgf8* subfamily of medaka consists of three genes: *mfgf8a*, *mfgf8b* and *mfgf18* with 26.8% amino acid identity and 59 amino acid residues positionally conserved. The amino acid residues of these genes are more than 200 and all of them have signal peptides at the N-terminal end. The exons of these genes are separated by 4 introns. Among them *mfgf8a* and *mfgf8b* are two homologs with 43.6% amino acid identity. The human *FGF8* showed 59.3% amino acid identity with *mfgf8a* and 49.1% identity with *mfgf8b*. The conserved gene order analysis of *mfgf8a* and *mfgf8b* also show some commonly linked genes with human *FGF8* locus; however, the gene orders in *mfgf8a* locus is inconclusive (not an identical or inverse order) with human *FGF8* locus and genes at *mfgf8b* locus show an inverse order with human *FGF8* (Table 1). Therefore, from the amino acid identity, *mfgf8a* is considered to be the medaka ortholog of human *FGF8* and *mfgf8b* is the paralog of *mfgf8a* potentially generated from *mgf8a* by genome duplication and gene translocation. The other member of the *fgf8* gene subfamily of medaka is *mfgf18* which shows 60.2% amino acid identity with human *FGF18*. The conserved synteny of *mfgf18* is identical with human *FGF18* locus. Therefore, *mfgf18* is the medaka ortholog of human *FGF18*.

Subfamily *fgf9*

The members of *fgf9* subfamily are *mfgf16*, *mfgf20a* and *mfgf20b* which showed 45.2% amino acid identity with ~210 amino acid residues of which 99 are positionally conserved. All of them have signal peptides at the N-terminal end and the exons are separated by 4 introns. The *mfgf16* shows 42.2% amino acid identity with human *FGF16*. The conserved synteny of *mfgf16* locus is in inverse order with human *FGF16*. Therefore, *mfgf16* is considered to be the medaka ortholog of human *FGF16* with regard to phylogenetic analysis. The *mfgf20* gene has two homologs: *mfgf20a* and *mfgf20b* which show 64.4% amino acid identity. The chromosomal locus of *mfgf20a* is yet to be determined (ultracontig 23); however, *mfgf20b* locus is identified on chromosome 10. The amino acid identity of *mfgf20*a and *mgf20b* with human *FGF20* is 66.1 and 69.2%, respectively. Moreover, the conserved synteny of human *FGF20* locus is in the same order as in *mfgf20b* and in inverse order with *mfgf20a*. Therefore, *mfgf20b* is considered to be the medaka ortholog of human *FGF20* and *mfgf20a* is the paralog of *mfgf20b* potentially generated from *mgf20b* due to genome duplication and gene translocation.

Subfamily *fgf11*

The *fgf11* subfamily of medaka consists of 7 genes: *mfgf11a*, *mfgf11b*, *mfgf12*, *mfgf13a*, *mfgf13b*, *mfgf14a* and *mfgf14b*. These genes showed 30.8% amino acid identity with 80 amino acid residues are positionally conserved. The amino acid residues are ~250 in number. All these genes except *mfgf14b* have no signal peptides at the N-terminal end. The exons are separated by 4 introns; however, in *mfgf14b*, which has transmembrane domains, has 8 introns. All the members of *fgf11* subfamily except *mfgf12* have two homologs. The amino acid identity of *mfgf11a* and *mfgf11b* is 64.4%; however, human *FGF11* shows 66.1% amino acid identity with *mfgf11a* and 69.2% identity with *mfgf11b*. The conserved synteny of *mfgf11a* and *mfgf11b* shows some commonly linked genes with human *FGF11* locus; however, *mfgf11a* locus has an inverse gene order and *mfgf11b* locus has an inconclusive gene order (N) with human *FGF11* locus (Table 1). Therefore, from phylogenetic analysis *mfgf11b* is considered as the medaka ortholog of human *FGF11* and *mfgf11a*, which was potentially originated from *mfgf11b* as a result of genome duplication and gene translocation, is the paralog of *mfgf11b*. The *mfgf12* shows 87.7% amino acid identity with human *FGF12*, and the conserved synteny analysis also indicate an identical order with human *FGF12* locus. Therefore, *mfgf12* is to be considered as the medaka ortholog of human *FGF12*. The *fgf13* gene in medaka has two homologs, *mfgf13a* and *mfgf13b*, which showed 80.3% amino acid identity. Moreover, *mfgf13a* shows 95.5% and *mfgf13b* shows 79.5% amino acid identity with human *FGF13*. Conserved gene order analysis identified some commonly linked genes in *mfgf13a* and *mfgf13b* loci which are arranged in an inverse order with regard to human *FGF13* locus (Table 1). Therefore, from phylogenetic analysis, *mfgf13a* is considered as the medaka ortholog of human *FGF13*, and *mfgf13b* is the paralog of *mfgf13a* potentially originated from *mfgf13a* by genome duplication and gene translocation. Medaka *fgf14* gene has also two homologs, *mfgf14a* and *mfgf14b*. The amino acid identity between these two genes is 64.0%; however, *mfgf14a* showed 61.5% and *mfgf14b* showed 89.6% amino acid identity with human *FGF14*. The conserved synteny analysis of *mfgf14a* and *mfgf14b* shows some commonly linked genes with human *FGF14* locus; however the gene order in *mfgf14a* locus is in inverse direction and *mfgf14b* locus is in identical direction with human *FGF14* locus. Therefore, on the basis of phylogenetic identity and conserved synteny analysis *mfgf14b* is considered as the medaka ortholog of human *FGF14* and *mfgf14a*, which was originated from *mfgf14b* by genome duplication and gene translocation, is the paralog of *mfgf14b*.

Subfamily *fgf19*

The *fgf*19 subfamily of medaka has two genes, *mfgf19* and *mfgf23*, with only 19% amino acid identity and 51 amino acid residues positionally conserved. The number of amino acid residues is 209 (*mfgf19*) and 268 (*mfgf23*), and both of them have signal peptides at the N-terminal end. The exons of these genes are separated by 2 introns. The chromosomal locus of *mfgf19* is on chromosome 3 which is linked with *mfgf4* and *mfgf3*. The amino acid identity of human *FGF19* and *mfgf19* is only 30.3%, however, as in medaka, human *FGF19* is also linked with *FGF4* and *FGF3*. Therefore, from conserved gene order analysis *mfgf19* is considered to be the medaka ortholog of human *FGF19*. The *mfgf23* shows only 27.3% amino acid identity with human *FGF23*. The conserved synteny analysis of *mfgf23* identified some commonly linked genes which are arranged in an inverse order with human *FGF23* locus. However, both in medaka and human *FGF23* gene is closely linked with *FGF6* (in medaka *fgf6b*) genes. Therefore, *mfgf23* is considered as the medaka ortholog of human *FGF23*.

Discussion

From the data base searching we have identified 28 *fgf* genes in medaka of which 9 are paralogs. In human and mouse there are 22 *FGF* gene family members; however, in zebrafish there are 27 reported *fgf* gene family members are reported (Itoh, 2007, Itoh and Konishi, 2007). We have also searched the *fgf* genes of fugu (*Takifugu rubripes*) and found that this fish species also contains 27 *fgf* genes of which 6 are paralogs (data not shown). Although medaka has a smaller genome size than zebrafish, the *fgf* gene numbers are more or less equal in these three fish species (zebrafish, medaka and fugu). Moreover, several of the *fgf* genes identified in zebrafish and fugu are not found in medaka. We expect that these *fgf* genes are probably lost from the medaka genome through evolution.

From the conserved synteny analysis it was observed that medaka *fgf19*, *fgf4* and *fgf3* loci and *fgf6b* and *fgf23* loci are closely linked on chromosome 3 and 23, respectively, which is also observed in human (in chromosomes 11 and 12) and zebrafish (chromosomes 7 and 4) (Itoh and Konishi, 2007). Therefore, these genes need to be included in one subfamily. Accordingly, *fgf3/19/23* genes are included as a subfamily member of *fgf4*. Therefore, the subfamily numbers of medaka *fgf* genes as in human and zebrafish (Itoh, 2007) are reduced to six, and they are *fgf1* (*fgf1a/1b/2*), *fgf4* (*fgf3/4/5a/5b/6a/6b/19/23*), *fgf7* (*fgf7/10a/10b/22*) *fgf8* (*fgf8a/8b/18*), *fgf9* (*fgf16/20a/20b*) and *fgf11* (*fgf11a/11b/12/13a/13b/14a/14b*).

Most of the *FGF*s mediate their biological response by binding with *FGFR*s; however, several of them act as intracellular proteins and as a *FGFR*-independent manner (Goldfarb, 2005). These *FGF*s belong to the *FGF11* subfamily and are also referred to iFGFs (Itoh and

Ornitz, 2008). In contrast to iFGFs, those that bind with *FGFR* are known as canonical *FGFs* (Itoh and Ornitz, 2008). In medaka, 7 *fgf* genes (*fgf11a/11b/12/13a/13b/14a/14b*) are the members of *fgf11* subfamily and except for *mfgf14b,* they are devoid of signal sequences. Therefore, it is assumed that these *fgfs* are *iFGFs* of medaka and are most probably functioning inside the cells. Moreover, these genes show only 30.8% amino acid identity within the family members with 80 amino acid residues positionally conserved, but their amino acid identity with corresponding human genes are more than 60%, which suggests that these genes are the probable ancestor of canonical *fgfs*. The current concept of the origin of canonical *fgfs* indicates that iFGFs which have four introns might be the ancestor of canonical fgfs. In medaka, fgf genes are classified into four groups with regard to their intron numbers. The maximum numbers of introns (eight) are found in *fgf14b*. Four introns in *fgf8a/8b/11a/11b/12/13a/13b/14,* three in *fgf1a/10b/16/16/20a/22,* and two in *fgf1b/2/3/4/5/6a/6b/7/19/23* are found. Therefore, it appears that the origin of all *fgf* genes occur either from *fgf14b* or similar genes that have more introns as well as signal peptides at the N-terminal end. During evolution two types of genes were originated from that ancestral gene(s) (*fgf14b*); one group with four introns and the signal peptides at the N-terminal end (members of *fgf8*: *fgf8a/8b/18*) and the other groups with four introns and no signal peptides (*fgf10a* and all the members of *fgf11* or *iFGFs*). Later, possibly from the members of *fgf8* or *fgf11* all other *fgfs* (three introns with no signal peptides: *fgf1a/10a/16/20a/22*; two introns with signal peptides: *fgf3/5a/5b/6a/6b/7/19/23*, and two introns with no signal peptides: *fgf1b/2/4/20b*) were evolved.

Therefore, it is evident from this review that although medaka were separated from zebrafish line of evolution ~ 150 millions years ago (Wittbrodt et al., 2002), the evolution of *fgf* gene families are identical in both species. It also supports the concept that fish genomes underwent partial or full genome duplication after the fish line separated from mammalian line.

5. Acknowledgements

We are grateful to Bhabak Shariat-Madar for his generous help during the preparation of the manuscript. The work was supported by funding from the Environmental Toxicology Research Program, National Center for Natural Product Research, School of Pharmacy, and a small grant from the Office of Research and Sponsored program, University of Mississippi. This publication was made possible by NIH grant number RR016476 from the MFGN INBRE Program of the National Center for Research Resources.

References

Bally-Cuif L; Wassef M. Determination events in the nervous system of the vertebrate embryo. *Curr. Opin. Gen. Dev.* (1995), 5:450-458.

Chen L; Deng CX. Roles of FGF signaling in skeletal development and human genetic diseases. *Front Biosci.* (2005), 10:1961-1976.

Chen GJ; Forough R. Fibroblast growth factors, fibroblast growth factor receptors, diseases, and drugs. *Recent Patents Cardiovasc Drug Discov*. (2006), 1:211-224.

Carl M; Wittbrodt J. Graded interference with FGF signaling reveals its dorsoventral asymmetry at the mid-hindbrain boundary. *Development* (1999), 126:5659-5667.

Cotton LM; O'Bryan MK; Hinton BT. Cellular signaling by fibroblast growth factors(FGFs) and their receptors (FGFRs) in male reproduction. *Endo. Rev.* (2008), 29: 193-216.

Cox WG; Hemmati-Brivanlou A. Caudalization of neural fate by tissue recombination and bFGF. *Development*, (1995), 121:4349-4358.

Dvorak P; Hampl A. Basic fibroblast growth factor and its receptors in human embryonic stem cells. *Folia Histochem. Cytobiol.*(2005), 43:203-208.

Emori Y; Yasuoka A; Saigo K. Identification of four FgF receptor genes in medaka fish (Oryzias latipes). *FEBS Lett.*(1992), 314:176-178.

Ford-Perriss M; Abud H; Murphy M. Fibroblast growth factors in the developing central nervous system. *Clin.Exp. Pharm. Physiol.* (2001), 28:493-503.

Furutani-Seiki M; Wittbrodt J. Medaka and zebrafish: an evolutionary twin study. *Mech Dev* 121:629-637 (2004).

Goldfarb M. Fibroblast growth factor homologous factors: evolution, structure, and function. *Cytokine Growth Factor Rev.* (2005), 16:215-220.

Itoh N. The Fgf families in humans, mice and zebrafish: their evolutional processes and roles in development, metabolism and disease. *Biol Pharm Bull.* (2007), 30:1819-1825.

Itoh N; Ornitz DM. Functional evolutionary history of the mouse fgf gene family. *Dev. Dyn.* (2008), 237:18-27.

Johnson DE; Williams LT. Structural and functional diversity in the EGF receptor multigene family. *Adv Cancer Res.* (1993), 60:1-41.

Kengaku M; Okamoto H. Basic fibroblast growth factor induces differentiation of neural tube and neural crest lineages of cultured ectoderm cells from xenopus gastrula. *Development*, (1993),119:1067-1078.

Kengaku M; Okamoto H. bFGF as a possible morphogen for the anteroposterior axis of the central nervous system in xenopus. *Development*, (1995),121, 3121-3130.

Lamb TM; Harland RM. Fibroblast growth factor is a direct neural inducer, which combined with noggin generates anterior-posterior neural pattern. *Development*,(1995),121:3627-3636.

McFarlane S; Cornel E; Amaya E; Holt CE. Inhibition of FGF receptor activity in retinal ganglion cell axons causes errors in target recognition. *Neuron* (1996), 17:245-254.

Robinson ML. An essential role for EGF receptor signaling in lens development. *Semin Cell Dev Biol*(2006), 17:726-740.

Saitou N; Nei M. The number of nucleotides required to determine the branching order of three species, with special reference to the human-chimpanzee-gorilla divergence. *J. Mol. Evol.* (1986), 24:189-204.

Shimada A; Yabusaki M; Niwa H; Yokoi H; Hatta K; Kobayashi D; Takeda H. Maternal-zygotic medaka mutants for fgfr1 reveal its essential role in the migration of the axial mesoderm but not the lateral mesoderm. *Development*, (2008),135:281-290.

Takaishi S; Sawada M; Morita Y; Seno H; Fukuzawa H; Chiba T. Identification of a novel alternative splicing of human FGF receptor 4: soluble form of splice variant expressed in human gastrointestinal epithelial cells. *Biochem Biophys Res Commun.* (2000), 267:658-662.

Terasaki H; Murakami R; Yasuhiko Y; Shin-I T; Kohara Y; Saga Y; Takeda H. Transgenic analysis of the medaka mesp-b enhancer in somitogenesis. *Dev. Growth Differ.*(2006), 48:153-168.

Wassef M; Joyner AL. Early mesencephalon/metancephalon patterning and development of the cerebellum. *Perspect Dev Neurobiol.*(1997), 5:3-16.

Watanabe A; Hatakeyama N; Yasuoka A; Onitake K. Distributions of fibroblast growth factor and the mRNA for its receptor, MFR1, in the developing testis of the medaka, Oryzias latipes. *J Exp. Zool.*(1997), 279:177-184.

Watanabe A; Kobayashi E; Ogawa T; Onitake, K. Fibroblast growth factor may regulate the inhibition of oocyte growth in the developing ovary of the medaka, Oryzias latipes. *Zool. Sci.*(1998), 531-536.

Wittbrodt J; Shima A; Schart M. Medaka- a model organism from the far east. *Nature Rev. Genetics*, (2002), 3:53-64.

Wuechner C; Nordqvist AC; Winterpacht A; Zabel B; Schalling M. Developmental expression of splicing varients of fibroblast growth factor receptor 3 (FGFR3) in mouse. *Int J Dev Biol.* (1996), 40:1185-1188.

Xu X; Weinstein M; Li C; Deng C. Fibroblast growth factors (FGFRs) and their roles in limb development. *Cell Tissue Res*, (1999), 296:33-43.

In: Development Gene Expresión Regulation
Editor: Nathan C. Kurzfield

ISBN: 978-60692-794-6
©2009 Nova Science Publishers, Inc.

Chapter XII

The Role of Internal Fluid Invironment for Regulation of Germ Genes Expression in Early Development of Some Cyprinid Fishes and their Intergeneric F1 Hybrids

A.M. Andreeva

Institute of Biology of Inland Waters, Borok, Yaroslavl region, Russia

Abstract

Connection between dynamics of yolk lipovitellin degradation and specific features of germ genes expression was studied in early development of intergenetic reciprocal F1 hybrids of the bream, roach and blue bream.

According to the modern view lipovitellin is the main protein of oocyte and embryo yolk. Its main function is reserve, nutritional and structural. But moreover lipovitellin is active component of internal fluid embryo invironment in wich embryo cells and germ genes are developing and expressing. There is a view that synchronous expression of parental alleles of genetic loci is related, as a rule, to kindred fish crossing, and asynchronous expression – to remote fish crossing. But when we analysed character of expression of some loci of intergeneric F1 hybrids (*6-Pgd, Ldh-B, β–Est-1, 2, 3, Aat-1, Me-1, 2* and others) with different expression time in embryogenesis, we show that character of loci expression is related to stage of development also.

As objects we used zygotes, embryos, larvas and frysof bream, roach, blue bream and intergeneric reciprocal F1 hybrids. Identification and analysis of enzymes activity and lipoviteelline properties were performed using methods of disk-, gradient and SDS-electrophoresis in polyacrylamide gel.

There are some arguments for regulation function of lipovitellin : first argument is connected with different expressions character (synchronous and asynchronous) of germ genes in early and late stages of embryogenesis; second - with different distribution of isoenzymes activity in early and late stages of embryogenesis; and third – with biocatalytic activity of lipovitelline because of oogenesis isoenzymes connected with lipovitelline by weak connection.

When first locus expression was timed to the early stages (blastodisk – gastrula), the gene parental alleles were activated asynchronously according to the maternal types. When the first expression was timed to later stages (yolk sac resorption), parenteral alleles were activated synchronously. In early development lipovitellin and oogenesis enzymes form the maternal (by origin) metabolic medium, which preferential activation of maternal alleles. At later developmental stages, when the yolk reserves are partially or fully resorbed and maternal proteins don' t play importance role, and germ proteins form new (germ) internal fluid invironment, in wich the embryonic genes are activated synchronously.

Introduction

The internal fluid organism invironment consist of some external fluids (interstitial fluid, lymph, plasma) which are connected by blood system. The main internal invironment proteins are plasma proteins wich can permeate into tissue fluid in process of transcapillary change. The internal fluid embryo invironment is formed on the basis of nonfertility egg cytoplasm, after fertilization perivitelline space and blastocel fluids are added to it. Main stock of fish egg external proteins are kept into yolk sac and sac contents amalgamate with internal fluid embryo invironment also.

The main yolk protein component of embryo is lipovitelline, its contents is more than 90% from general yolk protein. Its precursor vitellogenin is synthesized into fish female liver before spawning and then across blood gets into oocyts where disintegrates to lipovitelline and fosvitin. Traditionally lipovitelline is characterized as reserve (in oocyte) and storage and nutritional (in embryo) protein; nutritional function is realized in the fish embryos only after the gastrula stage (Нейфах, Тимофеева, 1977; Hartling, Kunkel, 1995; Babin et al., 1999; Новиков, 2000; Wang et al., 2000; Naoshi et al., 2002).

In this work we attempt to find out which functions apart from nutritional and storage are carried out by lipovitelline at the earlier development of fish (from egg fertilization to the stage of yolk sac dissapearence in the larva and to stages of fry and underyearling). The aim of this work is study of lipovitelline properties as possible regulator in fish embryogenesis. We analyse space-temporal connections between lipovitelline degradation and character of expression of parental alleles of embryo genes in early development. Also we analyse biocatalytic activity of lipovitelline wich is explained by oogenetic ezymes "set" on the Lv surface in different stages of early development of some cyprinid fishes and their intergenetic F1 hybrids.

Materials and Methods

Experimental materials such as the progenies of intraspecific and intergenetic reciprocal crosses of the bream *Abramis brama* L., roach *Rutilus rutilus* L . and blue bream *Abramis ballerus* L. were obtained during spring-summer periods of 1999-2006 in seven groups: 1. bream x bream, 2. roach x roach, 3. blue bream x blue bream, 4. roach x blue bream, 5. blue bream x roach, 6. bream x roach, 7. roach x bream. Intraspecific crosses involved: roach and bream four pairs of spawners for each, and blue bream one pair. Intergeneric crosses involved: roach and bream four pairs of spawners, bream and roach four pairs, roach and blue bream two pairs, and blue bream and roach one pair (Андреева и др., 2000; Слынько, 2000; Лапушкина и др., 2002; Андреева 2005*a*, 2005*b*, 2007).

Mature fishs were caught during the spawning period in the region of Volzhskii ples (Rybinsk Water Reservoir). The eggs were fertilized using the dry method (Ryabov, 1981) and standart technology of rearing F1 hybrids (Слынько, 2000).

The embryos and larvae were sampled at the following stages: 1. formation of perivitelline space and blastodisk, 2. blastula, 3. gastrula, 4. three segments, 5. peak of body segmentation (18-23 segments), 6. end of segmentation and onset of blood circulation (43-48 segments), 7. hatching, 8. stage of mixed feeding, 9. stage of full yolk sac resorption and transition to external feeding, 10. fry and 11. underyearling (Крыжановский 1949, 1968). Half of the larvae, analyzed at the stages 5-7, were prepared in membrane, embryo and yolk sac contents.

Unfertilized eggs from ripe females and the muscles of spawners were used as a control (Андреева, 2001, 2005) (Figure 1). Separate eggs, larvae and pieses of spawner muscle tissue were poured with a fixed volum of 20% saccharose and freezed. Before work the samples were centrifuged for 20 min at a rate of 18000 rev/min and a temperature of 4^0C. The supernatant was used in the work.

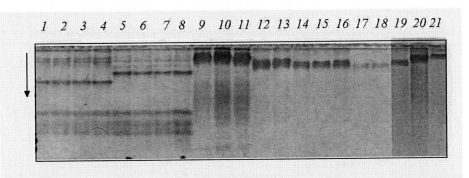

Figure 1. Disk-electrophoresis of watersoluble proteins from spawners muscles : bream (*1-4*) and roach (*5-8*); unfertilization eggs : bream (*9-11*) and roach (12-18); embryo : roach (19), blue bream (20) and bream (21) on stage of blastula.

The fractionation of water-soluble proteins was carried out in the buffer system tris-borate-EDTA (Davis, 1964; Ornstain, 1964), gradient of PAAG concentrations under nondenaturing conditions (Kopperschlander et al., 1969), SDS-electrophoreses under denaturing conditions (Laemmly, 1970). Molecular masses (MM) were calculated using program package Scion-95 and OneDscan.

The isozyme spectra of 6-phosfogluconate dehydrogenase 6-PGD (1.1.1.44); malat NADP-dehydrogenase or malic-enzyme ME (1.1.1.40); lactate dehydrogenase LDH (1.1.1.27); esterases β- (β–EST), acetylholin- AHE (3.1.1.7), butyrilholinesterase BuHE (3.1.1.8), aspartataminotransferase AAT (2.6.1.1), izocytrat dehydrogenase IDH (1.1.1.42) and general protein electpophoretic spectra were obtained by gel staining according to the histochemical procedures (Бернстон, 1965; Корочкин и др., 1977).

Activation of embryo genes was marked by expression of paternal alleles wich were shown at the isozyme spectra when paternal and maternal enzymes differed by electrophoretic mobility R_f.

For genetic analysis of enzyme loci was obtained for all experimental embryo groups (Table 1) we take into account information about structure of analysed loci.

Aat-1. Structural locus *Aat-1* controls soluble enzyme s-AAT. AAT is dimer, its MM is 93 kDa (Корочкин и др., 1977). Three different alleles of *Aat-1* locus from bream, roach and blue bream were described (Slynko, 1997). In using crosses (Table 1) parental blue bream fish were gomozygous by slow allele (*Aat-sl*), parental bream fish were gomozygous by fast allele (*Aat-f*), corresponding gomodimeric enzymes we mark as AAT-sl and AAT-f. Roach had allele *Aat-med* and product of this allele (AAT-med) was localized at the media position on the electrophoregramm : between AAT-f и AAT-sl (Андреева, 2007). Parental roach fishs were represented twoo types: gomozygous by *Aat-sl* and geterozygous by *Aat-sl/med*. Geterozygous fish was used only for intergenetic cross blue bream x roach (Table 1).

Ldh-B, Ldh-A. LDH is controlled by two loci *Ldh-B* and *Ldh-A*, tetramer, MM 135 kDa, consist of subunits of two types – A and B (Markert, 1963). Roach from natural population had polymorphism by *Ldh-B*, represented two alleles; very seldom polymorphism by *Ldh-A* occurs before bream also represented two alleles (Слынько, 1987; 1991; 1993; 1997). In using crosses (Table 1) parental bream fish were gomozygous by *Ldh-B* and *Ldh-A*; parental roach fish were gomozygous as by fast (B) as by slow (B$'$) alleles *Ldh-B*; parents were marked as R and R$'$ correspondingly.

β-Est-1, 2, 3. β-EST -1, 2, 3 – are monomeric proteins, they are controlled by three nonallele genes (Корочкин и др., 1977). Roach fishs from Rybinsk Water Reservoir population had polymorphism by all three loci : there are 5 alleles for *β–Est-1*, three alleles for *β–Est-2*, four alleles for *β-Est-3* (Слынько, 1993). Blue bream and bream had monomorphic loci *β–Est-1* and *3*, locus *β-Est-2* had two alleles (Слынько, 1993).

Me-1, 2, Idh-1, 6-Pgd, Ahe. Soluble tetrameric malik enzyme from mouse is controlled by locus *Mod-1* (Корочкин и др., 1977); enzyme from fish is controlled by locus *Me-1, 2* (Слынько, 1993) and perhaps is dimeric protein. Bream, blue bream and roach were monomorphous by loci *Me-1* and *Me-2*, only roach *Me-2* was represented by two alleles (Слынько, 1993). Soluble IDH is dimeric enzyme, controlled by pair kodominant alleles from autosomic locus *Idh-1* (Корочкин и др., 1977). Human 6-PGD is dimeric enzyme, MM 95 kDa; is controlled plural alleles from polymorphic locus *6-Pgd* (Корочкин и др., 1977). 6-PGD from white muscle of bream and roach had spacies-specific electrophoretic mobility R_f. AHE from human erythrocytes is controlled by two alleles autosomic locus (Корочкин и др., 1977).

Table 1. Isoenzymes and corresponding loci from spawners of roach, bream and blue bream participating in intraspecific and intergeneric crosses [*].

Enzyme	NN	1. Isoenzymes of spawners (♀ ♂) participating in crosses; 2. Loci, alleles; 3. Cross group (♀ ♂).			
LDH	1.	x	x	x	
	2.	Loci *Ldh-A* and *Ldh-B*, alleles *Ldh-A, B and B'*			
	3.	B x B, R x R	R x R, B x R	R x B	
AHE, BuHE	1.	x	x	x	x
	2.	Loci *Ahe, Buhe*			
	3.	B x B	R x R	B x R	R x B
ME	1.	x	x	x	x
	2.	Loci *Me-1, 2*			
	3.	B x B	B x R	R x B	R x R
β-EST	1.	x	x	x	
	2.	Loci *β-Est-1, 2, 3*			
	3.	B x B, R x R, B x R, R x B	BB x R, R x BB	BB x R, BB x BB	
AAT	1.	x	x	x	x
	2.	Locus *Aat-1*, alleles *Aat-fast, Aat-slow, Aat-media*			
	3.	B x R	R x B	R x BB, BB x BB, R x R	BB x R
6-PGD	1.	x	x	x	x
	2.	Locus *6-Pgd*, alleles *Pgd-A, Pgd-B, Pgd-C, Pgd-D*			
	3.	B x R	R x B	B x B	R x R
IDH	1.	x	x	x	x
	2.	Locus *Idh-1*			
	3.	B x B, R x B, B x R	R x R	R x R	R x BB

[*] R – roach, B – bream, BB – blue bream

Results and Discussion

Lipovitellin Identification

Lipovitellin the main protein of internal fluid invironment of fish embryo was identified by 1) high relative contents among soluble embryo proteins; 2) character of inheritance from reciprocal F1 hybrids; 3) molecular mass; 4) dynamic of degradation.

Thus, on the electrophoregam of soluble embryo proteins there is detected macrocomponent, wich portion of general protein is more than 95% (Figure 2). This macrocomponent had spacies-specific electrophoretic mobility: 0,093 for roach and 0,08 for bream (and blue bream). The macrocomponent from reciprocal hybrids had values R_f wich were submitted to "maternal effect" (Figure 2). All these properties characterize macrocomponent as lipovitellin.

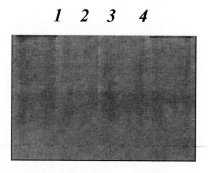

Figure 2. Lipovitellin from roach (*1*), bream (*2*) and hybrids F_1 roach x bream (*3*) and bream x roach (*4*) embryos on the stage blastula in disc-electrophoresis.

At gradient of PAAG concentrations lipovitellin was represented by two components with MM 200 and 335 kDa (Figure 3), at SDS-PAAG – by one component with MM 180 kDa, it is agreeing with other authors datas (Wang et al., 2000) and also characterize macrocomponent as lipovitelline.

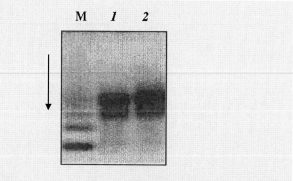

Figure 3. Definition of molecular mass MM of lipovitellin from bream embryo (*1*) and hybrid Bream x Roach (*2*) on the stage blastula in gradient of concentration PAAG (5-40%). M – marker protein human serum albumin.

Relative quantity of lipovitellin decreases and of proteins with small MM (PSM) – increase in embryo development (Figure 4), wich may be explayed by action of yolk proteases to the lipovitellin (Нейфах, Давидов, 1964; Немова, 1992).

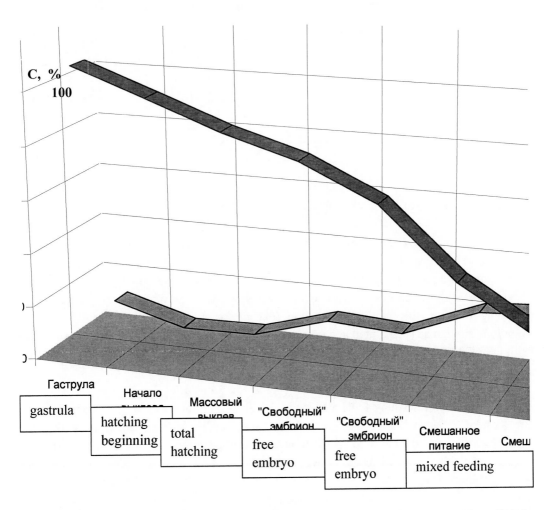

Figure 4. Dynamic of lipovitellin degradation (dark line) and increase of content of proteins with small MM (light line) from roach embryo on different developmental stages. C – protein content in %.

Lipovitellin degradation at all experimental fish group was detected after blastula. Research yolk degradation from salmon fish shows that sac rezorbtion begins on stage of blastodisk (Новиков, 2000); other authors notice the lipovitellin degradation after gastrula (Naoshi et al., 2002). Complete disappearance of macrocomponent was detected on the stage of sac rezorbtion. Parallels with increase contents of PSM there were change of qualitative composition of PSM-molecules perhaps because of consecutive degradation of products of lipovitellin proteolys to more small fragments (Table 2).

All recapitulating peculiarities show that this macrocomponent takes part in plastic metabolism and identify it as lipovitellin.

Table 2. Molecular mass (kDa) of water-soluble proteins from roach embryo on stage 2 blastomers (2bl) and then by 12-108 hours after hatching.

Stage	2bl	blastula	blood circulation.	hatching.	12 h.	36 h.	60 h.	108 h.
MM	330	335	300	305	301	300	301	280
	260	279	235	248	239	235	237	209
	200	202	159	166	153	151	156	158
			78	123	118	112	111	116
				95	90	84	83	70
				76	73	72	70	59
							53	38

Lipovitellin as Regulator of Early Development

In order to analyze the regulatory characteristics of lipovitellin its biocatalytic features were analyzed as well as the link between the time and character of embryonic gene expression in stages "enriched" by the lipovitellin and those when its supply was depleted.

1. Biocatalytic characteristics of lipovitellin. Activity of all analyzed enzymes including those in the lipovitellin fraction (except for AAT) was found in the yolk of embryons and larvae (preparated and not). Identification of oogenetic (including those linked with lipovitellin) and embryonic isoenzymes on the electrophoregramms will be shown using LDH of roach embryos as an example.

LDH-activity in roach embryos homozygous by slow allel Ldh-B. On the stage of blastodisc and until hatching LDH activity in the disc-electrophoresis was stably revealed in the form of 3 bands (Figure 5). Two of them matched B'_4 и B'_3A isoenzymes from the unfertilized eggs of the ready to spawn females that allowed considering them as oogenetic isoenzymes; third band matched lipovitellin and disappeared at the stage of exogenous feeding (Figure 5).

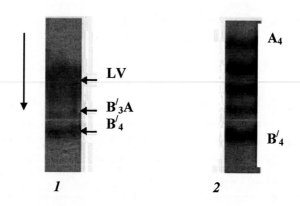

Figure 5. LDH-isozymes by roach homozygous by slow allele (B') of *Ldh-B* –locus on blastula-stage (*1*) and external feeding (*2*). LV – lipovitellin; B'_4, B'_3A and A_4 – isozymes of LDH.

Oogenetic LDH (linked and not linked with lipovitellin) on the stages of two blastomeres accounted for up to 79% of overall activity, on the stage of mixed feeding (60 hours after hatching) – up to 43,7%; then until full extinction of the activity on the stage of exogenous feeding (Figure 6). At that, oogenetic LDH linked with lipovitellin accounted for up to 69% of overall LDH activity on the stage of blastodisc, 59,9-63% in blastula and 47,6-50% in gastrula. During the development of the embryo share of LDH activity in the zone of lipovitellin motility dropped until the stage of the exogenous feeding inclusive (Figure 7).

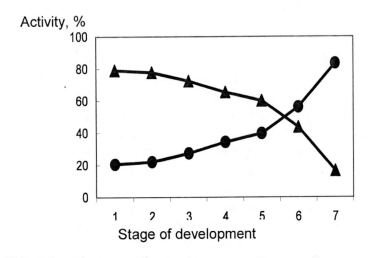

Figure 6. Distribution of LDH-activity from embryos between fractions of oogenetic (▲) and embryo (●) enzymes. 1-7 – stages of development correspondingly: 2 blastomers; blastula; hatching; 12, 36, 60 and 108 hours after hatching by hybrid F_1 Roach x Bream.

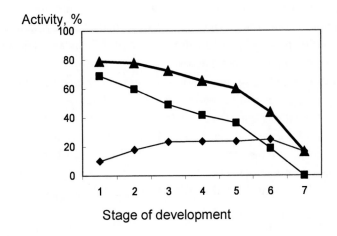

Figure 7. Distribution oogenetic LDH ( ) between fractions of enzymes, connected ( ) and doesn' t connected ( ) with lipovitellin. 1-7 – stages of development correspondingly: 2 blastomers; blastula; hatching; 12, 36, 60, 108 hours after hatching by hybrids F_1 Roach x Bream.

LDH activity in the yolk fraction in mictopreparated larvae (until hatching) was localized in the zone of lipovitellin and in the embryonic fraction – in the B'_4 and B'_3A zones. Therefore, oogenetic isoenzymes B'_4 and B'_3A from the cytoplasm of the embryo and oogenetic LDH linked with lipovitellin of the yolk were the most active ones. Fractioning of LDH isoenzymes by MM value in the PAAG concentrations' gradient did not lead to separation of oogenetic LDH from lipovitellin. This fact gives evidence for the presence of weak bonds between lipovitellin and yolk LDH.

Oogenetic isoenzymes loaded into the yolk during the mother's oogenesis and "sitting" on the surface of the lipovitellin condition the biocatalytic characteristics of lipovitellin. Disc-electrophoregrammes of water-soluble proteins of embryos of all stages up until complete depletion of the yolk showed the activity of analyzed enzymes (except for AAT) to be within the motility zones of either lipovitellin or corresponding oogenetic isoenzyme not linked with the lipovitellin or in both zones of motility simultaneously. Oogenetic isoenzymes not linked with the lipovitellin located in accordance with the values of their MM (β-EST) in the PAAG concentration gradient, while oogenetic isoenzymes associated with the lipovitellin (LDH, 6-PGD, ME, IDH, AHE, BuHE) stayed in the lipovitellin's zone of motility which was due to their interaction with the lipovitellin, probably. Yolk and the embryo itself in the micropreparated larvae contained different fractions of the enzymes: thus LDH linked associated with lipovitellin was found in the larval yolk, while isoenzyme not associated with the lipovitellin was found in the embryo (without yolk).

Exept LDH oogenetic β-EST, IDH and ME were differentiated on two fractions also: connected and disconnected with lipovitellin; the activity of AHE and BuHE was revealed exclusively in a zone of mobiliry of lipovitellin (Picture 8). An activity of β-EST in a lipovitellin zone of mobility was revealed only from a gastrula stage inclusive, and in gradient of conce+tntration of PAAG enzyme moves separately from lipovitellin.

2. Distribution of biocatalytic activity between isozymes at different stage of embryo development. Distribution of biocatalytic activity between isozymes, varying during embryo development, reflects, first of all, processes of a tissue differentiation of a germ. However, such redistribution of activity occurs also on a background of a varying internal environment of a embryo and indirectly confirms regulatory aspect of influence of this environment and its components on expression of germinal genes.

3. Time and character of expression of germinal genes. The time of the first expression of germinal genes in all experimental groups of germs was defined for the analysis of space-temporal conections between degradation of lipovitellin and character of expression of of parental alleles of germinal genes. The moment of activation of germinal genes was registrated at time of the first expression of paternal alleles (Андреева, 2005а, 2005в, 2007), considering the fact of a simultaneity of inclusion of parental alleles of structural genes in early development (Корочкин, 1976, 1999; Чадов, 2002) in posterity from intraspecific crossing.

Ldh :

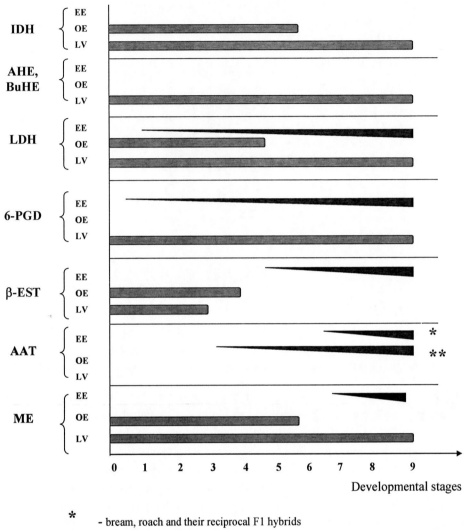

* - bream, roach and their reciprocal F1 hybrids

** - blue bream and Blue bream x Roach

Figu
re 8. Distribution of biocatalytic activities of ME, AAT, □-EST, 6-PGD, LDH, AHE, BuHE, IDH in embryos between oogenetic (connected with lipovitellin (LV) and nonconnected with lipovitellin OE) and embryo enzymes (EE) by bream, roach, blue bream and their F1 hybrids on stages from zygote (0) and blastodisk (1) to yolk sac resorption (9).

The time of expression of germinal genes Ldh-A and Ldh-B on an example of a roach heterozygous by Ldh-B. Mother of heterozygotic posterity was homozygous on a fast allele B, and the father was homozygous on a slow allele B' , embryos should have only two parental oogenetyc isozymes B_4 and B_3A at the earling stages of development down to hatching. However, already at stage of blastodisk they had four isozymes, one of which was the same in mobility with isoenzyme B_4 , and three others settled down between B_4 and B'_4 (Figure 9).

These three isoenzymess could be the variants of combinations of B- and B' - subunits, or the result of noncovalent modifications B_4 and B_3A, the presence of wich was considered during the early development for *Danio rerio* (Клячко, Озернюк, 2001). The comparison of these three isozymes at the stages of cleavage, segmentation, and before and after hatching demonstrated the invariability of their electrophoretic mobility. This fact suggests that they are variants of combinatorial combination of B- and $B^{/}$- subunits, and not the modification of oogenetic B_4 and B_3A-isoforms.

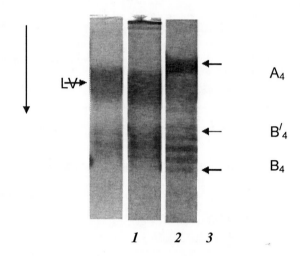

Figure 9. Isoenzyme LDH-spectrums LDH by roach heterozygous by *Ldh-B*-locus on developmental stages of cleavage (*1*), gastrula (*2*) and after hatching (*3*). LV – lipovitellin; B_4 , $B^{/}_4$ и A_4 – isoenzyme of LDH.

The fifth isozyme with Rf coinciding with the Rf of paternal $B^{/}_4$-isozyme appeared in the electrophoregram after gastrula. Theoretical calculation of the number of all $B/B^{/}$ –isozymes gives the value 5 (B_4, $B_3B^{/}$, $B_2B^{/}_2$, $BB^{/}_3$, $B^{/}_4$), which is the case. In heterozygous roach the oogenetic isozyme B_3A could possibly coincide in Rf with one of the $B/B^{/}$ - isozymes (Аттер и др., 1991), or could eplace one such isozyme because of its inactivity. The activation of tissue-specific locus *Ldh-A* coincides with appearance of myotomes (Кирпичников, 1987), and because of this, the derivates of interaction of the products of *B* and $B^{/}$ alleles with the products of locus *Ldh-A* added in sequence to the existing isozymes of $B^{/}$ / B – composition after the beginning of segmentation. At the hatching stage the isozyme spectrum was presented by 10 bands, at the stage of endogenic feeding by 11, and exogenous feeding by 13 bands (Figure 10).

Time of Ldh-A and B expression in progeny of reciprocal bream and roach F1 hybrids. Comparative analysis of dynamica of formation of isoenzyme spectrums in progeny of intraspecific and intergeneric crossing shows the lateness of expression time of paternal alleles *Ldh-B* in hybrids compared to "clean" species.

Thus, studying identic combination of parental LDH genotypes (*Ldh-B* x *Ldh-B* $^{/}$) in groups Roach x Roach ("clean" roach) and Bream x Roach (hybrid) we revealed that paternal *Ldh-B* alleles by hybrids were expressed only after hatching while paternal *Ldh-B* alleles by "clean" roach were expressed on blastodisk stage (Figure 10).

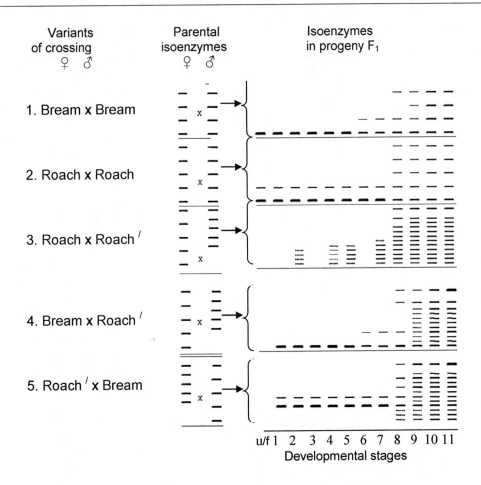

Figure 10. Formation of isoenzyme LDH-spectrums in progeny F1 from bream and roach crossing. Roach and Roach ' - are parents of roach homozygous by fast (*B*) and slow (*B'*) alleles of locus *Ldh-B*. u/f – unfertilizationary eggs, 1-11 – stage of development from blastodisk (1) toyolk sac resorption (11).

The early *Ldh-B* expression in progeny from intraspecific fish crossing allows to classify LDH as enzyme with early activation. The lateness of expression time of paternal subunits of B-type compared to maternal ones in hybrids characterizes the expression of parental alleles *Ldh-B* as asynchronous according maternal type.

ME. In all experimental embryo groups the *Me-1* and *Me-2* – expression was on later stage (after hatching and by underyearling) (Figure 11). The character of *Me-1* expression was synchronous, the character of *Me-2* expression was not detected because of coincidence of isoenzyme ME-2 spectrums by crossing fishs.

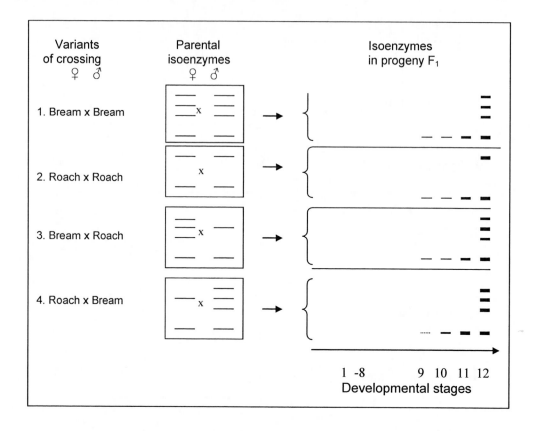

Figure 11. Formation of isoenzyme ME-spectrums in progeny F1 from bream and roach crossing. 1-11 – stage of development from blastodisk (1) to yolk sac resorption (10) and underyearling (12).

6-PGD. Heterozygous isoenzyme spectrums in progeny from reciprocal crossing of bream and roach counted 11 bands (Figure 12). In Roach x Bream hybrid first expression of parental (maternal) alleles *6-Pgd* was at the beginning of segmentation, paternal alleles were expressed after hatching. In Bream x Roach hybrid first expression of maternal alleles *6-Pgd* was at the gastrula, paternal alleles were expressed after full yolk sac resorption. Thus 6-PGD is belongs to the enzymes with early activation. The character of *6-Pgd* expression by hybrid was asynchronous according to the maternal type.

β–EST. The early expression of embryo genes *β-Est-1*is was revealed by B.bream and B.bream x Roach (gastrula), in other embryo group – later: by roach and bream genes *β-Est-1, 2* and *3* was revealed on the end of segmentation and hatching stages correspondingly. The full isoenzyme *β*-EST spectrums in all embryo were formed on hatching stage. After stage of full yolk sac resorption there were a re-distribution *β*-EST–activity between *β*-EST-1, *β*-EST-2 and *β*-EST-3 by larvas (Figure 13).

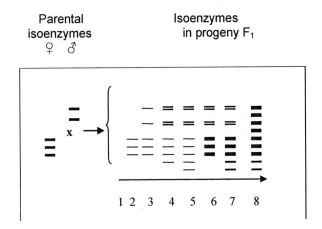

Figure 12. Formation of 6-PGD isozyme spectrums by hybrid Roach x Bream on stages: 1 – blastodisk - gastrula, 2 – beginning of segmentation, 3 – 12 hours after hatching, 4 - 48 hours, 5 – 60 hours, 6 – hours, 7 - 132 hours after hatching, 8 –full resorption of yolk sac.

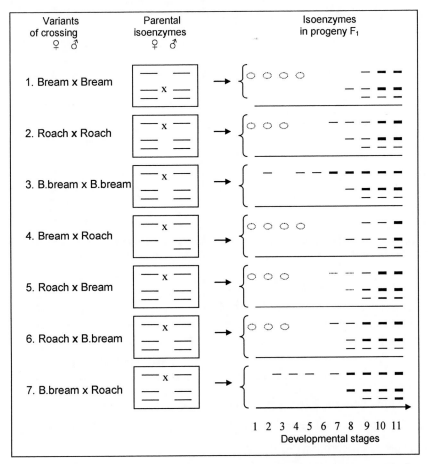

Figure 13. Formation of isoenzyme β-EST-spectrums in progeny F1 from eam, roach and blue bream crossing. 1-11 – stage of development from blastodisk) to yolk sac resorption (10).   – oogenetic β-EST.

As coincidence R_f from corresponding β-EST isoenzymes of parents parting in crossing, the analysis of implement of isoenzymes to maternal or paternal type was not possible. But to take into consideration early β-EST-1 expression by blue bream and more later expression by roach and bream we can surmise that appearance of β-EST-1 on gastrula stage by B.bream x Roach is expression of maternal (B.bream) gene β-Est-1. In this case the character of expression β-Est-1 by B.bream x Roach hybrid we consider asynchronous according to the maternal type passed into synchronous on hatching stage.

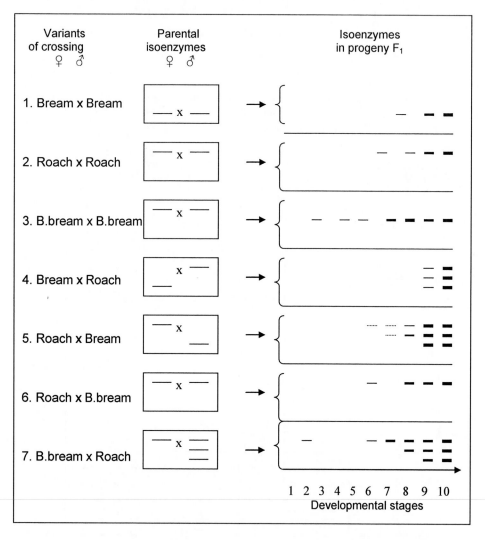

Figure 14. Formation of isoenzyme AAT-spectrums in progeny F1 from m, roach and blue bream crossing. 1-11 – stage of development from :odisk (1) to yolk sac resorption (10).

AAT. First expression of Bream *Aat-1* as AAT-f homodimer enzyme was on hatching stage, of Roach *Aat-sl* as AAT-sl homodimer was on peak of body segmentation stage, of Blue bream AAT-sl homodimer were revealed on gastrula stage (Figure 14).

In hybrid groups: first expression of Blue bream x Roach *Aat-1* locus by one band of maternal homodimer AAT-sl was on gastrula stage until mixed feeding stage; heterozygous isozyme spectrum consisting of homodimer AAT-sl and heterodimer AAT-sl/med was on mixed feeding stage; later homodimer AAT-sl, heterodimer AAT-sl/med and homodimer AAT-med was formed at the mixed feeding-fry stage (Figure 14). In hybrids Roach x Blue bream isozyme spectrum was presented by one band of homodimer AAT-sl on the end of segmentation stage and later. In hybrids Roach x Bream isozyme spectrum was presented at first by two bands: homodimer AAT-sl and heterodimer AAT f/sl (the end of segmentation); later – by three bands: AAT-sl and AAT-f homodimers and heterodimer AAT f/sl (full yolk sac resorption). The appearance of paternal homodimer AAT-f in Roach x Bream hybrids was apparently related to an increased activities of paternal allele. In hybrids Bream x Roach at different temperature variation (Andreeva, 2007) the full three-band AAT spectrum was formed on hatching stage (Figure 14).

Our result suggest that in the Blue bream and Blue bream x Roach F1 hybrids, embryonic *Aat-1* genes were, apparently, activated during early period (gastrula) of embryonic development. In other experimental grous of fishes, these genes were expressed much later. This allowed us to consider AAT in the Blue bream and Blue bream x Roach hybrids as an enzyme with early activation and in the Bream, Roach, and Bream x Roach, Roach x Bream, and Roach x Blue bream hybrids as an enzyme with late activation. Analyses of character of *Aat-1* expression suggest that in hybrids with early and late *Aat-1* expression parental *Aat-1* alleles were activated, relatively, asynchronously and synchronously. This connection was preserved at different temperatures of incubation of the embryos (Андреева, 2007).

Conclusion

The result of studing of biocatalytic activity of maternal and embryo proteins, and time and character of expression of parental alleles of embryo genes also, allow to consider that change of internal fluid invironment in early development of intergeneric F1 hybrids plays important role for regulation of parental genoms coordination in hybrid organism equally to factor of a degree of relationship of crossing species, according it synchronous expression of parental alleles of genetic loci is related to kindred crossing but asunchronous – to remote crossing (Глушанкова и др., 1973; Champion, Whitt, 1976; Neyfakh et al., 1976; Кирпичников, 1987).

Congruousness of processes of lipovitellin degradation and diminution its biocatalytic function on the one hand, gene expression and increase of embryo enzymes activity on the other hand, also adaptation of these prosesses to definite developmental stages, allow to think their intimate connection which had space-temporal character (Figure 8).

In hybrid experimental groups these connection are realized at time and character of parental alleles of embryo genes expression, and determined, first of all, by mechanizms of regulation of intergenomic interaction (Корочкин, 1999), and also by functional expedience of gene activation (Андреева, 2005b).

In case of early embryo genes expression asynchronous activation of parental alleles to the maternal type for hybrids was revealed, wich latter develops into synchronous. In case of late embryo genes expression synchronous activation of parental alleles for hybrids was revealed (Table 3). This connection is storaged for different temperature regimes of embryos incubation (Andreeva, 2007). We consider that It based exclusively on the maternal effect expressed in species specific timing of locus activation.

Yolk and oogenetic enzymes connected with maternal (by origin) lipovitellin form the maternal (by origin) metabolic invironment which enhances preferential activation of maternal alleles of embryo genes (Figure 15). At later developmental stages the yolk reserves are partially or fully resorbed and maternal proteins don't play basic role in forming of properties of embryo internal invironment, parental alleles expression was synchronous. Describing space-temporal connections between biocatalytic activity of lipovitellin and embryo genome expression allow to consider lipovitellin not only inert plastic material but as a active component of internal invironment, a link in system of regulation (initiation) of embryo genes activity and therefore as metabolic regulator in early development.

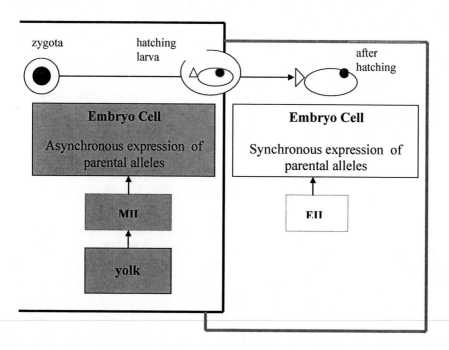

Figure 15. Influence of yolk to character of expression of parental alleles of embryo genes генов from intergeneric F_1 hybrids. MII, EII – maternal internal invironment and embryo internal invironment correspondingly. Compartments marked by grey color contain "maternal" yolk proteins.

References

Андреева А.М. Сывороточные □-глобулины рыб // Вопр. ихтилогии. 2001. Т. 41. N[0]4. С. 550-556.

Андреева А.М. Особенности формирования изоферментных спектров лактатдегидрогеназы в раннем развитии плотвы Rutilus rutilus (L.)// Вопр.ихтиол. 2005a. Т.45. N[0] 2. С.277-282.

Андреева А.М. Особенности проявления генов лактатдегидрогеназы в раннем развитии леща Abramis brama (L.), плотвы Rutilus rutilus (L.) и их реципрокных гибридов F$_1$ // Вопр. Ихтиол. 2005b. Т.45. N[0] 3. С.411-417.

Андреева А.М. Особенности проявления генов аспартатаминотрансферазы в раннем развитии некоторых видов карповых рыб и их межродовых гибридов F$_1$ // Онтогенез. 2007. Т.38. N[0] 1. С.44-51.

Андреева А.М., Басова Е.Е., Слынько Ю.В. Особенности временной экспрессии изоферментов лактатдегидрогеназы, аспартатаминотрансферазы и β–эстеразы у плотвы, леща и их гибридов //Материалы международной конференции студентов и аспирантов по фундаментальным наукам «Ломоносов» / М.: Изд-во МГУ, 2000. Вып. 4. С. 9-10.

Аттер Ф.М., Эберсолд П., Уингнс Г. 1991. Интерпретация генетической изменчивости, выявляемой методом электрофореза // Риман Н., Аттер Ф. (ред.). Популяционная генетика и управление рыбным хозяйством. М.: Агропромиздат. 353с.

Берстон М. Гистохимия ферментов. М.: Мир, 1965. 464 с.

Глушанкова М.А., Коробцова Н.С., Кусакина А.А. 1973. Использование теплоустойчивости белка при исследовании проявления генов альдолазы у гибридных зародышей рыб // Биохимическая генетика рыб Л: Наука. С.76-84.

Кирпичников В.С. Генетика и селекция рыб. Л: Наука, 1987. 520 с.

Клячко О.С., Озернюк Н.Д. 2001. Изменение функциональных и структурных особенностей изоферментов лактатдегидрогеназы на разных стадиях онтогенеза Danio rerio // Онтогенез. Т.32. № 5. С.374-376.

Корочкин Л.И. Взаимодействие генов в развитии. М.: Наука, 1976. 276 с.

Корочкин Л.И. Введение в генетику развития. М.: Наука, 1999. 253 с.

Корочкин Л.И., Серов О.Л., Манченко Г.П. Понятие об изоферментах. Номенклатура изоферментов и соответствующих им генов. Молекулярные механизмы возникновения множественных форм ферментов // Генетика изоферментов. М.: Наука, 1977. С. 5-18.

Крыжановский С.Г. Эколого-морфологические закономерности развития карповых, вьюновых и сомовых рыб // Тр.Ин-та морфол.живот. АН СССР. 1949. N[0] 1. С. 3-332.

Крыжановский С.Г. Закономерности развития гибридов рыб различных систематических категорий. М.: Наука, 1968. 220 с.

Лапушкина Е.Е., Андреева А.М., Слынько Ю.В. Особенности раннего развития и анализ устойчивости к действию пестицидов леща Abramis brama L., плотвы Rutilus rutilus L. и их гибридов F$_1$ // Сб. статей: Биология Внутренних вод: проблемы экологии и биоразнообразия. Изд-во: ЯрГУ. 2002. №3. С.132-144.

Нейфах А.А., Тимофеева М.Я. Молекулярная биология процессов развития. М.:Наука, 1977. 310 с.

Нейфах А.А., Давидов Е.Р. Отсутствие ядерного контроля над увеличением активности катепсина в раннем эмбриональном развитии// Биохимия, 1964.Т.29. С.273-282.

Немова Н.Н. Внутриклеточные протеиназы в эколого-биохимических адаптациях у рыб: Автореф.дис. ...д-ра биол.наук.М., 1992. 42с.

Новиков Г.Г. Рост и энергетика развитиякостистых рыб в раннем онтогенезе.М.: Эдиториал УРСС, 2000. 296с.

Рябов И. Н. Методы гибридизации рыб на примере семейства карповых // Исследование размножения и развития рыб. М.: Наука, 1981. С. 195-215.

Слынько Ю.В. 1987. Коэффициенты имбридинга и структура вида Abramis brama L. // Информ. Бюлл. Биол.внутрен.вод. №75. Л.: Наука. С.39-43.

Слынько Ю.В. 1991. Полиморфизм мышечных изоферментов карповых рыб СССР. I Лактатдегидрогеназа // Информ. Бюлл. Биол.внутрен.вод. №90. Л.: Наука.С.75-84.

Слынько Ю.В. Распределение рыб в масштабе крупных географических областей. // Экологические факторы пространственного распределения и перемещения гидробионтов. С.-Пб.: Гидрометеоиздат, 1993. Гл. IV. С. 259-279.

Слынько Ю.В. Генетическая структура и состояние рыб Рыбинского водохранилища // Современное состояние рыбных запасов Рыбинского водохранилища. Ярославль: Изд-во ЯрГТУ, 1997. С. 153-178.

Слынько Ю.В. Система размножения межродовых гибридов плотвы (Rutilus rutilus L.), леща (Abramis brama L.) и синца (Abramis ballerus L.) (Leuciscinae: Cyprinidae) // Дис. ... к.б.н. Санкт-Петербург, 2000.186 с

Чадов Б.Ф. «Образ» регуляторного гена в опытах на дрозофиле // Генетика. 2002. Т. 38. N^0 7. С.869-881.

Babin P.J., Bogert J., Kooiman F.P., Van Marrewijk W. J.A., Van der Horst D.J. 1999. Apolipophorin II/I, apolipoprotein B, vitellogenin, and microsomal triglyceride transfer protein genes are derived from a common ancestor // J.Mol.Evol. Vol. 49. $N^0$150.P.1-9.

Champion M.J., Whitt G.S. 1976. Synchronous allelic expression at the glucosephosphate isomerase A and B loci in interspecific sunfish hybrids // Biochem. Genet. 1976. V.14. № 9-10. P.723-738.

Davis B.J., Disk-electrophoresis. II. Method and application to human serum proteins // Ann.N.Y.Acad.Sci. 1964.V.121. P.404-427.Frankel J.S., Hart N.H. Lactate dehydrogenase ontogeny in the genus Brachydanio (Cyprinidae) // J. Hered. 1977. V. 68. №2. P. 81-86.

Hartling R.C., Kunkel J.G. Proteolytic cleavage of yolk protein during flounder (Pleuronectes americanus) development: characterization of lipovitellin from eggs and embyos// Mol.Biol.Cell. 1995. V. 6. P. 321a.

Laemmli U.K. Cleavage of structural proteins during the assembly of the head of bacteriophage // Nature. 1970. V.4. № 5259. P. 680-685.

Markert C.L. 1963. Lactate dehydrogenase isozymes: dissociation and recombination of subunits // Science. № 140. P.1329-1330.

Kopperschlander G., Dicrel W., Bierwagen B., Hofman E. Molecular weiths bestium yongen durch polyacrylamid gel electrophorese unter verwendurg lines linearen gel gradienter //FEBS Lett. 1969. V.5. P.221-227.

Naoshi et al., Akihiko. Identification and characterization of proteases involved in specific proteolysis of vitellogenin and yolk proteins in salmonids // J.Exp.Zool. 2002. V.292. $N^0$1. P.11-25.

Neyfakh A.A., Glushankova M.A. Kusakina A.A.1976. Time of function of genes controlling aldolase activity in loach embryo development // Devel.Biol. V.50. № 2. P.502-510.

Ornstein L. Disc-electrophoresis. I. Background and theory // Ann.N.Y.Acad.Sci. 1964. N^0 121. P.321-349.

Wang H., Yan T., Tan J.T.T., Gong Z. A zebrafish vitellin gene (vg3) encodes a novel vitellogenin without a phosvitin domain and may represent a primitive vertebrate vitellogenin gene // Gene. 2000. V. 256. P. 303-310.

In: Development Gene Expresión Regulation
Editor: Nathan C. Kurzfield

ISBN: 978-60692-794-6
©2009 Nova Science Publishers, Inc.

Short Communication

Cloning and Expression Analysis of the Homeobox Gene *Abdominal-A* in the Isopod *Asellus Aquaticus*

Philipp Vick[4], Axel Schweickert and Martin Blum

Institute of Zoology, University of Hohenheim,
Garbenstr. 30, 70593 Stuttgart, Germany

Abstract

Regulation of embryonic axis patterning by Hox genes has been shown to be widely conserved among metazoans. In *Drosophila melanogaster* the Hox gene *abdominal-A* (*abd*-A) is important for the development of the legless abdomen. In contrast to the clear tagmata-correlated activity in insects the analysis of Hox expression patterns during crustacean development has turned out to be more diverse. While in the branchiopod brine shrimp *Artemia franciscana* the posterior genes show a more ancestral overlapping arrangement, this is not the case for the malacostracan isopod *Porcellio scaber*. In this more modern species *abd-A* is mainly restricted to the developing pleon. Here we present the cloning and expression pattern of the *abd-A* gene from the freshwater crustacean *Asellus aquaticus*. In contrast to the related isopod *Porcellio scaber*, *Asellus aquaticus* differs in the regulation of posterior segment patterning. While *Porcellio scaber* displays distinct and separate segments in the pleon, posterior segments are partially fused in *Asellus aquaticus* to yield a pleotelson. The *abd-A* signal was significantly reduced or absent in the pleotelson of *Asellus aquaticus*, while *abd-A* was expressed in the free segments of *Porcellio scaber*. An additional correlation between *abd-A* gene expression and patterning in these two species was found in that the orientation of walking legs was towards the posterior pole in segments expressing *abd-A*. *Asellus aquaticus* thus may provide a highly interesting and novel arthropod model organism to study evolution of segment identity and patterning.

---

[4] email:pvick@uni-hohenheim.de

Introduction

Homeotic mutations, in which one body segment adopts the identity of another one, have fascinated naturalists for more than a century (Bateson 1894). For example, mutants in Drosophila are known in which antennae are changed into legs, or halteres into wings (Duncan and Montgomery 2002). Homeotic mutations have also been described in vertebrates, such as changes in identity of vertebrae (Jegalian and De Robertis 1992). The identification of homeotic genes as the targets of such mutations has opened the molecular analysis of body segment identity, and - as segment identity in many cases underlies changes of body plans during evolution - provided an entry point into the investigation of macroevolution at the molecular level. All homeotic genes encode transcription factors which are characterized by the presence of a conserved DNA binding motif, the homeobox (Gehring 1987).

Segment identity in Drosophila is the best studied paradigm to date. Clusters of homeotic genes in the Antennapedia and the Bithorax complex specify segment identity along the anterior-posterior body axis (Lewis 1978; Carroll 1995). The Antennapedia complex contains the five genes *labial, proboscipedia, Deformed, Sex combs reduced (Scr)* and *Antennapedia (Antp)*, and the Bithorax complex comprises the three genes *Ultrabithorax (Ubx), abdominal-A (abd-A)* and *Abdominal-B (Abd-B)*. In *Drosophila* five of these genes are involved in defining the trunk segments. *Scr, Antp* and *Ubx* are specifically expressed in the developing thoracic segments, whereas *abd-A* is important for conferring abdominal segments their identity in concert with *Abd-B* (Carroll 1995).

Differences in body plans between the different classes of the arthropods concern for example the subdivision into head, thorax and abdomen in insects as compared to prosoma and opisthosoma in spiders. Analysis of Hox expression revealed that genes active in the *Cupiennius salei* prosoma were exclusively found in the *Drosophila* head (Damen et al. 1998). In addition, mRNAs of these genes were also found in the head of crustaceans and myriapods (Abzhanov and Kaufman 1999; Abzhanov and Kaufman 1999; Hughes and Kaufman 2002). These findings have led to the conclusion that the chelicerate prosoma is homologous to the head tagma of other arthropods (Damen et al. 1998; Abzhanov and Kaufman 1999; Abzhanov et al. 1999; Hughes and Kaufman 2002; Carroll 2004). In general, these and other analyses have helped to resolve the evolution of arthropod lineages more clearly, at times confirming or rejecting morphology-based phylogeny (Akam 2000; Cook et al. 2001; Averof 2002; Schram and Koenemann 2004).

The emerging hypothesis from such studies has been that although Hox genes are highly conserved between different arthropod classes and species, changes in expression patterns were the driving force for alteration in body plan (Akam 2000; Cook et al. 2001; Averof 2002; Schram and Koenemann 2004). Crustaceans offer a particularly revealing case. Specifically, Hox genes have been used to visualize the differentiation of the major groups. Averof and Akam showed that in the ancestral branchiopod species *Artemia franciscana* the trunk Hox genes *Antp, Ubx* and *abd-A* were expressed in a broad overlapping fashion (Averof and Akam 1993; Averof and Akam 1995). Additionally Kaufman and colleagues showed in several studies differences in Hox gene expression patterns in two crustaceans, the isopod *Porcellio scaber* and the decapod *Procambarus clarkii* as well as in the myriapod *Lithobius atkinsoni* (Abzhanov and Kaufman 2000; Abzhanov and Kaufman 2000; Hughes and Kaufman 2002). The crustacean studies revealed that in contrast to the obviously more primary Hox expression patterns of *Artemia*, those in *Porcellio* and *Procambarus* were more tagmata-confined. *Ubx* was mostly expressed in the pereon whereas *abd-A* was mostly restricted to the pleon. *Antp* showed a strong signal in the pereon, which was much weaker in the pleon. These studies demonstrated the differences of Hox expression between major crustacean groups and how these corresponded to the diverse adult morphologies.

Interestingly, Hox expression patterns could also be used to describe variations in external morphology within a class of arthropods. In crustaceans and especially the Malacostraca, anterior Hox gene comparisons were used to analyze the formation of maxillipeds. Maxillipeds originally represented anterior thoracic walking legs which have been converted into additional feeding appendages during evolution. In several crustacean groups these transformations occurred at different positions along the anterior-posterior body axis. In all cases analyzed the anterior expression domains of *Ubx* and *abd-A* correlated with the transition of maxillipeds to walking legs (Averof and Patel 1997; Abzhanov and Kaufman 1999).

The above mentioned examples demonstrated the power of Hox gene analysis for evolutionary studies at the level of phyla and classes. Examples of studies at lower taxonomic levels are scarce and restricted to dipterans (Yoder and Carroll 2006). We thus wondered whether such analyses could be extended to closely related species, which display significant differences in segment morphology. In the present study we chose two isopod species, namely the water louse *Asellus aquaticus* (LINNAEUS, 1758) and the wood louse *Porcellio scaber* (Abzhanov and Kaufman 2000).

The freshwater isopod *Asellus aquaticus* is widely distributed in standing and slow flowing freshwaters of the Palearctic region. Like in all higher crustaceans (Malacostraca) its body is divided into different functional regions, so-called tagmata. Isopods in general have a head fused to the first thoracic segment forming a cephalothorax, as well as seven free thoracic segments presenting the walking leg bearing pereon (Fig.1A+B). The posterior pleon consists of six segments bearing the taxon-specific limbs. The sixth pleonic segment is always fused with the telson and bears the uropods (Gruner 1965; Gruner 1993).

Differences in pleonic limb morphology and degree of fusion of pleonic segments reflect phylogenetic relationships inside the isopoda. Specifically, members of the Asellidae and the genus *Asellus* have two very small free pleonic segments, and the last four segments are fused to the telson forming a pleotelson (Fig.1A). The third pair of pleopods builds up an operculum covering the fourth and fifth pair which function as gills. In addition, although thoracomeres are quite similar in morphology, only the last three are orientated posteriorly. Consistent with this notion the posterior three pairs of thoracopods differ from the remaining one in that they are significantly longer and directed posteriorly while the anterior ones point to the front (Fig. 1A+B; von Haffner 1937; Gruner 1965; Gruner 1993).

In contrast, Porcellio scaber, a member of the Oniscidea differs significantly from the Asellota design. Porcellio also possesses seven free thoracomeres, but these and the thoracopods are nearly identical in morphology. The pleon shows no further fusion of segments, resulting in five free pleomeres of equal size and shape (Gruner 1965). Kaufmann and colleagues have described expression patterns for various Hox genes in developing

Porcellio embryos (Abzhanov and Kaufman 2000), and found a clear correlation of abd-A and pleonic segments. We therefore hypothesized that differences should exist in Hox gene expression between Porcellio and Asellus and started out to characterize Hox genes in the water louse.

Here we present the expression analysis of *abd-A* during development of *Asellus aquaticus*. We show by whole-mount *in situ* hybridization that beside very similar expression characteristics compared to *Porcellio* there are differences between these two isopod species which reflect differences in their family-specific posterior morphology. In *Asellus* the anterior and the posterior expression boundary each were shifted more anteriorly. Additionally we detected an expression domain in the epithelial cells of the developing early endodermal digestive gland tissue. This study therefore provides an example of how modest but highly significant differences in expression patterns of Hox genes reflect morphological differences between families of the same arthropod order.

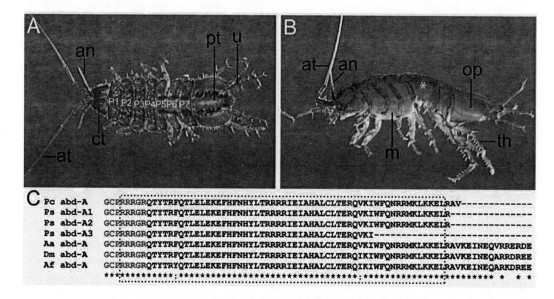

Figure 1. Morphology of the water louse *Asellus aquaticus* and amino acid sequence alignment of arthropod *abdominal-A* (*abd-A*) homeodomains.(A) Dorsal view of an adult female displaying a cephalothorax (ct) with a pair of antennules (an) and antennae (at), seven free pereonic segments (P1-P7), two free pleonic segments and the pleotelson (pt) bearing the uropods (u).(B) Lateral view of same specimen displaying the seven pairs of thoracopods (th), marsupium (m) and operculum (op). Please note the lost 5th pereopod marked by an asterisk. (C) Amino acid sequence alignment of the homeodomain and flanking sequences of the heomeobox gene *abd-A* from *Asellus aquaticus* (Aa), *Procambarus clarkii* (Pc), *Porcellio scaber* (Ps), *Drosophila melanogaster* (Dm) and *Artemia franciscana* (Af). Primer sequences are indicated in grey; the boundary of the homeobox is marked by the dotted box; asterisks mark identical amino acids, gaps indicate missing or un-matched residues, colons highlight conserved amino acid changes.

Results and Discussion

In order to correlate gene expression patterns with body segment differentiation, a number of Hox genes and other homeobox genes were selected for analysis. Following molecular cloning of cDNA fragments and preliminary expression analysis, *abd-A* was chosen for an in-depth study, as this gene proved to be the most promising candidate. Cloning and expression patterns of further genes will be described elsewhere (Vick and Blum, in preparation).

Cloning and Sequence Analysis of the Homeobox Gene *Abd-A* from *Asellus Aquaticus*

Known arthropod sequences were retrieved from the databases and aligned to select degenerate primers for cloning of the *Asellus aquaticus* ortholog of *abdominal-A* by RT-PCR. First, a standard PCR was used to obtain a short stretch of *Asellus*-specific sequences. In a second step 3'-RACE PCR was performed with two *Asellus*-specific nested and one sequence-tagged unspecific primer targeted to the poly-A tail of messenger RNAs. Following PCR reactions, electrophoresis and sequencing we obtained a 238 base pair (bp) fragment of *abd-A* spanning the whole homeodomain and parts of the 3' region. Surprisingly, the 3' sequences recovered did not contain the poly-A tail or UTR sequences, indicating that the unspecific oligo dT anchor primer annealed to A-rich stretches within the coding region, which we found to be particularly AT-rich in coding regions of *Asellus* as compared to *Drosophila melanogaster*. In *Porcellio*, three different splice variants of *abd-A* were described (Abzhanov and Kaufman 2000). We only recovered one form, which does however not exclude that additional isoforms exist. We did not attempt to isolate additional variants, as the one we cloned should be able to detect all three potential isoforms in *in situ* hybridization experiments.

The alignment of the deduced amino acid sequence with those of *Drosophila melanogaster*, *Artemia franciscana*, *Procambarus clarkii* and *Porcellio scaber* revealed a very high degree of conservation inside the homeodomain as well as in the neighboring region (Fig. 1C). A single change was found to *Drosophila abd-A*, namely a conserved amino acid replacement from valine to isoleucine at position 45 of the homeodomain. Within the crustaceans, identity in all positions was detected except for *Artemia franciscana*, in which in addition to the change in position 45 a second conserved change from phenylalanine to tyrosine in position 11 of the homeodomain was found.

Embryonic Development and Gene Expression Pattern of Abd-A in *Asellus Aquaticus*

The development of isopods in general and of *Asellus aquaticus* in particular has been described in several studies in the 19th and 20th century (Rathke 1834; Dohrn 1867; McMurrich 1895; Dejdar 1930; Länge 1958; Weygoldt 1959). As with all Peracarida, *Asellus* displays direct development. The female lays its eggs into the marsupium, a ventral brood pouch located between the sternites and the oostegites of the anterior thoracic segments, in which the whole development takes place (Fig. 1B; Gruner 1965). Following superficial cleavage and gastrulation, segments and limbs form from anterior to posterior, creating a gradient of mature body regions towards the posterior end (McMurrich 1895; Weygoldt 1959). These early stages of development were not analyzed for *abd-A* gene transcription, as embryos were fragile, and because development of the pleotelson had not yet been initiated at that stage.

With increasing length, head and telson meet inside the chorion and vitelline membrane. Because isopods have no ventral groove they form a dorsal curvature (Fig. 2A). When the embryo hatches from the chorion it has built up most segments as well as the cephalothoracic and pereonic limb buds. Shortly thereafter the pleonic limbs occur. At that stage *abd-A* was expressed in a defined domain at the posterior pole of the embryo (Fig. 2A). Discrete segment boundaries could not be distinguished at that stage in whole-mounts. A parasagittal section confirmed posterior-specific expression, and revealed signals to be present in pleonic segments 1-4 as well as pereonic segment P7 (Fig. 2A''). An additional expression domain was found in the paired primordia of the endodermal epithelia. Signals were faintly visible in ventral view of whole-mounts (Fig. 2A'), as the ectodermal epithelium was translucent at that stage. The histological analysis in Fig. 2A'' confirmed the specificity of this expression, which is further supported by the absence of signal in sense control hybridized specimens (Fig. 2B-B''). Probe trapping seems highly unlikely, as signals were clearly located inside the epithelial cells lining the endodermal cavity. This pattern came as a surprise, as no endodermal mRNA signals of *abd-A* have been reported in any model organism so far. However, *abd-A* was shown to be expressed in the midgut-surrounding visceral mesoderm and its loss of function influenced *labial* expression in the endodermal midgut of *Drosophila* (Manak et al. 1994; Staehling-Hampton and Hoffmann 1994).

As development progresses and the embryonic axis lengthens the inner membrane ruptures and sheds. This process is induced by the downward bending of the pleon (cf. Fig. 2A and C). At that stage (75% of development; Fig. 2C) signal intensity had markedly increased. Expression had begun to shift anteriorly, and was now predominantly found in pereonic segments P5-7, and – progressively fading – in the anterior-most pleonic segments (Fig. 2C). Additional signals were detected in transverse histological sections at the level of the fifth (Fig. 2C') and seventh (Fig. 2C'') pereonic segments. Staining was seen in the ventral nerve cord in a bilaterally symmetrical fashion (Fig. 2C', C''), likely representing differentiated neurons. This neural expression of *abd-A* correlates well with studies of *abd-*A expression in *Porcellio,* where neural signals were described to be localized specifically to pereonic segments P5-7 (Abzhanov and Kaufman 2000), in perfect agreement with our study. In addition to the segmental signals, transcription was also clearly present in pereonic limbs, with stronger staining in the proximal areas, fading towards the distal ends (Fig. 2C, C', C''). Strikingly, expression was restricted to pereonic limbs P5 and P6, i.e. limbs which orient themselves toward the posterior pole in the adult. In *Porcellio*, expression in P6 and P7 was also reported, however, much fainter signals compared to *Asellus*, which correlated with the only moderate posterior bending of pereopods in *Porcellio* (Abzhanov and Kaufman 2000).

Strongest signals in *Porcellio* were restricted to more posterior segments, namely pleonic segments 1-5 (Abzhanov and Kaufman 2000).

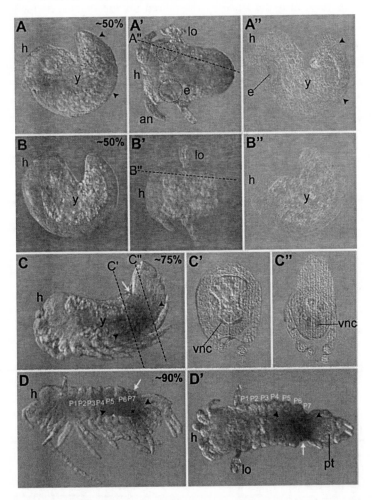

Figure 2. *Abdominal-A* mRNA expression during development of *Asellus aquaticus* embryos.Staging of embryos was according to Whitington et al. (1993). **(A-A'')** Embryo at ~50% of development showing *abd-A* mRNA expression mainly in the pleon and in the paired endodermal anlagen of the midgut (e); in (A) only the posterior expression domain can bee seen. (A') Ventral view showing lateral organs (lo), and endodermal primordia visible through the translucent ectoderm (dotted rings). A sagittal section in (A'') highlights expression domains.**(B-B'')** Control embryo at ~50% of development hybridized with a sense probe showing no expression in any tissue.**(C-C'')** *abd-A* mRNA expression in a ~75% embryo. Expression was located between the 5th pereonic segment and the anterior pleon with decreasing intensity towards the posterior pleon. Transversal sections (C'+C'') showed strong expression in the ventral nerve cord (vnc) and weaker signals in the proximal limbs and the remainder of the mesodermal and ectodermal trunk tissue, which gradually faded towards the dorsal side. No expression was detected in the dorsal most part and in the distal limbs.**(D+D')** Expression of *abd-A* in an embryo at ~90% of development. The signal was clearly restricted between P5 and pleonic segment 3 with strongest expression in P6+P7 (D'), the developing 7th pereopod (*) and the proximal limbs of the anterior pleonic segments.A, A'', B, B'', C, D lateral views, anterior to the left, dorsal to the top; A', B' ventral views, anterior to the left; D' dorsal view, anterior to the left; C', C'' transversal sections, dorsal to the top. White arrows mark tagmata boundary between pereon and pleon. Black arrowheads indicate approximate anterior and posterior expression boundaries. Asterisk marks 7th pereopod. Dashed lines indicate approximate planes of section. an, antennule; h, head; pt pleotelson; y, yolk.

Between 75% and 90% of development, the embryo hatches again and thereby finalizes limb development (Dohrn 1867). At ~90% the embryo shows already movements, and the remaining yolk has been fully absorbed by the paired midgut glands. *abd-A* gene expression now was most clearly restricted to P5-7 and to pleonic segments 1-3, with no remaining signals in the more posterior region, where the pleotelson had begun to form (Fig. 2D'). Expression in the neural tissue persisted until this stage (not shown), while signals in pereonic limbs, which by now had completed their posterior orientation, were hardly detectable. The expression in pleonic segment 3 was restricted to the ventral aspect of the segment (not shown). Strikingly, this expression again correlated with a morphogenetic event, namely the fusion of segments at the dorsal aspect in pleonic segments 3-5, resulting in the formation of the pleotelson. In contrast, expression was clearly present in anterior pleopods (Fig. 2D), and in the newly developing pereopod 7 (asterisk in Fig. 2D).

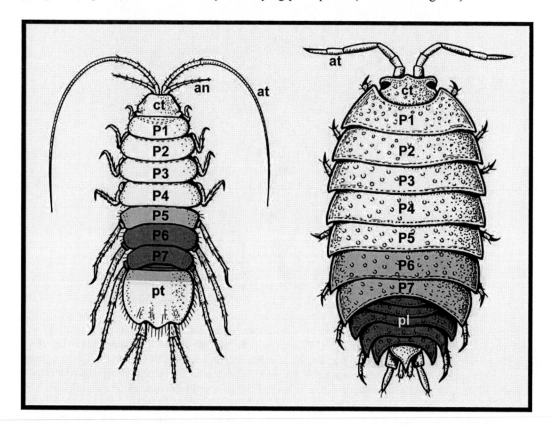

Figure 3. Schematic comparison of embryonic *abdominal-A* mRNA expression domains in *Asellus aquaticus* (left) and *Porcellio scaber* (right) in relation to adult morphology. In both cases, dark blue shading marks strong expression and light blue coloring indicates weak expression. *Asellus* showed strong expression of *abd-A* in P6, P7 and the free pleonic segments 1 and 2, and weaker expression in P5 and pleonic segment 3. In contrast, strong expression in *Porcellio* was restricted to the pleonic segments 1-5. In P6 and P7 a weak expression was reported (Abzhanov and Kaufman 2000). at, antennae; an, antennule; ct, cephalothorax; pl, pleon; pt, pleotelson.

Conclusions

In summary, our *abd-A* expression analysis in *Asellus aquaticus* demonstrated a clear correlation of expression domains with distinct morphogenetic units in two cases (Fig. 3): pleotelson fusion (absence of *abd-A* transcription) and pereonic limb orientation (strong *abd-A* expression during the process of posterior bending). Compared to *abd-A* expression in *Porcellio*, the adult morphology divergence is mimicked by a corresponding change in gene expression, i.e. *abd-A* activity throughout the pleonic segments (which do not fuse), and lack of activity in posterior pereopods which do not orient posteriorly (von Haffner 1937; Gruner 1965). Thus, even between highly related crustaceans, shifts of Hox gene expression boundaries correlate with adult morphology variations (Fig. 3). It could, therefore, be quite beneficial to analyze the promoter regions of *abd-A* in these two species, as these changes should be represented in a comparably small number of nucleotide changes. Further, it might be worthwhile to extend this analysis to the most basal isopods, i.e. the suborder Phreatoicidea, which show an even more pronounced dimorphism of leg orientation (von Haffner 1937). Finally, the hypothesis put forward here, namely that temporal-spatial changes in *abd-A* gene expression are responsible for morphogenetic variations in adult morphologies of *Asellus aquaticus* and *Porcellio scaber* might be testable in the future, provided techniques for genetic manipulation of embryos become available.

Material and Methods

Animals and Embryos

Embryos bearing adult females were collected from a local pond during the month of May to August. Typically, 15-40 embryos were recovered from females by dissecting the marsupium. Embryos were staged in analogy to the crustacean staging system described by Whitington et al. (1993), fixed in 4 % paraformaldehyde and stored in methanol at –20°C until further analysis.

Whole-Mount *in Situ* Hybridization and Histological Analysis

In situ hybridization followed standard procedures. For histological analysis embryos were embedded in a gelatine-albumin mix and sectioned at 30 μm using a vibratome. Sections were embedded and photographed using Nomarski optics.

Accession Number

The *abdominal-A* sequence was deposited at gene bank, accession number EU882729.

References

Abzhanov, A. and Kaufman, T. C. (1999). "Homeotic genes and the arthropod head: expression patterns of the labial, proboscipedia, and Deformed genes in crustaceans and insects." *Proc Natl Acad Sci U S A* **96**(18): 10224-9.

Abzhanov, A. and Kaufman, T. C. (1999). "Novel regulation of the homeotic gene Scr associated with a crustacean leg-to-maxilliped appendage transformation." *Development* **126**(6): 1121-8.

Abzhanov, A. and Kaufman, T. C. (2000). "Crustacean (malacostracan) Hox genes and the evolution of the arthropod trunk." *Development* **127**(11): 2239-49.

Abzhanov, A. and Kaufman, T. C. (2000). "Embryonic expression patterns of the Hox genes of the crayfish Procambarus clarkii (Crustacea, Decapoda)." *Evol Dev* **2**(5): 271-83.

Abzhanov, A., Popadic, A. and Kaufman, T. C. (1999). "Chelicerate Hox genes and the homology of arthropod segments." *Evol Dev* **1**(2): 77-89.

Akam, M. (2000). "Arthropods: developmental diversity within a (super) phylum." *Proc Natl Acad Sci U S A* **97**(9): 4438-41.

Averof, M. (2002). "Arthropod Hox genes: insights on the evolutionary forces that shape gene functions." *Curr Opin Genet Dev* **12**(4): 386-92.

Averof, M. and Akam, M. (1993). "HOM/Hox genes of Artemia: implications for the origin of insect and crustacean body plans." *Curr Biol* **3**(2): 73-8.

Averof, M. and Akam, M. (1995). "Hox genes and the diversification of insect and crustacean body plans." *Nature* **376**(6539): 420-3.

Averof, M. and Patel, N. H. (1997). "Crustacean appendage evolution associated with changes in Hox gene expression." *Nature* **388**(6643): 682-6.

Bateson, W. (1894). *Materials for the Study of Variation Treated with Especial Regard to Discontinuity in the Origin of Species*. London, Macmillan and co.

Carroll, S. (2004). *From DNA to Diversity*, Blackwell Science Inc.

Carroll, S. B. (1995). "Homeotic genes and the evolution of arthropods and chordates." *Nature* **376**(6540): 479-85.

Cook, C. E., Smith, M. L., Telford, M. J., Bastianello, A. and Akam, M. (2001). "Hox genes and the phylogeny of the arthropods." *Curr Biol* **11**(10): 759-63.

Damen, W. G., Hausdorf, M., Seyfarth, E. A. and Tautz, D. (1998). "A conserved mode of head segmentation in arthropods revealed by the expression pattern of Hox genes in a spider." *Proc Natl Acad Sci U S A* **95**(18): 10665-70.

Dejdar, E. (1930). "Die Funktion der "blattförmigen Anhänge" der Embryonen von *Asellus aquaticus* (L.)." *Zeitschrift für Morphologie und Ökologie der Tiere* **19**.

Dohrn, A. (1867). "Die embryonale Entwicklung von *Asellus aquaticus*." *Zeitschrift für Morphologie und Ökologie der Tiere* **17**.

Duncan, I. and Montgomery, G. (2002). "E. B. Lewis and the bithorax complex: part I." *Genetics* **160**(4): 1265-72.

Gehring, W. J. (1987). "Homeo boxes in the study of development." *Science* **236**(4806): 1245-52.

Gruner, H.-E. (1965). *Die Tierwelt Deutschlands und der angrenzenden Meeresteile - 51. Crustacea, V. Isopoda*, Gustav Fischer Verlag Jena.

Gruner, H.-E. (1993). *Lehrbuch der speziellen Zoologie - Band 1, Teil 4: Arthropoda*, Gustav Fischer Verlag.

Hughes, C. L. and Kaufman, T. C. (2002). "Exploring the myriapod body plan: expression patterns of the ten Hox genes in a centipede." *Development* **129**(5): 1225-38.

Jegalian, B. G. and De Robertis, E. M. (1992). "Homeotic transformations in the mouse induced by overexpression of a human Hox3.3 transgene." *Cell* **71**(6): 901-10.

Länge, H. (1958). "Bau und Entwicklung der blutbildenden Organe von *Asellus aquaticus* (L.)." *Zeitschrift für wissenschaftliche Zoologie* **161**: 144-208.

Lewis, E. B. (1978). "A gene complex controlling segmentation in Drosophila." *Nature* **276**(5688): 565-70.

Manak, J. R., Mathies, L. D. and Scott, M. P. (1994). "Regulation of a decapentaplegic midgut enhancer by homeotic proteins." *Development* **120**(12): 3605-19.

McMurrich, J. P. (1895). "Embryology of Isopod Crustacea." *Journal of Morphology* **11**.

Rathke, H. (1834). "Recherches sur la formation et le développement de l'Aselle d'eau dpuce." *Annal. des Sci. Nat., 2me sér.* **ii.**

Schram, F. R. and Koenemann, S. (2004). Developmental genetics and arthropod evolution: On body regions of crustacea. *Evolutionary Developmental Biology of Crustacea.* G. Scholtz, Aa Balkema.

Staehling-Hampton, K. and Hoffmann, F. M. (1994). "Ectopic decapentaplegic in the Drosophila midgut alters the expression of five homeotic genes, dpp, and wingless, causing specific morphological defects." *Dev Biol* **164**(2): 502-12.

von Haffner, K. (1937). "Untersuchungen über die ursprüngliche und abgeleitete Stellung der Beine bei den Isopoden." *Zeitschrift für wissenschaftliche Zoologie* **149**: 513-536.

Weygoldt, P. (1959). "Beitrag zur Kenntnis der Malakostrakenentwicklung. Die Keimblätterbildung bei *Asellus aquaticus* (L.)." *Zeitschrift für wissenschaftliche Zoologie* **163**: 342-354.

Whitington, P. M., Leach, D. and Sandeman, R. (1993). "Evolutionary change in neural development within the arthropods: axonogenesis in the embryos of two crustaceans." *Development* **118**(2): 449-61.

Yoder, J. H. and Carroll, S. B. (2006). "The evolution of abdominal reduction and the recent origin of distinct Abdominal-B transcript classes in Diptera." *Evol Dev* **8**(3): 241-51.

Reviewed by Dr. Matthias Gerberding, Max-Planck-Institute of Developmental Biology, Tübingen, Germany.

Index

B

C

D

E

F

G

H

I

J

K

L

M

N

O

P

S

T

U

V

W

X

Y

Z